INDUSTRIAL WATER ANALYSIS HANDBOOK

N.MANIVASAKAM M.Sc.,
Formerly **CHIEF WATER ANALYST**
PRINCIPAL PUBLIC HEALTH LABORATORY
COIMBATORE-641018 (T.N.)
INDIA

Chemical Publishing Company

INDUSTRIAL WATER ANALYSIS HANDBOOK

ISBN: 978-0-8206-0040-6

Chemical Publishing Company:
www.chemical-publishing.com
www.chemicalpublishing.net

Originally published 2005

Printed in the United States of America

DEDICATED

TO

THE LOTUS FEET OF

LORD MURUGA

AND TO MY PARENTS

C.K.NATARAJ & ARUNAGIRI AMMAL

I gratefully acknowledge,

1) American Public Health Association, Washington

2) American Society for Testing and Materials, Philadelphia

3) International Organization for Standardization, Switzerland

4) Hach Company, Loveland, Colorado, U.S.A

For their kind permission to reproduce some of the tables, and paragraphs from their publications,

- N.MANIVASAKAM

PREFACE

Water analysis has become extremely important for many reasons. Not only to design suitable water treatment plant during commissioning of the industry but it is essential for continuous monitoring of the quality of treated water. Without a proper water analysis program, the efficacy of water treatment cannot be ascertained. A proper water analysis would also reduce the cost on treatment significantly. Knowing this fully well, every industry is setting up their own water analysis laboratory as a part of their industry. Abundant literature is available on the analysis of water, but this book is especially written to suit the needs of industries. All the test procedures for the parameters pertaining to industrial water analysis are provided in this book and discussed in detail. Another salient feature of this book is that the analytical procedures are given in a step by step way so that it becomes easier for the analyst to carryout analysis of water. Chapters are arranged in alphabetical order for easy access to the required chapter. Thus this book would serve as a valuable reference to carryout analysis of water for any intended industrial purpose. This apart, it is extremely useful to carryout examination of waster water too.

This book is divided into four parts.

Part – I 'Introduction' provides a detailed introduction on analysis of water along with parameters to be determined for each industrial use, thus helps to reduce the time required for analysis and labor involved in analysis.

In Part – II 'Chemical Analysis', elaborate testing procedures for all the parameters necessary for industrial uses are given. Exclusive chapters in the beginning on 'Sampling of water' and 'Sampling of Boiler water' will provide the analyst a detailed sampling program and the important sampling points so that the analysis would be more meaningful and more useful. Detailed analytical procedures for 65 chemical parameters are given which makes this book a handy reference for carrying out analysis without any difficulty.

Part – III 'Microbiological Analysis' deals with the identification and determination of the density of microbial organisms that are likely to interfere in industrial processes. To aid the analyst, separate chapters starting from the Requirement of Chemicals, Glassware and Equipment, Technics involved in

Microbiological Analysis, Sampling of Water for Microbiological Examination and Preparation of Media and Reagents, to identification of specific organisms are provided with a detailed discussion.

In Part – IV 'Microscopical Examination', a brief account of microscopical organisms is given. The common organisms present in water along with their habitat and significance are also dealt.

This book would serve as a handy reference to all wet processing industries. This book would be helpful in multifarious ways to Analysts, Chemists, Engineers, and Managers of industries and Water Treatment Consultants, Firms engaged in Water Treatment and other personnel engaged in water analysis and water treatment. This apart, this book would also be a source book to students of Industrial Engineering, Chemical Engineering, Industrial Chemistry, Applied chemistry, Environmental Engineering and Environmental Science and other allied faculties of Colleges and Universities.

This book has been prepared with the help and assistance of a number of people. Their assistance made this book a most useful volume. I owe a great debt of gratitude to all of them. Some of the organizations have been most gracious in granting permission to reproduce certain illustrations, figures and portions from their publications. Their cooperation is gratefully acknowledged. A special debt of gratitude is due to M/S. Chemical Publishing Company, Gloucester, U.S.A., for their keen interest shown towards the publication of this book.

Comments and criticisms as well as suggestions for the improvement of the book are solicited.

Coimbatore – 42 (TN) **- N.MANIVASAKAM**
India.

This book is divided into the following parts.

Part I. ***INTRODUCTION***

Part II. ***CHEMICAL ANALYSIS***

Part III. ***MICROBIOLOGICAL ANALYSIS***

Part IV. ***MICROSCOPICAL EXAMINATION***

CONTENTS

PART I

INTRODUCTION

CHAPTER - 1
INTRODUCTION

Water analysis has become increasingly important for many reasons. Water is the material that is widely employed in industries in multifarious ways – water is the raw material, water is the ingredient, water is the medium of reaction, water is the conveying medium, water is used for washing, water is used to generate steam and it is the major cooling agent and so on. For each and every industrial use, the quality of water is very important and it is the judging factor of the quality of finished product as well as the criterion for protecting boilers, cooling systems, machineries and equipments from scale formation and corrosion. Water analysis is the key component in designing suitable water treatment plant and to maintain the water quality. Without a proper water analysis program, the efficacy of water treatment cannot be ascertained. A proper water analysis would also reduce the cost on treatment considerably. Since each industrial use needs a water of specific quality, a detailed analysis is essential to know the quality of water available and to apply suitable treatment to render the water fit for the specific purpose and to maintain the same.

Analysis of water includes four components namely,

1. *Chemical Analysis*
2. *Microbiological Analysis*
3. *Microscopical Examination*
4. *Testing for Radioactivity*

CHEMICAL ANALYSIS

Examination for physical characteristics and chemical composition together called as 'Chemical Analysis'. All the parameters that comes under chemical analysis are listed below.

Physical Properties	Metals	Non-metallics	Dissolved Gases	Organics	Other Tests
Color	Aluminium	Acidity	Carbon - dioxide	Hydrazine	BOD
Electrical Conductivity	Arsenic	Alkalinity	Chlorine (Residual)	Nitrogen (Ammonia)	Calcium carbonate Stability Test
Odor	Boron	Chloride	Oxygen (Dissolved)	Nitrogen (Albuminoid)	Chlorine demand
pH Value	Cadmium	Cyanide		Nitrogen (Organic)	Coagulant demand
Solids	Calcium	Fluoride		Nitrogen (Total Kjeldahl)	COD
- Total	Chromium	Iodide		Oil and grease	Equivalent Mineral Acidity
- Dissolved	Copper	Nitrate		Phenols	Langellier Index
- Suspended	Hardness	Nitrite		Surfactants	Particle Counting
- Settleable	Iron	Phosphate		Tannin and Lignin	Permanganate Value
Temperature	Lead	Silica		Volatile acids	Residual Sodium Carbonate
Turbidity	Magnesium	Sulfate			Silt Density Index
	Manganese	Sulfide			Sodium - Adsorption Ratio
	Mercury	Sulfite			Sodium Slip
	Nickel				
	Potassium				
	Selenium				
	Sodium				
	Silver				
	Zinc				

MICROBIOLOGICAL ANALYSIS

Microbiological analysis is carried out to detect the presence and density of microbial organisms in water. These organisms may render the water unsuitable in one or other way for the intended use, especially for food industries, pharmaceutical industries, fermentation industries and electronics industries. Hence is it essential that the water used in industries should also be free from microorganisms. In this book, procedures for the following organisms, are given.

1. *Standard Plant Count (To detect the number of viable micro-organisms present in water)*
2. *Coliform bacteria (To detect fecal contamination)*
3. *Fecal coliforms (Escherichia coli) (To detect fecal contamination)*
4. *Proteolytic Bacteria (Protein degrading bacteria. Essential test for Food Industries)*
5. *Lipolytic Bacteria (Lipid degrading bacteria. Essential test for Food Industries)*
6. *Thermophilic Bacteria (To be tested for water used in Dairy)*
7. *Gelatin Liquefying Bacteria*
8. *Iron Bacteria*
9. *Sulfate-Reducing Bacteria*
10. *Sulfur-Oxidizing Bacteria*
11. *Yeast and Mould*
12. *Slime Forming Organisms*

See Part – III, 'MICROBIOLOGICAL ANALYSIS' for more details

MICROSCOPICAL EXAMINATION

Water may contain numerous minute organisms that could be detected only under microscope. Many industrial processes virtually require water free of such microscopical organisms, as they impart color, odor and turbidity to water. Some of the higher forms may affect the health also. A detailed microscopical examination will reveal the presence and concentration of phytoplanktons such as algae, diatoms, flagellates and zooplanktons such as protozoans, ciliated protozoans, rotifers, crustaceans and other forms. For details, refer Part – IV, 'MICROSCOPICAL EXAMINATION'.

TESTS FOR RADIO ACTIVITY

Radio active materials find their way into water courses mainly from the discharge of waste water from nuclear power plants and due to extensive use of radio tracers. They are highly undesirable in processing industries. Stringent limits for radio active materials are stipulated for drinking water and water used for food processing and fermentation industries. Water used in these industries must comply with these requirements. Radio active measurements are helpful to detect the radio active contamination and contaminant levels in water.

ANALYSIS OF RAW WATER & TREATED WATER

The primary objective of the analysis of raw water in industries is to find out its quality to design a suitable treatment plant and to plan a treatment program. Detailed analysis is necessary to design an effective and comprehensive treatment plant. Seasonal variations and corresponding fluctuations* in raw water quality have to be taken into account while designing a treatment plant. If two or more sources with plenty of water are available, raw water analysis would help to select the source to which minimum treatment is required. Such selection would result in considerable savings in treatment costs and in the subsequent control analysis.

After the treatment plant is installed, the treated water has to be tested periodically. The periodicity (hourly, daily, weekly etc.,) is dependent on the industrial purpose for which the water is used and the type of treatment plant installed. The treated waters are tested to find out the efficacy, uniformity and consistency of treatment. Any over treatment will lead to unnecessary expenditure and under treatment would result in poor quality water and consequently lead to equipment failure and / or inferior product.

* Throughout the world, rivers are the most commonly used water sources for industrial purposes. Since they are subject to pollution by industries, sewage, agricultural run-off and other man made activities, considerable variation in the concentration of many constituents should be anticipated along with variations due to seasonal changes. Whereas ground waters are essentially constant in composition and there may not be much variation in quality characteristics.

ANALYTICAL PROGRAM

Drawing an analytical program is highly imperative to attain the objective of the best performance of the treatment system and securing water of desired quality for the particular industrial use and maintaining the treatment. The basic components of an analytical program are,

i. Fixing the sampling points [(i.e) the stages of treatment system that need control]
ii. Frequency of sampling
iii. Parameters to be determined [(i.e) constituents that affect the process at the particular stage]
iv. Method of analysis.

All these components are entirely dependent on the quality requirement of water for the intended industrial use and the type of treatment plants installed.

While planning the program of sampling and analysis, care should be exercised that the program should not be too extensive incorporating unnecessary data which involve then more labour and also expensive and wastage of time. Similarly an inadequate program would lead to improper conclusions and improper treatment resulting in poor quality products and damage to equipments. The best program is the one with minimum sampling points and tests, at the same time providing necessary information. Regular analyses for raw water, treated water used in process have to be carried out. Analysis of condensate returns and recirculating cooling water is also necessary.

Most of the industries require testing of water, only to asses the physical characteristics and chemical composition. Some other industrial processes need water of satisfactory bacteriological quality also (eg) food industries, fermentation industries and electronics industries. In such cases, both chemical and bacteriological analysis need to be carried out. When the water is susceptible to radio active contamination, it has to be tested for radio activity also.

In Chapter - 2, the parameters that need to be determined are given industry wise and may be referred for further information. For fixing the right sampling points refer Chapter - 5, 'SAMPLING OF WATER FOR CHEMICAL ANALYSIS' and Chapter - 6 'SAMPLING OF BOILER WATER'. For detailed analytical procedures refer Part – II and Part – III.

CHAPTER - 2

PARAMETERS TO BE DETERMINED

(INDUSTRY WISE)

Given below is the list of parameters to be tested on a regular basis. To aid the analyst the list is given industry wise. These are the minimum tests to be carried out regularly which are also termed as 'Control Tests'. Even among these, if some of the parameters remain constant and exhibit no variation, they need not be done. It is reiterated however, while commissioning the industry the water has to be tested for all parameters to assess its suitability for the particular industrial use and to design a suitable treatment plant. If surface waters are used, their characteristics have to be studied seasonally prior to designing treatment plant. In addition to the control parameters, they have to be checked for other parameters also when they are suspected to vary in character due to seasonal variations and/or pollution from upstream side.

The present day softeners, demineralizers, reverse osmosis plants and steam generation systems are equipped with fully automatic (or semiautomatic) continuous analyzers which measure the pH, temperature, turbidity conductivity, hardness, dissolved oxygen and silica. A schematic diagram of water treatment plants and steam generating equipments along with continuous analyzers are depicted in Fig. 2 – 1. Needless to say that in addition to such measurements, a testing schedule has to be established to make sure that sufficient information is obtained with the minimum number of tests. Such schedules and control parameters are described in the following sections.

CONTROL PARAMETERS FOR TREATMENT PROCESSES

Lime – Soda Softening

1. **pH value :** Continuous analysis is desirable
2. **Hardness:** May be tested once per shift
3. **Phosphate:** For a significant reduction of hardness, phosphate is usually added in hot lime-soda softening. Maintenance of a residual of 6 – 8 ppm is necessary. May be tested once per shift.

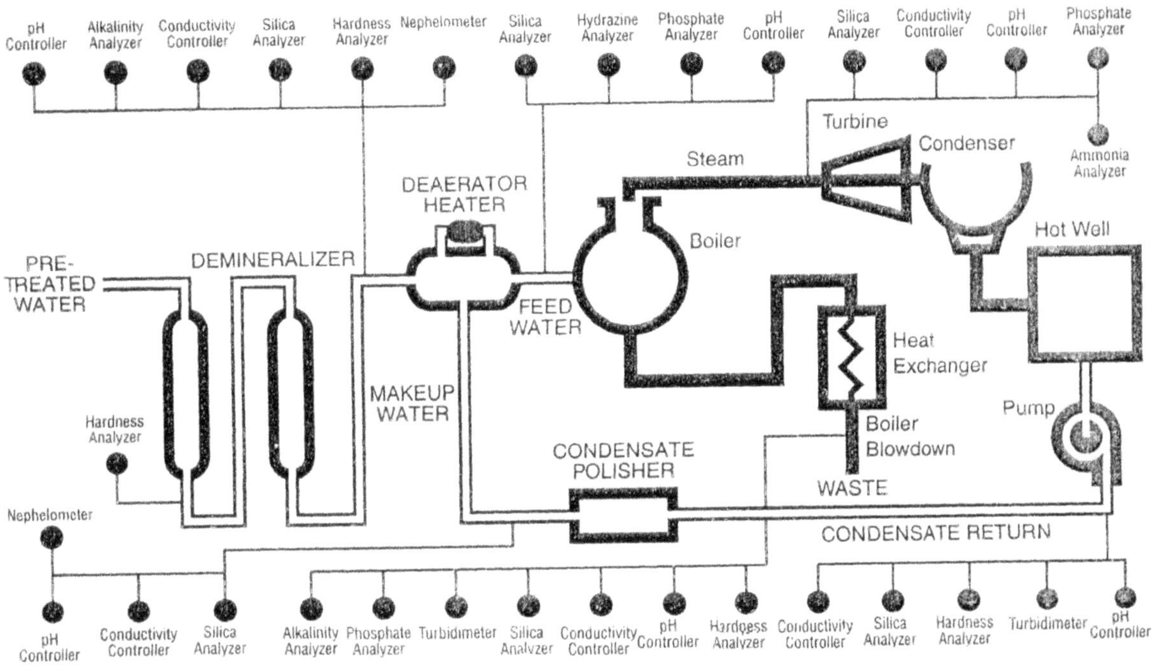

Fig. 2 – 1. Water Treatment Plants and steam Generation Systems Continuous Analyzers and Strategic Points at which they are installed*.

Cation – exchange Softening

1. **pH value:** Continuous analysis is desirable.

2. **Hardness:** A continuous analyzer for hardness is convenient for monitoring the performance of cation exchangers. The instrument is normally installed on the softener outlet header. When it alarms, the softened water must be checked for leakage of hardness. EDTA titric metric method is recommended for checking. (In the absence of a continuous analyzer, the softened water should be checked for hardness every 2 hr.)

* Fig. 2 – 1 is reproduced with the kind permission from M/s. Hach Company, Colorado, U.S.A., from their publication, "Hach Water Analysis Hand Book" 4th Edition (Revision – 2) – 2003.

Demineralization

1. Conductivity
2. pH value
3. Silica
4. Hardness
5. Sodium
6. Alkalinity
7. Turbidity.

The Demineralizers are usually installed with continuous analyzers for monitoring pH and conductivity. Both conductivity and pH measurements are used to indicate breakthrough in demineralization. As the resin becomes exhausted, the pH falls and the conductivity rises.

In normal operation, the pH of the anion exchanger effluent is about 8. If the pH becomes 9 or more it is an indication of excessive sodium ion leakage from the cation exchanger. The present day trend is to install a continuous analyzer for sodium in the cation exchanger outlet so that sodium leakage could be immediately known.

Automatic silica analyzers are also installed at the anion exchanger outlet. When demineralizers approach exhaustion, silica is the first impurity to breakthrough into the effluent. Due to its low ionic strength, silica cannot be detected readily by conductivity meters. However automatic analyzers are capable of detecting increasing levels of silica that occur before complete exhaustion. Regeneration can then be initiated.

CONTROL PARAMETES FOR STEAM GENERATION SYSTEMS

Feed Water

Boiler feed water is a mixture of condensate and makeup water. Usually condensate will be in high proportion (ranging from 95% to 99% - sometimes total feed water is condensate) and make up will be in a lower proportion. Before feeding, the quality must be known in order to protect the boiler system from scaling and corrosion. Detailed analysis may be carried out at suitable intervals. However, all the parameters need not be done daily. A few control tests alone have to be done which clearly give a picture on the quality of feed water. Necessary control parameters are tabulated (See Table 2 – 1)

Boiler Water

The analysis of boiler water is extremely useful in detecting the conditions inside the boiler and to detect under or over treatment. Based on the analysis of boiler water, dosage of chemicals can be controlled. In this case also certain chemical parameters have to be checked frequently which are given in Table 2 – 2.

Steam

It is extremely important to determine the purity of steam frequently. Needless to say that steam purity is affected by certain operating variables and therefore it is best to sample and monitor continuously the quality of steam. Unlike with other samples, obtaining a representative sample of steam is very difficult and hence several precautions need to be observed. Purity of steam is determined by measuring either the conductivity or sodium concentration in a condensed sample. For a detailed analytical procedure refer, chapter – 76.

Condensate

Condensate is nothing but the condensed steam and is a water of extreme purity. The condensate return system is a revealing sampling point to monitor total boiler water treatment performance. The parameters, turbidity, electrical conductivity and hardness are used to find out in-leakage through heat exchangers and condensers. If the conductivity of the combined condensate rises, individual returned streams should be checked for hardness. Test for silica also may be carried out on such return streams to detect contamination. Regular measurement of the pH of the condensate is necessary to regulate the dosage of amines. (After the addition of neutralizing amines the pH should be in the range of 8 – 9). Tests for iron, copper and dissolved oxygen are carried out to detect corrosion problems. Silica and sodium are important parameters which indicate carryover problems and / or condenser leakage (see Table 2 – 3). As indicated already, continuous analyzers are available for the measurement of sodium as low as 0.1 ppb. The major advantage with these analyzers is that these instruments suffer no interference from hydrazine, morpholine, cyclohexylamine or ammonia.

Blow Down

Blowing down is an essential operation in any boiler water treatment program. Blowdown removes precipitated sludge and dissolved solids from the boiler before harmful levels are reached. Blowdown can be continuous or intermittent. In either case, analysis is required to maintain the critical balance between solids removal and the loss of heat and treatment chemicals that must be replaced. The control parameters are listed below. (See Table 2 – 4).

Raw Water

Raw waters have also to be tested occasionally to detect any variations in quality. Ground waters normally do not show any variations. Surface waters, however are subject to fluctuations in quality due to seasonal variations and/or pollution from upstream side and may require frequent checking.

TABLE 2–1

BOILER FEED WATERS – CONTROL PARAMETERS AND TESTING SCHEDULE

Sl.No.	Parameter	Frequency
1.	Electrical Conductivity	Continuously
2.	pH value	- do -
3.	Total Hardness	Daily
4.	Total Alkalinity	- do -
5.	Chloride	Once per shift (i.e., once in 8 hrs)
6.	Silica	- do -
7.	Dissolved Oxygen	- do -
8.	Residual Sodium sulfite / Hydrazine	- do -
9.	Phosphate	Need to be done only if the water is softened by Lime – Soda process. Once per day.
10.	Iron	Occasionally
11.	Copper	- do -
12.	Oil	- do -

TABLE 2–2

BOILER WATER - CONTROL PARAMETERS AND TESTING SCHEDULE

Sl.No.	Parameter	Frequency
1.	pH value	Continuously
2.	Total Hardness	Once per shift (once in 8 hrs)
3.	Total Alkalinity	- do -
4.	Caustic Alkalinity	- do -
5.	Chloride	- do -
6.	Silica	- do -
7.	Dissolved Oxygen	Continuously
8.	Residual Sodium sulfite / Hydrazine	Once per shift
9.	Phosphate	- do -
10.	Iron	- do -
11.	Copper	- do -
12.	Oil	Occassionally
13.	Residual Chelant	Once per day

TABLE 2–3

CONDENSATE–CONTROL PARAMETERS AND TESTING SCHEDULE

Sl.No.	Parameter	Frequency
1.	Turbidity	Once per day
2.	Electrical conductivity	Continuously
3.	pH value	- do -
4.	Total Hardness	Once per day
5.	Silica	- do -
6.	Sodium	- do -
7.	Iron	- do -
8.	Copper	- do -
9.	Oil	- do -

TABLE 2–4
BOILER BLOWDOWN
CONTROL PARAMETERS AND TESTING SCHEDULE

Sl.No.	Parameter	Frequency
1.	Electrical Conductivity	Once per shfit (once in 8 hrs)
2.	Total Dissolved Solids	- do -
3.	pH value	- do -
4.	Total Alkalinity	- do -
5.	Caustic Alkalinity	- do -
6.	Silica	- do -
7.	Residual Sodium sulfite / Hydrazine	- do -
8.	Chloride	- do -
9.	Phosphate	- do -
10.	Residual Chelant	- do -

CONTROL PARAMETERS FOR COOLING WATER AND WATER FOR HUMIDIFICATION

COOLING WATER

- Turbidity
- Hardness
- Alkalinity
- Iron
- Manganese
- Residual chlorine
- Chromates / Polyphosphates / Nitrites
- Iron Bacteria

HUMIDIFICATION

- Turbidity
- Suspended Solids
- Hardness
- Corrosivity Test
- Iron Bacteria

CONTROL PARAMETES (INDUSTRYWISE)

ALUMINIUM INDUSTRY

- Tests for Boiler Feed Water
- Tests for Cooling Water

AUTOMOTIVE INDUSTRY

- Tests for Boiler Feed
- Tests for Cooling Water
- For Pickling, Electroplating and Anodizing
    - Hardness
- For all purposes
    - Color
    - Turbidity
    - Iron
    - Manganese
    - Biological growths such as Iron Bacteria

BAKERY

- Color
- Turbidity
- Taste
- Odor
- Iron
- Manganese
- Alkalinity
- Hardness
- Hydrogen sulfide

BEVERAGE PRODUCTION

- Color
- Turbidity
- Odor
- Taste
- Alkalinity
- Total Dissolved Solids
- Hardness
- Iron
- Manganese
- Chloride
- Sulfate
- Fluoride
- Hydrogen sulfide
- Permanganate Value (Oxygen absorbed) - to find out organic matter
- Microscopical Examination (to detect algae and protozoa)

BOTTLE / UTENSIL / FOOD EQUIPMENT WASHING

- Color
- Turbidity
- Odor
- Taste
- Hardness
- Iron
- Manganese
- Fluoride
- Permanganate Value (Oxygen absorbed) - to find out organic matter
- Bacteriological Examination
    - Standard Plate Count
    - Coliform Bacteria

BREWING

- Color
- Turbidity
- Taste
- Odor
- Total Dissolved Solids
- pH value
- Alkalinity
- Iron
- Manganese
- Nitrate
- Nitrite
- Chloride
- Fluoride
- Silica
- Hydrogen sulfide
- Calcium
- Calcium chloride
- Magnesium
- Magnesium sulfate
- Magnesium chloride
- Sodium chloride
- Sodium sulfate
- Sodium carbonate

CERAMIC INDUSTRY

- Tests for Boiler Feed water
- Color
- Iron
- Turbidity
- Manganese
- Hardness
- Total Dissolved Solids
- Iron bacteria
- Sulfate-reducing Bacteria

CHLOR-ALKALI MANUFACTURE

- Color
- Turbidity

- Iron
- Manganese
- Calcium (in brine) has to be monitored continuously to prevent fouling of membranes.

CITRUS FRUIT INDUSTRY

- Color
- Turbidity
- Taste
- Odor
- Total Hardness

CORN-PRODUCTS MANUFACTURE

- Color
- Turbidity
- Taste
- Odor
- Hardness
- Magnesium
- Iron
- Manganese
- Bacteriological Examination
    - Standard Plate Count
    - Coliform Bacteria

DAIRY

- Color
- Turbidity
- Taste
- Odor
- Total Dissolved Solids
- Hardness
- Copper
- Iron
- Manganese
- Sulfate
- Chloride
- Nitrate
- Nitrite
- Ammonia
- Permanganate value (Oxygen absorbed) - to find out organic matter
- Tests for Boiler Feed Water
- Bacteriological Examination
    - Standard Plate Count
    - Coliform Bacteria
    - Proteolytic Bacteria
    - Lipolytic Bacteria
    - Thermophilic Bacteria

DISTILLERY

See "Fermentation Industries".

EDIBLE OIL REFINERY

- Tests for Boiler Feed Water
- Color
- Turbidity
- Odor
- Hardness
- Iron
- Manganese
- Bacteriological Examination
  - Standard Plate Count
  - Coliform Bacteria

ELECTRONICS INDUSTRY

- Conductivity
- Silica
- Particle Count
- Particle Size
- Total Organic Carbon
- Copper
- Chloride
- Nitrate
- Nickel
- Phosphate
- Potassium
- Sodium
- Sulfate
- Zinc
- Bacteriological Examination
  - Standard Plate Count
  - Endotoxins

ELECTRO PLATING

- Color
- Turbidity
- Hardness
- Sodium
- Iron
- Bicarbonates
- Chloride
- Permanganate Value (Oxygen absorbed) - to find out organic matter

FERMENTATION INDUSTRIES

- Color
- Turbidity
- Odor
- Alkalinity
- Sulfate
- Chloride

- Total Dissolved Solids
- Hardness
- Calcium
- Magnesium
- Iron
- Manganese
- pH Value
- Nitrate
- Fluoride
- Tests for Boiler Feed Water
- Tests for Cooling Water
- Bacteriological Examination
    - Standard Plate Count
- Coliform Bacteria

In addition to the above parameters, while commissioning the industry, analysis for other parameters including heavy metals such as Arsenic, Cadmium, Chromium, Copper, Lead, Mercury, Selenium along with Cyanide which are considered to be toxic have to be carried out. Zinc and Phenols have also to be tested. If surface water is used, tests for these substances have to be made at regular intervals.

FOOD CANNING

- Tests for Boiler Feed Water
- Tests for Cooling water
- Color
- Turbidity
- Taste
- Odor
- pH value
- Total Dissolved Solids
- Hardness
- Iron
- Manganese
- Alkalinity
- Nitrate
- Fluoride
- Sodium chloride
- Ammonia
- Bacteriological Examination
    - Standard Plate Count
    - Coliform Bacteria
    - Test for Pathogens

FOOD PROCESSING

- Tests for Boiler Feed Water
- Tests for Cooling water
- Tests as per Food Equipment Washing
- Color
- Turbidity
- Taste
- Hardness
- Iron
- Manganese
- Fluoride
- Bacteriological Examination
    - Standard Plate Count

- Odor
- Total Dissolved Solids
    - Coliform Bacteria
    - Test for Pathogens

Also see the control parameters under

- Confectionary
- Dairy
- Ice Manufacture
- Edible Oil Refinery
- Meat Packing
- Soft Drinks Manufacture
- Sugar Industry

In addition to the above parameters, while commissioning the industry, analysis for other parameters including heavy metals such as Arsenic, Cadmium, Chromium, Copper, Lead, Mercury, Selenium along with Cyanide which are considered to be toxic have to be carried out. Zinc and Phenols have also to be tested. If surface water is used, tests for these substances have to be made at regular intervals.

GLUE MANUFACTURE

Hard waters retards the breaking up of the tissue. Therefore soft waters are preferable for the glue manufacture. Hence the most important parameter is hardness.

ICE MANUFACTURING

- Color
- Turbidity
- Taste
- Odor
- pH value
- Total Dissolved Solids
- Total Suspended Solids
- Hardness
- Calcium
- Magnesium
- Sodium
- Iron
- Manganese
- Alkalinity
- Sulfate
- Chloride
- Fluoride
- Bacteriological Examination
    - Standard Plate Count
    - Coliform Bacteria

MEAT–PACKING

- Tests for Boiler feed Water
- Tests for Cooling water
- Tests as per Food Equipment Washing
- Color
- Turbidity
- Taste
- Odor
- Hardness
- Iron
- Manganese
- Permanganate Value (Oxygen absorbed) - to find out organic matter

PULP AND PAPER INDUSTRY

- Color
- Turbidity
- Total Dissolved Solids
- Total Suspended Solids
- Hardness
- Calcium
- Magnesium
- Alkalinity
- Iron
- Manganese
- Chloride
- Silica
- Carbon dioxide
- Residual chlorine
- Permanganate Value (Oxygen absorbed) - to find out organic matter
- Microscopical Examination to detect algae and fungi

RAYON MANUFACTURE

- Color
- Turbidity
- pH value
- Total Dissolved Solids
- Total hardness
- Iron
- Manganese
- Alkalinity
- Silica
- Aluminum
- Copper
- Other heavy metals (lead, zinc, calcium, selenium etc.,)
- Bacteriological Examination
    - Standard plate count
- Microscopical Examination (to detect algae etc.,)

SEMI CONDUCTOR MANUFACTURE

- Electrical Conductivity
- Total Dissolved Solids
- Particle Count
- Dissolved Silica
- Total Organic Carbon
- Aluminum
- Ammonium salts
- Chromium
- Copper
- Iron
- Manganese
- Potassium
- Sodium
- Zinc
- Bromide
- Chloride
- Nitrate
- Nitrite
- Phosphate
- Sulfate
- Bacteriological Examination
    - Standard Plate Count

SILK REELING AND PROCESSING

- Color
- Turbidity
- Total Suspended Solids
- Hardness
- Iron
- Manganese
- Soluble alkalinity
- Nitrate
- Nitrite

SILVER PLATING OF GLASS

- Color
- Turbidity
- Total Suspended Solids
- Iron
- Manganese

SOAP MANUFACTURE

- Color
- Turbidity
- Total Dissolved Solids
- Total Suspended Solids

- Hardness

SOFT DRINKS MANUFACTURE

- Color
- Turbidity
- Taste
- Odor
- Total Dissolved Solids
- Hardness
- Iron
- Manganese
- Alkalinity
- Permanganate Value (Oxygen absorbed) - to find out organic matter
- Bacteriological Examination
    - Standard Plate Count
    - Coliform Bacteria

See also parameters under the section, “Bottle Washing“, “Boiler Feed Water” and “Cooling Water’

STARCH MANUFACTURE

- Color
- Turbidity
- Taste
- Odor
- Total Dissolved Solids
- Total Suspended Solids
- Total Hardness
- Iron
- Manganese
- Permanganate Value (Oxygen absorbed) - to find out organic matter

SUGAR INDUSTRY

Cane Sugar Industry

- Color
- Turbidity
- Total Dissolved Solids
- Total Suspended Solids
- Iron
- Manganese
- Permanganate Value (Oxygen consumed) - to find out organic matter

Beet Sugar Industry

Parameters as per cane sugar industry. In addition, the following tests have to be conducted.

- Alkalinity
- Sodium chloride
- Hydrogen sulfide

- Bacteriological Examination
    - Standard Plate Count
    - Coliform Bacteria

For cane sugar and beet-sugar industries, water has to be checked for boiler feed and cooling. The condensate should also be tested for sugar and ammonia that are likely to enter.

TANNERY

- Color
- Turbidity
- pH value
- Total Hardness
- Alkalinity
- Iron
- Manganese
- Chloride
- Carbon dioxide
- Bacteriological Examination
    - Standard Plate Count

TEXTILE PROCESSING (COTTON & SYNTHETIC TEXTILES)

- Color
- Turbidity
- Odor
- pH value
- Total Dissolved Solids
- Total Suspended Solids
- Hardness
- Iron
- Manganese
- Alkalinity
- Sulfate
- Sulfide
- Chloride
- Nitrate
- Phosphate
- Aluminum
- Copper
- Lead (or other heavy metals)
- Silica
- Dissolved Oxygen
- Carbon dioxide
- Chlorine
- Ammonia
- Oils & Grease
- Corrosivity Test
- Microscopical Examination (to detect algae etc.,)

THERMAL POWER PLANTS

Extensive sampling and continuous analysis of makeup, condenser hot well water, deaerated water, feed water at the economizer inlet and boiler water need to be done. It is also essential to examine super heated steam. (The total solids concentration in the steam should be below 50μg/l (ppb) and silica concentration should not exceed 10μg/l to avoid turbine fouling. The steam solids should be free from sodium hydroxide to minimize stress corrosion cracking of turbine blades and other parts connected to turbine).

The important parameters to be tested are:

1) Electrical conductivity
2) pH value
3) Corrosivity test
4) Dissolved Oxygen
5) Sodium
6) Silica (on Demineralized makeup water).

These measurements have to be made continuously.

In addition to the above parameters, Iron, Copper and Nickel have to be tested on feed water, heater drains and condensate. Hydrazine is also to be tested for feed water. Ammonia has also to be tested before and after condensate polisher.

ULTRA PURE WATER

- Color
- Turbidity
- Electrical conductivity
- pH value
- Silica
- Total Organic carbon
- Particle count
- Particle size
- Bacteriological examination
  - Standard Plate Count
- Endotoxin Test

WOOL SCOURING & PROCESSING

- Color
- Turbidity
- pH value
- Total hardness
- Iron
- Manganese
- Aluminium
- Copper
- Alkalinity
- Nitrate
- Nitrite

PART II

CHEMICAL ANALYSIS

CHAPTER - 3

EQUIPMENTS, GLASSWARE AND CHEMICALS REQUIRED FOR CHEMICAL ANALYSIS

In this chapter, equipments, glassware, miscellaneous items and chemicals required for chemical analysis are given as separate lists.

Equipments must be purchased from renowned concern with good service back up conforming to specifications.

Only high quality glassware to be purchased and they should withstand thermal shock and should not leach any chemical substances while heating or when used for other purposes. Pyrex, Corning and Borosil brands are recommended. While purchasing, glassware should be specified clearly according to the requirements.

For distillation , use of quick – fit assembly is recommended.

It is advised to purchase and use only analytical reagent grade (AR Grade) chemicals manufactured by reputed firms.

EQUIPMENTS

1. Electronic Analytical Balance.

Accuracy 0.1 mg (for accurate weighings)

2. Analytical Balance.

Accuracy 1 mg – Electrically operated (for weighing higher quantities of chemicals).

3. Electrical Conductivity meter.

4. pH meter.

(with single combined electrode).

5. Turbidity meter.

Turbidity - 4 ranges.
Starting from 0.1 to 400 NTU.

6. Spectrophotometer.

Complete set with cuvettes etc.,

7. Flame Photometer.

With oil free pump and other necessary accessories.

8. Cold Vapor Atomic Absorption Spectrometer.

With all accessories for mercury determination.

9. Water Bath.

Made up of stainless steel – 6 holed with concentric rings.

10. Hot Plate.

Rectangular Size.

11. Hot Air Oven.

With thermostatic controls.
Approximate size of chamber: 35 × 35 × 35 cm.

12. Magnetic Stirrer.

With speed adjuster, 4 magnets and retriever.

13. Centrifuge : (Electrical)

14. Distilled Water Still.

Made up of stainless steel. Capacity 2 litres / hour.

15. Residual Chlorine Comparator.

With Chlorine Discs.
Two ranges , 0.1 to 1.0 mg/l chlorine.
0.2 to 2.0 mg / l chlorine.

16. Voltage Stabilizers

17. LPG Gas Burners

Gas cylinder with Regulator and Pressure Tubing.

GLASS WARE

1. Beakers, glass , Griffin, Low form (Graduated / Non graduated) with spout.

- Capacity 2000 ml.
- Capacity 1000 ml.
- Capacity 400 ml.
- Capacity 250 ml
- Capacity 150 ml.

2. Beakers, Polythene / Polycarbonate with spout.

- Capacity 250 ml.
- Capacity 150 ml.

3. Volumetric Flasks with interchangeable glass stopper. (High Accuracy)

- Capacity 1000 ml
- Capacity 500 ml
- Capacity 250 ml
- Capacity 100 ml
- Capacity 50 ml

4. Conical flask (Erlenmeyer), wide mouth with rim

- Capacity 500 ml
- Capacity 250 ml
- Capacity 100 ml

5. Conical flask (Erlenmeyer), Narrow mouth, stoppered with, interchangeable stopper.

- Capacity 250 ml

6. Burette, Straight bore, Stopcock, 50 ml capacity with graduation interval of 0.1 ml.(High Accuracy)

7. Nessler cylinder, for color comparison, Flat bottom

– 100 ml capacity.

With graduation markings at 100 ml and 50 ml.

App Height × OD = 180 mm × 35 mm.

8. Nessler Cylinder for Color Comparison Flat Bottom

– 50 ml capacity

With graduation ,marking at 50 ml and 25 ml

App height x OD = 150 mm x 25 mm

9.Comparison Tubes Matched.

– 150 mm × 16 mm with markings at 2 ml and 10 ml.

10. Measuring Cylinder Graduated Glass with pour out with hexagonal base.

- Capacity 1000 ml. with Graduation interval of 10 ml.
- Capacity 500 ml. with Graduation interval of 5 ml.
- Capacity 250 ml. with Graduation interval of 2 ml.
- Capacity 100 ml. with Graduation interval of 1 ml.
- Capacity 50 ml. with Graduation interval of 1 ml.
- Capacity 10 ml. with Graduation interval of 0.2 ml.

11. Measuring cylinder, Graduated, Polythene / Polycarbonate.

- Capacity 500 ml.
- Capacity 250 ml.
- Capacity 100 ml.

12. Pipet – Bulb (Transfer – Volumetric) (High Accuracy)

- Capacity 50 ml
- Capacity 25 ml
- Capacity 20 ml
- Capacity 10 ml
- Capacity 5 ml

13. Pipet – Graduated (Measuring – Mohr type) (High Accuracy)

- Capacity 25 ml with graduation interval at 0.2 ml
- Capacity 10 ml with graduation interval at 0.1 ml
- Capacity 5 ml with graduation interval at 0.05 ml
- Capacity 1 ml with graduation interval at 0.01 ml
-

14. Bottle, Glass, Reagent, Plain narrow mouth, with interchangeable Flat Head Stopper.

- Capacity 1000 ml.
- Capacity 500 ml.
- Capacity 250 ml.
- Capacity 125 ml.

15. Bottle, Glass, Reagent, Plain narrow mouth with interchangeable Flat Head Stopper – Amber.

- Capacity 250 ml.
- Capacity 125 ml.

16. Bottle, Reagent, Polythene / Polycarbonate

- Capacity 500 ml.
- Capacity 250 ml.
- Capacity 125 ml.

17. Bottles, Weighing with interchangeable stopper, Squat form.

- Capacity 40 ml. App height × OD – 40 x 60 mm.
- Capacity 20 ml. App height × OD – 35 x 50 mm.

18. Dishes, Glass, Evaporating, Flat bottom with pour out.

- App height x OD - 45 x 80 mm.
- App height x OD - 55 x 105 mm.

19. Dishes made up of Vycor Glass Evaporating, Flat bottom with pour out (for Boron determination).

- Capacity about 50 ml
  - App height – 27 mm.

- Bottom dia - 35 mm.
- Upper dia - 70 mm.

- Capacity about 20 ml

20. Funnel, Glass, 60°

- Dia 10 cm
- Dia 7.5 cm
- Dia 5.0 cm

21. Funnel, Separting, Conical (Pear shape) with stopcock and interchangeable stopper. (short stemmed)

- Capactiy 1000ml
- Capacity 500 ml
- Capacity 250 ml
- Capacity 125 ml.

22. Desiccator, large, with cover, Knob Top.

- App chamber depth – 110 mm
- App overall height – 300 mm
- App ID of Ground Flange – 200 mm.

23. Wash bottles, Glass, Complete set with interchangeable stopper.

- Capacity 1000 ml (Size of interchangeable stopper 29/32)
- Capacity 500 ml (size of interchangeable stopper 24/29)

24. Wash bottles, LDPE, Squeeze type, Screw cap fitted with stopper and delivery tube.

- Capacity 500 ml.

25. Watch Glass.

- Dia 150 mm
- Dia 100 mm.
- Dia 75 mm.

26. Test tube with rim (for general use)

- App. length × OD - 150 × 25 mm.
- App. length × OD - 150 × 15 mm.
- App. length × OD - 100 × 12 mm.

27. Glass rod with a button at one end (for stirring)

- App length 200 mm.
- Rod dia 7 mm.
- Button dia 10 – 12 mm.

28. Glass tube (for dispensing indicators etc.,)

- App length 200 mm.
- Dia 5 mm.

29. Modified Gutzeit Apparatus (for Arsenic determination)

(as per the Figure given in Chapter - 13 'ARSENIC' and as per the specifications mentioned therein)

30. Distillation Assembly – complete set with the following components (for cyanide and phenol determination)

- Flask, distillation, 1000 ml capacity with 24/29 interchangeable joint.
- Adapter with 24/29 interchangeable joint
- Condenser Graham, coiled type, with 24/29 interchangeable joint.
- App. length of jacket 300 mm
- App. overall height – 450 mm

MISCELLANEOUS ITEMS

Whatman Filter Paper - No.4
Whatman Filter Paper - No 42.
Qualitative Filter Paper , Grade-1
(in circles)

- dia 11.0 cm.
- dia 12.5 cm.

Porcelain Casserole.

Capacity 1000 ml , dia about 160 mm

China dish , porcelain

With pourout , capacity 80 ml , dia 80 mm.

China dish , Porcelain

With pourout , capacity 180 ml , dia 100mm.

Silica Crucible with Lid.

Burette Stand.

Steel , with burette holder.

Pipet Holder.

Wooden / HDPE (White)

Nessler Rack

Wooden / HDPE (white) to hold 10 Nos of 50 ml nessler tubes.

Wooden / HDPE (white) to hold 5 Nos of 100 ml nessler tubes.

Funnel Stand.

Wooden / HDPE (white)

Tripod stand.

Mortar and Pestle , Agate

Tongs for crucible.

Tongs for Beakers.

Tongs for flasks.

Spatula

stainless steel – different sizes.

Wire gauze with asbestos centre.

Pipe clay triangle.

Porcelain tile.

Pure white - 150 mm × 150 mm.

Polythene tubing.

Different dia.

Rubber Tubing.

Different dia.

Rubber corks of various sizes.

Cotton

Absorbent.

CHEMICALS

Unless otherwise specified, purchase only AR grade chemicals.

ACIDS

1. Sulfuric acid , Conc (sp gr. 1.84)
2. Hydrochloric acid Conc (sp gr. 1.18)
3. Nitric acid , Conc (sp gr. 1.42)
4. Acetic acid , Conc , glacial
5. Sulfuric acid (Fuming)
6. Phosphoric acid (Ortho) (H_3PO_4)

ALKALIS

1. Sodium hydroxide, pellets.
2. Potassium hydroxide, pellets.
3. Ammonia solution , Conc (sp. gr 0.88 or 0.91)
4. Sodium Carbonate

pH buffers

1. pH 4 buffer tablets
2. pH 7 buffer tablets
3. pH 9.2 buffer tablets

INDICATORS

1. Bromocresol green (high quality)
2. Methyl orange (high quality)
3. Methyl red (high quality)
4. Phenolphthalein (high quality)
5. Eriochrome black - T
6. Eriochrome cyanine - R
7. Patton- Reeders Indicator (2-hydroxy -1- (2-hydroxy – 4 sulfo naphthyl azo-3 - naphthoic acid))

8. Rhodamine- B
9. Rhodanine
10. Thymol blue
11. Starch (soluble)

OTHER CHEMICALS AND FINE CHEMICALS

1. Alizarin Red - S
2. Aluminium Metal AR
3. Aluminium potassium sulfate
4. 4- Amino antipyrine (also called 4-Amino Phenazone)
5. p-amino dimethylaniline oxalate (also called N,N dimethyl p-phenylene diamine)
6. Ammoinum chloride
7. Ammonium persulfate
8. Ammonium phosphate
9. Ammonium thiocyanate (or Potassium thio cyanate)
10. Ascorbic acid
11. Arsenic trioxide
12. Barium chloride
13. Bis - pyrazolone
14. Boric acid
15. Bromine
16. Cadmium metal
17. Calcium carbonate
18. Carbon tetra chloride
19. Chloroform
20. Copper foil (electrolytic)
21. Copper sulfate
22. Cuprous chloride (Cu_2Cl_2)
23. Curcumin
24. 3-3' – Diamino benzidine
25. Diammonium hydrogen phosphate
26. Dimethyl glyoxime
27. S - diphenyl carbazide
28. Disodium hydrogen phosphate

29. Dithizone
30. Ethylene Diamine Tetra Acetic acid – disodium salt (EDTA – disodium salt)
31. Ethyl alcohol
32. Ferrous ammonium sulfate
33. Ferric chloride
34. Formic acid
35. Glycerol
36. Hydrogen peroxide
37. Hydrazine sulfate
38. Hexa methylene tetramine (Hexamine)
39. Hydroxyl amine hydrochloride (Hydroxy ammonium chloride)
40. Iodine
41. Isopropyl alcohol(Propane - 2 - ol)
42. Lead acetate
43. Lead nitrate
44. Linear alkyl sulfonate(Sodium dioctyl sulfo succinate preferred)
45. Magnesium metal
46. Magnesium sulfate
47. Manganese metal (High purity 99.9 %)
48. Manganous sulfate
49. Mercuric chloride
50. Mercuric sulfate
51. Methylene blue
52. α - Naphthyl amine hydrochloride
53. Neocuproine(9 – dimethyl – 1 ,10 , phenanathroline hemihydrate)
54. Nickel sulfate
55. Orthotolidine hydrochloride
56. Oxalic acid
57. 1 - phenyl 3-methyl 5-pyrazolone
58. Phenol
59. Potassium bromate
60. Potassium bromide
61. Potassium carbonate (anhydrous)
62. Potassium chloride
63. Potassium chromate

64. Potassium cyanide
65. Potassium dichromate
66. Potassium ferri cyanide
67. Potassium iodide
68. Potassium nitrate
69. Potassium permanganate
70. Potassium persulfate
71. Pyridine
72. Silver diethyl dithiocarbamate
73. Sodium acetate
74. Sodium bicarbonate
75. Sodium chloride
76. Sodium citrate
77. Sodium hexameta phosphate
78. Sodium hydrogen phosphate (NaH_2PO_4)
79. Sodium hydrogen selenite
80. Sodium nitrite
81. Sodium selenite
82. Sodium sulfide (if AR grade not available , sodium sulfide flakes – fused may be purchased)
83. Sodium thio sulfate
84. SPADNS (Sodium 2 – (parasulfophenyl azo) -1, 8 hydroxy – 3,6 – naphthalene disulfonate)
85. Stannous chloride
86. Sulfanilic acid
87. Tartaric acid
88. Toluene
89. Triethanol amine
90. Zeocarb – 225 (ion exchange resin)
91. Zinc metal (AR) , 20 to 30 meth (arsenic free)
92. Zirconium oxychloride

CHAPTER - 4

WATER FOR ANALYTICAL LABORATORY USE

Water used for analytical laboratory purposes must be of high purity. Such high purity is essential as the purified water is used for the preparation of reagent solutions, for making dilution, to prepare standards and blanks. Whatever may be the method of production of pure water it is important that the water should not contain any detectable amount of the element or compound under investigation at the detection limit of analytical procedure. Water intended for inorganic analysis is usually prepared either by distillation* or demineralization (deionization). The recent methods for the preparation of pure water includes reverse osmosis, micro filtration, ultra filtration and electrodialysis. The requirements of pure water for organic analysis are more stringent. Hence a combination of the above methods are chosen to produce water of high purity that its trace contaminants are measured in the concentration range of parts per billion (ppb) or parts of per trillion (ppt).

Technical standards on water quality for analytical uses have been established by a number of professional organizations including the International Organization for Standardization (ISO). The American Chemical Society (ACS), ASTM International, the U.S. National Committee for Clinical Laboratory Standards (NCCLS which is now CLSI) and U.S.Pharmacopeia (USP). In this book, requirements specified by ISO are given. ISO has specified in its bulletia ISO 3696:1987(E), the quality requirements of water for laboratory use for the analysis of inorganic chemicals. These requirements are applicable to water for inorganic analysis only and not applicable to water for organic trace analysis, to water for the analysis of surface active agents or to water for biological or medical analysis. ISO further classifies the purified water into Grade – 1, Grade – 2 and Grade – 3, a brief account of which are given below.

All the grades of water shall be clear and colorless as assessed by visual inspection.

* Generally metal stills, or stills made of borosilicate glass or vitreous silica are employed for distillation. The product water obtained from metallic stills are of inferior quality when compared to that obtained from glass or silica stills.

Grade–1

Essentially free from dissolved or colloidal ionic and organic contaminants and suitable for the most stringent analytical requirements including those of high performance liquid chromatography; should be produced by further treatment of grade 2 water (for example reverse osmosis or deionization followed by filtration through a membrane filter of pore size 0.2 μm to remove particulate matter or redistillation from a fused silica apparatus).

Grade–2

Very low in inorganic, organic or colloidal contaminants and suitable for sensitive analytical purposes, including atomic absorption spectrometry (AAS) and the determination of constituents in trace quantities; should be produced, for example, by multiple distillation, or by deionization or reverse osmosis followed by distillation.

Grade–3

Suitable for most laboratory wet chemistry work and preparation of reagent solutions; should be produced, for example, by single distillation, by deionization, or by reverse osmosis. Unless otherwise specified, it should be used for ordinary analytical work.

NOTE : It is assumed that the initial feed stock water is potable and reasonably pure. If it is heavily contaminated in any respect, some pretreatment may be necessary.

Requirements

The material shall comply with the appropriate requirements of the Table 4 – 1. Testing for compliance shall be carried out by means of the methods specified in clause 7.

TABLE 4–1
REQUIREMENTS

Parameter	Grade 1	Grade 2	Grade 3
pH value at 25°C inclusive range	Not applicable (see note 1)	Not applicable (see note 1)	5.0 to 7.5
Electrical conductivity µs/cm at 25°C, max	0.1 (see note 1)	1.0 (see note 2)	5.0
Oxidizable matter Oxygen (O) content mg/L, max	Not applicable (see note 3)	0.08	0.4
Absorbance at 254 nm and 1 cm optical path length, absorbance units, max	0.001	0.01	Not specified
Residue after evaporation on heating at 110°C Mg/kg, max	Not applicable (see note 3)	1	2
Silica (SiO_2) content mg/l, max	0.01	0.02	Not specified

NOTES

1) Because of the difficulties associated with the measurement of the pH value of high-purity water, and the doubtful significance of the value obtained, limits for the pH of grade 1 and grade 2 water have not been specified.

2) The values of conductivity for grade 1 and grade 2 water apply to freshly prepared water; during storage, it is possible for contaminants such as atmospheric carbon dioxide and alkalis from glass containers to be dissolved, leading to changes in conductivity.

3) A limit for oxidizable matter and residue after evaporation for grade 1 water is not specified because of the difficulty of testing for compliance at this level of purity. The quality of grade 1 water is, however, assured by compliance with the other requirements and by its method of preparation.

CHAPTER – 5

SAMPLING OF WATER FOR CHEMICAL ANALYSIS

(For Sampling Boiler Water - See Chapter - 6)

This chapter describes the method for collection of water samples for physical and chemical examination only. Sampling procedures for microbiological examination are dealt elaborately in Part – III, Chapter – 81.

Sampling is the first and foremost step in the analysis of water. Proper sampling to obtain a representative sample is extremely important. Before sampling, a program if drawn on the following lines would be immensely useful for effective sampling.

1) Sampling points have to be chosen carefully.
2) Number of samples to be collected should be decided first.
3) A forethought on the type of sample- whether it is a grab sample or composite - is necessary
4) Additional samples, along with principal samples for parameters such as dissolved oxygen, oils & grease etc., have to be collected in separate containers, in addition to the main samples. In the case of high purity waters, separate samples for iron, copper, sodium and silica are also to be collected. Such sampling points need to be identified first.
5) The type of container (glass, polyethylene or PTFE) depending on the nature of the sample have to be chosen.
6) Preservation methods, and if chemical preservatives are added, the right type of preservatives have to be decided.
7) Similarly, if any special type of sampling apparatus is required, it should also be decided first.

Due care and attention is necessary to obtain a representative sample. The following points should be strictly adhered to.

1) The samples must represent the large or small mass sampled and the conditions existing at the point taken and at the time of sampling as far as possible. Even the most careful analysis becomes useless unless the sample is representative.

2) The sampling point should be chosen carefully so that a representative sample of the water to be tested is obtained. The sampling method, location, and time of sampling must be coordinated so that the results obtained would serve the purpose more correctly , for which the sample is collected. Raw waters are not fixed in their composition. They show appreciable changes depending upon their sources, seasonal variations, temperature, pollution load, etc., Process waters are of uniform quality. These factors should be taken as guidelines in determining the frequency of sampling, appropriate sampling points and sampling devices.

3) The sample must be of sufficient volume for subsequent laboratory examination.

4) It must be taken periodically at specified intervals to monitor its quality to the required objective.

5) If analysis is not carried out immediately it has to be preserved properly.

GRAB SAMPLE AND COMPOSITE SAMPLE

Liquid samples are broadly classified into two major groups, namely, the one called instantaneous, spot, snap or grab sample and the other termed as composite sample. A grab sample is a manually collected single portion of sample of water (or waste water). When the grab samples of a particular water (or waste water) is collected at regular intervals (1hr, 2hr, 3hr etc.,) for a specified period say 12 or 24 hrs, and mixed, then the integrated sample is known as 'Composite sample'.

Grab samples are useful for spot checking and in investigating deviations from normality. A series of grab samples have to be collected and examined separately for tracing out constituent variations. (The volume of individual samples depend on the parameters to be tested).

Composite samples are normally required for waste waters only, as batch operations are carried out in an industry. Composite samples correlated with flow, show the average conditions existing. The most serious error that arise in composite sample is the suspended solids concentration. To avoid such errors, the grab samples have to be mixed thoroughly while transferring from grab sample container to composite sample container.

Separate individual samples have to be collected for the determination of unstable constituents (ammonia, carbon dioxide, chloride, oxygen, hydrogen sulfide etc.,)

CONTINUOUS SAMPLING

The grab and composite samples described above are collected manually. If a continuous record of conditions at a given sampling point is necessary, installation of automatic continuous sampling equipment is advantageous. They provide more frequent samples and eliminate any human error in sampling.

QUANTITY OF SAMPLE

Generally 2.5 litres of sample is sufficient for most of the physical and chemical examination. For certain special determinations, however larger volumes of sample are necessary.

A sample volume of 5 litres is preferable when the concentration of the constituents are much less. Where microgram concentrations of heavy metals are determined, a volume up to 20 litres is necessary. The number of tests and the quantity required for each test determines the volume of the sample to be taken. It is always desirable to have some of the sample left over for checking the analysis or sometimes for further analysis at a later day.

Note: The same sample should not be used for chemical, microbiological and biological examinations since the methods of collection and handling are quite different.

SAMPLE CONTAINER

For most of the purposes, the ordinary stoppered Winchester Quart bottle of 2.5 litres capacity is enough. Generally glass containers are preferable to those made of polythene or other plastic materials. However they do have the disadvantage of being very fragile. Polythene containers are least expensive and are preferable in some circumstances. (e.g) when the sample being examined extracts substances (such as boron, silica) from the glass or some of the constituents of the sample adhere to the glass. Quartz containers and PTFE (Poly tetra fluoro ethylene - Teflon) containers are the best, but are expensive.

Note-1: Soda-glass containers should not be used for collection of samples intended for heavy metal determination.

Note-2: For the determination of Silver, store the sample in amber bottle.

When composite samples are to be collected, it is better to use wide mouthed bottles of capacity 300 to 500 ml for the sub-samples. They are particularly useful in transferring the contents without leaving any settleable matter.

New glass bottles should be treated either with dil.hydrochloric acid (1+1) or dil.nitric acid* (1+1) and then repeatedly washed with distilled water. Subsequently they may be cleaned using detergents.

Note: If chromium or sulfate are to be determined, do not use acids for cleaning. If trace amounts of phosphates are to be determined, do not use detergents for cleaning.

POINTS OF SAMPLING & FREQUENCY OF SAMPLING

Water Bodies

Great care and discretion are necessary in the selection of the sampling point so as to obtain a sample that represents as fairly as possible the supply under examination.

Because of a wide variety of conditions found in the streams, lakes, reservoirs, and other bodies of water it is not possible to prescribe the exact point of sampling. The choice of location, depth and frequency of sampling should be based on local conditions and upon the purpose of the investigation.

* Nitric acid is the most preferred acid when tested for metals especially for lead.

A sample taken at any point in the cross-section is satisfactory in the case of stream where water is mixed to uniformity. This uniform mixing cannot be expected in large rivers and streams and three or more samples are taken at the mid point of equal cross-sectional areas of such rivers and streams. These samples are then combined together to obtain a composite sample. If only a grab sample is taken, it is better to collect in the middle of the stream and at mid depth.

When a tributary or an industrial waste or sewage is joining the river, it is best taken in the downstream sufficiently away to allow through mixing. In general, a distance of 1 to 4 Km below the tributary is sufficient.

Collect sample about 0.5 to 1 Km below dams or waterfalls such that the sample would be free from the entrained air. It is desirable to take a series of samples from any source of water to determine whether differences in composition are likely to exist before final selection of the sampling point.

Water Treatment Plants

All types of water treatment plants generally require sampling points for raw water and treated water.

The quality of the raw water and treated water is the governing factor in deciding the number of samples, frequency of sampling and the physical and chemical tests to be carried out.

If the raw water is subject to wide variations in quality, the frequency of sampling and tests have to be increased accordingly. Sampling and testing of the treated water should be frequent enough to show the incorrect treatment, excess / lower dosage of treatment chemicals and variations in raw water quality.

The sampling points in filters, water softeners, demineralizers, reverse osmosis plants, evaporators, surface condensers, condensate return lines, tanks, vats, pipelines and conduits have to be fixed with respect to the individual equipment containing the water to be tested, the character and changes occurring in the inlet and outlet water and the rate of passage through the equipment.

The variability in concentration of the constituents is used as the basis for determining the frequency of sampling. For a water supply of fairly constant composition, samples drawn at 8-hours intervals are usually satisfactory. However with certain supplies, whose composition changes rapidly, or a process that is more sensitive it may be necessary to collect samples at less than 1 hr intervals.

Ion exchange softeners and demineralizers and reverse osmosis systems may require several sampling points depending on the stages of treatment for satisfactory control over the treatment and to terminate the operation before treated water quality deteriorates significantly. The present day treatments are mostly fully automatic or semi automatic and the system itself contains instruments to record pH, temperature, turbidity, conductivity, hardness, residual chlorine, dissolved oxygen and silica. Despite such instrumental reading which help to monitor the quality continuously, manual sampling at the end of operating run and again following regeneration are necessary.

TAKING THE SAMPLE

When the liquid is homogeneous, it is suffice to take the grab sample only. Liquids flowing along a pipe or channel or over a weir can be taken by means of a bucket, a wide-mouthed vessel or if possible directly in to the sample bottle. Needless to say that the bottle should first be rinsed with the liquid being sampled and then filled.

Note: Samples collected in such a manner are not suitable for the determination of dissolved gases such as dissolved oxygen and carbon dioxide. As indicated earlier, additional separate samples have to be collected for the determination of such dissolved gases and substances that are affected by air contact. For details of collection of the samples for such determination, refer respective chapters.

When water is sampled from cocks or valves, insert the sample line or a thoroughly cleaned extension (usually glass or sulfur free rubber tube) of the sample line in to the sample bottle touching the bottom. Allow the water to flow into and overflow from the container before the sample is taken. If any constituent* to be determined is affected by air contact, take the sample out of contact with air and completely fill the container. When samples are to be taken

* Dissolved gases such as oxygen, free carbon dioxide, ammonia, free chlorine, and pH, iron, acidity / alkalinity etc.,

at depths from rivers, lakes or other water bodies, depth integrating samplers or point samplers have to be used. (Refer American Public Health Association, 'Standard Methods for the Examination of Water and Waste Water', 20th edition 1998 and American Society for Testing and Materials, 'Annual Book of ASTM Standards', Part – 23, 1972)

When hot samples are taken they must be cooled using cooling coils. (Refer Chapter – 4, "SAMPLING OF BOILER WATER" for more details)

When the liquid to be sampled is heterogeneous such as sewage or industrial waste, composite sample is necessary.

LABELLING OF SAMPLES

Immediately after sampling, the containers have to be labelled with the following information.

- (i) Sample Number
- (ii) Date and Time of sampling
- (iii) Source of sample
- (iv) Exact point of sampling (in detail) and
- (v) Name and designation of the sample collector.

INFORMATION TO BE SUBMITTED ALONG WITH THE SAMPLES

During sample collection, various points have to be observed and recorded separately. They should be submitted along with the sample.

If the sample is from a river or stream, the following points should be recorded.

- (i) Exact location and name of the place
- (ii) Depth below surface
- (iii) Rate of flow
- (iv) Whether the water level is above or below average
- (v) Recent rainfall or flood conditions
- (vi) Possible sources of pollution
- (vii) Temperature of water
- (viii) Results of field tests.

If the sample is taken from a well, the following points need to be recorded.

(i) Depth of water in the well
(ii) Depth at which the sample is taken
(iii) Yield of water
(iv) Any indication of pollution such as discoloration etc.,
(v) Nature and depth of subsoil, impervious layer and water bearing stratum.
(vi) Nature of surroundings and proximity to drains, sewers, stables or other possible sources of pollution.
(vii) Mode of construction of well.

TIME INTERVAL BETWEEN COLLECTION AND ANALYSIS

In general, the shorter the time between collection and analysis, the more accurate will be the results. No specific time interval between collection and analysis can be generalized; this depends upon the nature of the sample, and constituents to be determined. It is advisible to keep the sample out of light (to retard the growth of algae and other organisms) and at low temperature (4° C), if analysis gets delayed.

Make the determination of pH and of dissolved gases, such as oxygen, carbon dioxide, chlorine at the field itself.

PRESERVATION OF SAMPLES

No single method of preservation is entirely satisfactory, and the preservative should be so chosen with due regard to the determinations that are to be made. It should be noted that the preservative added for one constituent should not affect the determination of other constituent. As described above, keeping the sample in the dark at low temperature is the best possible preservation. However some of the specific preservation methods are described below. See Table 7.1.

SAMPLING INDUSTRIAL EFFLUENTS

The sampling and the analysis of industrial effluents require greater care and attention. Obtaining representative sample only is not the criterion and it is necessary to study during sample collection, the waste load, major sources of wastes within a plant, character of wastes, potential recovery of valuable materials at profit and the effect of discharged wastes on the receiving body of water.

Wastes discharged by industries are of great variety by nature. They are subject to rapid changes within a few minutes, owing to breakdowns, spillovers, floor washing and numerous other causes. In sampling industrial wastes, suitable sampling points as well as the frequency and type of sample to be collected must be decided first. Generally, sampling points are located where flow conditions encourage a homogeneous mixture.

To obtain a representative sample, it is necessary to collect individual samples at uniform intervals, say 10 or 15 minutes. The degree of variation in the rate of flow will determine the time interval for sampling. Care should be taken in collecting a sample containing oil. It may be helpful to collect a larger volume of such samples and withdrawing a suitable proportion using a narrow diameter sectional sampling tube. Collection of additional separate sample exclusively for the determination of oils may prove most useful.

Table 5–1

RECOMMENDED METHODS OF PRESERVATION OF SAMPLES

Determination	Recommended container	Required volume (ml)	Method of preservation	Holding time
I. Physical Properties				
1. Color	P/G	100	Cool to 4°C	48 hrs
2. Turbidity	P/G	100	Cool to 4°C. Store in dark	24 hrs
3. Odor	–	Sniff the odor in the field itself	–	–
4. Temperature	–	–	Measure immediately in the field itself	–
5. pH value	P/G (B)	50	Measure immediately in the filed itself	–
6. Electrical Conductivity	P/G	100	Cool to 4°C	48 hrs
7. Solids				
a. Total	P/G	100	Cool to 4°C	7 days
b. Dissolved	P/G	100	Cool to 4°C	7 days
c. Suspended	P/G	100	Cool to 4°C	7 days
d. Volatile	P/G	100	Cool to 4°C	7 days
e. Settleable	P/G	1000	Cool to 4°C	48 hrs.

Determination	Recommended container	Required volume (ml)	Method of preservation	Holding time
II. Inorganic (A) Non metallics				
8. Acidity	P/G (B)	100	Cool to 4°C	2 weeks
9. Alkalinity	P/G (B)	100	Cool to 4°C	2 weeks
10. Bromide	P/G	100	None required	1 month
11. Chloride	P/G	100	None required	1 month
12. Chloride	P/G	200	Determine in the field itself.	–
13. Cyanide	P/G	500	Add NaOH to pH > 12 Cool to 4°C (if any residual chlorine is present add ascorbic acid also)	2 weeks
14. Fluoride	P only	300	None required	1 month
15. Iodide	P/G	100	Cool to 4°C	24 hrs.
16. Nitrogen				
a. Ammonia	P/G	500	Add 0.8ml conc. H_2SO_4 per litre and refrigerate.	1 month
b. Nitrate	P/G	50	Cool to 4°C	2 days
c. Nitrite	P/G	50	Add 40 mg. $HgCl_2$/l and refrigerate.	2 days
d. Nitrate and Nitrite	P/G	100	Add H_2SO_4 to pH < 2 Cool to 4°C	1 month
e. Total (Kjeldahl)	P/G	500	Add H_2SO_4 to pH < 2 Cool to 4°C	1 month
17. Oxygen (Dissolved)	BOD Bottle (glass)	300	Determine immediately or fix DO (See Chapter – 50 Oxygen Dissolved)	–
18. Phosphate	(rinse the container with 1+1 HNO_3)			
a. Ortho Phosphate (dissolved)	P/G	50	Filter in the field itself. (See note 3) Cool to 4°C	2 days
b. Hydrolyzable	P/G	50	Add H_2SO_4 to pH < 2 Cool to 4°C	1 month
c. Total	P/G	50	Add H_2SO_4 to pH < 2 Cool to 4°C	1 month
d. Total and dissolved	P/G	50	Filter in the field itself Add H_2SO_4 to pH < 2 Cool to 4°C	24 hrs.
19. Silica	P only	50	Cool to 4°C	1 month.
20. Sulfate	P/G	50	Cool to 4°C	1 month.
21. Sulfide	P/G	500	Determine in the field or add 2ml 2 N zince acetate / 1 plus NaOH to pH > 9.	1 Week
22. Sulfite	P/G	50	Determine in the filed itself	–

Determination	Recommended container	Required volume (ml)	Method of preservation	Holding time
II. Inorganic (B) Metals				
23. All metals except chromium and mercury				
a. Dissolved	P/G	200 to 500	Filter in the field (See note 3) Add HNO_3 to pH < 2	6 months
b. Total	P/G	200 to 500	Add HNO_3 to pH < 2	6 months
24. Chromium (VI)	P/G	200	Cool to 4°C	24 hrs.
25. Mercury				
a. Dissolved	P/G	500	Filter in the field. Add HNO_3 to pH < 2	1 month
b. Total	P/G	500	Add HNO_3 to pH < 2	1 month
III. Organics				
26. BOD	P/G	300 to 1000	Add H_2SO_4 to pH < 2 Cool to 4°C	24 hrs.
27. COD	P/G	50 to 200	Cool to 4°C	1 month
28. Total organic carbon	P/G	50	Add H_2SO_4 or HCl to pH < 2 Cool to 4°C	1 month
29. Phenolics	G only	500	Add H3PO4 to pH < 4.0 and 1g. Cool to 4°C	1 month
30. Oils & grease	G only (wide mouth)	1000	Add H_2SO_4 to pH < 2 Cool to 4°C	1 month
31. Surfactants	P/G	250	Cool to 4°C	2 days

Notes:

1) G-Glass, P-Polythene or Plastic (The bottle cap should not contain any liner) G(B) – Glass, Borosilicate

2) Preserve the sample immediately after collection. During composite sample collection, the sub samples should be kept at 4°C until compositing and sample splitting.

3) If dissolved constituents are required, filter the samples through a 0.45 micron membrane filter and then add preservatives to the filtrate.

4) Samples should be analysed as soon as possible after collection. The holding time given is the maximum time limit upto which the sample may be hold for analysis and still considered valid.

ORDER OF TESTING

The sample after collection has to be analyzed immediately. A program of analysis has to be established as some of the characteristics are subject to change within minutes/hrs. Other parameters not affected by air contact or by other means can be analyzed afterwards.

Preservation and holding time for each parameter is given previously, which can also be taken as a rough guide in establishing the sequence of analysis. A systematic scheme of analysis is given below.

1) Immediately after the collection, pH, free carbon dioxide, dissolved oxygen, residual chlorine, hydrogen sulfide, sulfite, nitrite, ammonia and alkalinity have to be determined. The physical parameters such as color, turbidity, and odor have also to be carried out simultaneously.

2) Secondly, temperature should be measured during collection itself. Suspended solids, settleable solids, iron, manganese and hardness need to be determined.

3) In the third phase, analysis for sulfate, chloride, nitrate, phosphate and silica is to be made.

4) In the fourth category comes the parameters such as metals, oils and grease, surfactants, tannin & lignin etc.,

CHAPTER - 6

SAMPLING OF BOILER WATER

Sampling and testing of boiler water at frequent intervals is essential to protect the boiler from scales and corrosion. Before going in to the details of sampling, the term 'boiler water' has to be understood fully. 'Boiler water' is the circulating water inside the boiler, after the generated steam has been separated, and before the incoming feed water / chemicals mix with it.

POINT OF SAMPLING

The exact sampling point depends upon the boiler design, location of feed lines (water & chemicals) and other local conditions.

Usually special sampling nozzles are provided in a boiler for the collection of boiler water. In the absence of such nozzles, sample may be collected from the continuous blow down line at the boiler. (In the absence of sampling connections, samples may be drawn from the water column, drain lines or other convenient outlet below the water level. However, samples collected from these points are not truly representative.) In all cases water has to be collected at normal operating level.

As significant variations in the composition of water exist throughout the boiler, simultaneous sampling at several points is desirable. However care should be exercised in avoiding the possibility of steam inclusion, incoming feed water and added chemicals.

APPARATUS

It is suggested by many authors that condensate return system is a revealing sampling point to monitor total boiler water treatment performance.

For sampling boiler water, cooling coils are necessary to bring down the sample temperature to below 35°C. Three types of cooling coils are shown in Figures 8.1, 8.2 and 8.3. These coolers should be made of materials that are resistant to corrosion both by sample on one side and cooling water on the other side.

Where good quality cooling water is available, any of the three types shown is acceptable. If a water containing algae or slime is used, they may cause rapid fouling so that the efficiency of cooler is impaired. The Double – Tube Helical Coil heat exchanger shown in Fig 8.1 is susceptible to such plugging and scaling. In the case of other coolers (as shown in Fig 8.2 and Fig 8.3), the scale may be dissolved while the coil is in service by scale cracking or by thermal shocking procedure. The scale will fall to the bottom and be removed mechanically or by chemical cleaning.

SAMPLE CONTAINER

Chemically resistant glass or polythene containers are suitable. For samples containing small amounts of hardness, silica and sodium, polythene containers should only be used. The containers have to be cleaned thoroughly before sampling.

TAKING THE SAMPLE

The first step in collecting boiler water sample is the cleaning of sample lines by purging, such that the stagnant water and sediment if any will be removed. Purging is carried out by a cooled regulated flow of water for sufficient time[†]. Care should be exercised to avoid dilution of the water by condensed steam or concentration due to "flashing".

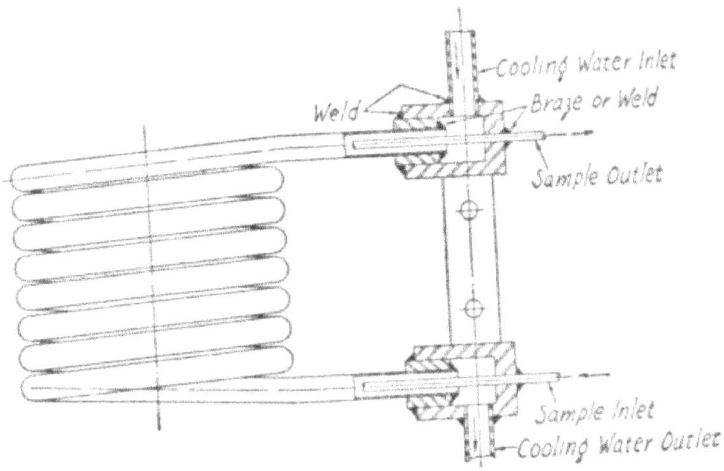

Fig. 6 - 1. Double - Tube Helical Coil Heat exchanger*.

[†] The purging time depends on the dia and length of pipeline varying from 5 seconds to 60 seconds per foot of pipeline.

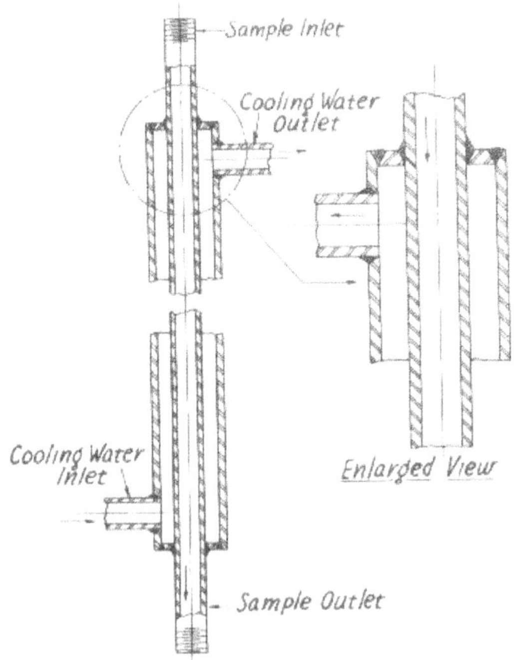

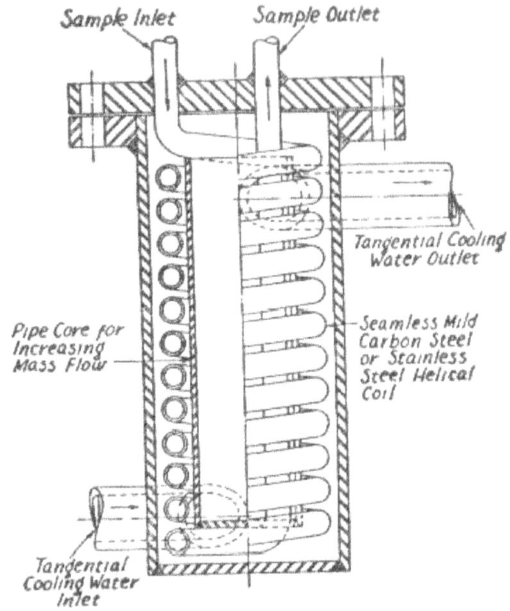

Fig. 6 - 2. Vertical-Tube Heat exchanger* **Fig. 6 - 3. Submerged Helical Coil Heat exchanger***

The second important criterion is flow of water. Before taking the samples, a continuous flow of not less than 500 ml per minute has to be established.

The sample has to be passed through cooling coils‡ and cooled down to less than 35°C and drawn in the specified container. Before collecting the sample, the container has to be rinsed atleast three times with the boiler water to be sampled.

While collecting the sample, the tube has to be extended to the bottom of the container. Allow the container to overflow atleast five times the volumes of the container and stop it with a stopper which is already washed with the boiler water sample.

---

* Figures 6 - 1, 6 - 2 and 6 - 3 are reproduced with the kind permission of American Society for Testing and Materials from their publication 'ANNUAL BOOK OF ASTM STANDARDS' Part – 23, 1972.

‡ If chemical equilibrium in the water at the boiler operating conditions is required, the water has to be filtered at the operating temperature and pressure of the boiler prior to passing through cooling coils, for which a pressure- type filter has to be installed between the boiler and cooling coil to permit the hot samples to be filtered as they leave the boiler.

Important note: Since boiler water concentration vary at different water levels and at different ratings, all the samples have to be collected only under normal operating level as indicated already.

LABELLING

Immediately upon taking samples, they have to be labelled clearly indicating the following details.

1) Sample Number
2) Date and Time of sampling
3) Point of sampling : (Should be given in detail)
4) Temperature and rate of flow in the equipment
5) Temperature of the sample
6) Results of Field Tests
7) Other relevant information

Immediately after collection, the samples have to be packed neatly and sent to laboratory for analysis.

For more details of sampling and sampling equipments for boiler water, readers are directed to refer, American Society for Testing and Materials, 'ANNUAL BOOK OF ASTM STANDARDS', Part – 23, 1972.

CHAPTER - 7

CHEMICAL ANALYSIS - GENERAL NOTES

The following points apply generally and they are not mentioned normally in individual determination.

Reagents

The quality of all reagents should be of "Analytical Grade "(AR Grade), unless otherwise stated.

Expression of concentration of Acid solutions

In addition to the terms, Normality (N), Molarity (M) the concentrations are sometimes expressed in additive volumes. In additive volume (a + b), the first number 'a' indicates the volume of the concentrated reagent and the second number 'b' denotes the volume of distilled water required for dilution. Thus 1 + 3 H_2SO_4 indicates that 1 volume of conc. H_2SO_4 has to be diluted with 3 volumes of distilled water.

Water for Preparation of solutions

The water used for preparing solutions should meet the requirements specified in the next Chapter - 4 "WATER FOR ANALYTICAL LABORATORY USE".

Significant Figures

The results should not be reported to a great number of figures than is justified by the accuracy of the method. It is sufficient to give three significant figures as maximum.

Unless specifically stated otherwise, sample should be shaken well before taking aliquots for individual tests. If necessary, the samples may first be filtered.

CHAPTER - 8

EXPRESSION OF RESULTS

As regards water and waste water analysis, the universal expression of analytical results is milligrams per litre (mg/l).

The term parts per million (ppm) – is a weight-to-weight ratio and is equal to mg/l when the density of water is exactly 1. However they are not equal when density differs. For example, seawater - the density of which exceeds normal waters. The density of industrial effluents also vary from normal water density. In these cases the expressions mg/l and ppm are not equal. It is better to express in terms of mg/l in such cases.

If the concentrations are less than 1 mg/l, it is convenient to express the result in terms of micro grams per litre (μg/l). This is also termed as parts per billion* (ppb).

The term 'percent' is preferred when the concentration is greater than 10000 mg/l, 1% being equal to 10000 mg/l.

Sometime results are also expressed in terms of milli equivalents per litre (meq/l). This is also termed as equivalents per million (epm). It is a convenient way of expression and facilitates any chemical calculations needed to determine the dosage of various chemicals used in water treatment. It is obtained by dividing the mg/l value by the equivalent weight of the element, ion or radical. For example, the calcium content of a water (500 mg/l calcium), may be expressed in terms of meq/l by dividing by 20, which is equivalent weight of calcium.

$$meq/l = \frac{mg/l}{Equivalent.wt} = \frac{500}{20} = 25$$

The factors for conversion of mg/l to meq/l and from meq/l to mg/l are given in Appendix (See Table - XIV in Appendix).

---

* Billion = 10^9

The unit meq/l is also useful to check the correctness of analysis. The sum of cations should be equal to the sum of anions, if the analysis is correct. See the following example.

***Example 1:** On analysis, a water is found to contain the following constituents in the concentrations specified. Find out whether the analysis is correct?*

Calcium (meq/l)	*= 1.5*	*Bicarbonate (meq/l)*	*= 1.2*
Magnesium (meq/l)	*= 0.4*	*Sulfate (meq/l)*	*= 0.4*
Sodium (meq/l)	*= 0.2*	*Chloride (meq/l)*	*= 0.4*
		Nitrate (meq/l)	*= 0.1*

Sum of the cations = 1.5 + 0.4 + 0.2 = 2.1
Sum of the anions = 1.2 + 0.4 + 0.4 + 0.1 = 2.1

The sum of the cations (2.1) equals the sum of the anions (2.1). Therefore the analysis is correct.

There are also some other units, namely, Grains per Gallon (US) – gpg (US) and Grains per Gallon (Imperial) – gpg (Imperial). For the conversion of mg/l value into gpg (US) and gpg (Imperial) and vice versa, see Table – XII in Appendix.

In certain determination such as color and turbidity, they are expressed as follows:

Color is expressed as the hue observed. Quantitatively it is expressed in terms of 'Hazen Units'.

Turbidity is expressed in 'Nephelometric Turbidity Unit' (NTU) or in the equivalent term 'Formazin Turbidity Unit' (FTU).

pH is expressed as a mere number and it has no unit for expression.

The results of analysis may conveniently be recorded and reported as in the following format.

RESULTS OF CHEMICAL ANALYSIS OF WATER

Date of collation: Collected by:	Sampling point:	Lab. No.

Physical Examination	Color-Hazen units ....
	Turbidity in NTU ....
	Odor ....
	Electrical conductivity Micro mhos/cm at 20°C ....
	Total Dissolved Solids (at 105°C), mg/l ....
	Total Suspended Solids (at 105°C), mg/l ....
	pH value ....
Chemical Examination	Total Hardness (as $CaCO_3$), mg/l ....
	Calcium Hardness (as $CaCO_3$), mg/l ....
	Magnesium Hardness (as $CaCO_3$), mg/l ....
	Sodium (as Na), mg/l ....
	Iron (as Fe), mg/l ....
	Manganese (as Mn), mg/l ....
Chemical Examination	Acidity (as $CaCO_3$), mg/l
	Methyl orange acidity ....
	Total acidity ....
	Alkalinity (as $CaCO_3$), mg/l
	Phenolphthalein alkalinity ....
	Methyl orange alkalinity ....
	Sulfate (as SO_4^{2-}), mg/l ....
	Sulfite (as SO_3^{2-}) mg/l ....
	Sulfide (as S^{2-}), mg/l ....
	Chloride (as Cl^-), mg/l ....
	Nitrate (as N), mg/l ....
	Nitrite (as N), mg/l ....
	Phosphate (as P), mg/l ....
Dissolved Gases	Dissolved Oxygen, (as O_2) mg/l ....
	Carbon dioxide, (as CO_2) mg/l ....
	Chlorine, (as Cl_2) mg/l; ....
Miscellaneous Determination	Hydrazine (as N_2H_4), mg/l ....
	Oils & Greases, mg/l ....
	Dissolved Silica (as SiO_2), mg/l ....
	Surfactants, mg/l ....
	Results of other Tests ....

CHAPTER - 9

PRELIMINARY ACID DIGESTION OF THE SAMPLE

(FOR HEAVY METALS DETERMINATION)

Since the metals have the tendency to form complexes readily with the organic constituents, it is necessary to destroy them by digestion with strong acids. This preliminary acid digestion not only destroys organic matter but also bring all metallic compounds in suspension to solution. Interferences due to nitrite, cyanide, sulfide, sulfite, thiosulfate and thiocyanate are also eliminated during digestion. The procedures for digesting the sample with nitric and sulfuric acid, and nitric and perchloric acids are given here. Certain metals, however, require special treatments which are described in the appropriate texts.

(i) Digestion with Nitric and Sulfuric acids

Place a suitable volume of the well mixed sample in an evaporating dish (porcelain or silica). Acidify to methyl orange with conc.H_2SO_4 and add 5 ml conc. HNO_3 and 2 ml 30% hydrogen peroxide to reduce chromate. Evaporate[1] by placing over a water bath or hot plate until white dense fumes characteristic of SO_3 just appear in the flask. (If the solution is still not clear, add a further 10 ml conc. HNO_3 and evaporate to fumes of SO_3) cool to room temperature, add 50 ml distilled water, boil to dissolve the solids if any present and filter through a sintered-glass crucible. Transfer the filtrate to a 100 ml volumetric flask and make upto the mark with distilled water. The resulting solution is 3N in H_2SO_4. Use aliquots of this made up solution for the determination of metals.

Note : If lead is present in the sample add 50 ml ammonium acetate solution (200 g/300 ml water) to the conical flask in which digestion was carried out and heat to incipient boiling. Pour through the crucible in which the digestate was filtered. Lead present as $PbSO_4$ remaining on the filter will be dissolved and carried over into solution. Transfer the filtrate to a 100 ml volumetric flask and make upto the mark with distilled water. Determine lead from this solution also.

[1] If necessary cover the evaporating dish with watch glass to avoid loss of material by spattering.

(ii) Digestion with Nitric and Perchloric acids

If it is difficult to oxidize the organic matter by digestion with nitric and sulfuric acids, digestion with nitric and perchloric acids may be employed.

CAUTION

Digestion with nitric and perchloric acids should be done carefully as the mixture of organic matter and perchloric acid may explode violently. Observe the following precautions.

1. Before the addition of perchloric acid ensure that the sample is pretreated with nitric acid.
2. Do not add perchloric acid to hot solutions.
3. Fuming should be done only under hoods of stone or asbestos cement.

Measure a suitable volume of the well mixed sample in an evaporating dish and acidify to methyl orange with conc. HNO_3. Add a further 5 ml of conc. HNO_3 and evaporate to about 10 ml by placing over a water bath or hot plate. Transfer it to a concial flask of 125 ml capacity and cool. Add 5 ml of conc. HNO_3 and 10 ml of perchloric acid $HClO_4$ (70%). Add a few glass beads or boiling chips and heat gently by placing on a hot plate until white dense fumes of $HClO_4$ just appear. (If even at this stage, the solution is not clear, add a further 10 ml of conc. HNO_3. Place a watch glass over the mouth of the conical flask and keep the solution just barely boiling until it clears).

Allow to cool and dilute to 50 ml with redistilled water. Boil to expel oxides of nitrogen and chlorine present if any. Filter through a sintered glass crucible and make up the filtrate to 100 ml in a volumetric flask. The solution now obtained is 0.8 in $HClO_4$. Use aliquots of the made up solution for the determination of heavy metals.

If lead is to be determined, the residue should be treated with ammonium acetate solution as given in Digestion with nitric and sulfuric acids.

CHAPTER - 10

ACIDITY

Acidity* of a water is the capacity of the water to neutralize hydroxyl ions and is expressed in terms of calcium carbonate.

Acidity is usually caused by the presence of free carbon dioxide, and weak organic acids such as acetic and tannic acids and strong acids such as sulfuric and hydrochloric acids. Iron and aluminium salts may hydrolyze in water to release mineral acidity. Surface waters and ground waters attain acidity from humic acids, from industrial wastes such as pickling liquors, and from acid mine drainage. However, the most common source of acidity in unpolluted waters is carbon dioxide only which is present in the form of carbonic acid.

Acidity is usually determined by titration with 0.02N sodium hydroxide solution using methyl orange and phenolphthalein indicators. The amount of sodium hydroxide required for the sample (having a pH below 4.5) to reach pH 4.5 (methyl orange end point) is a measure of "mineral acidity" and the amount to reach pH 8.3 (phenolphthalein end point) is a measure of the total acidity. (See figure 10.1).

Generally the pH of most of the samples is more than 7 and consequently CO_2 acidity is the only acidity present in them. Samples contaminated with acidic wastes only will have a pH below 4.5 and contain both mineral acidity and CO_2 acidity.

Color change titrimetric method or potentiometric titration may be adopted for the determination of acidity. Titrimetric method is described below.

* Acidity should not be confused with pH value for one water of pH 6.0 may have a lower acidity than a buffered water of pH 7 which may have a higher acidity.

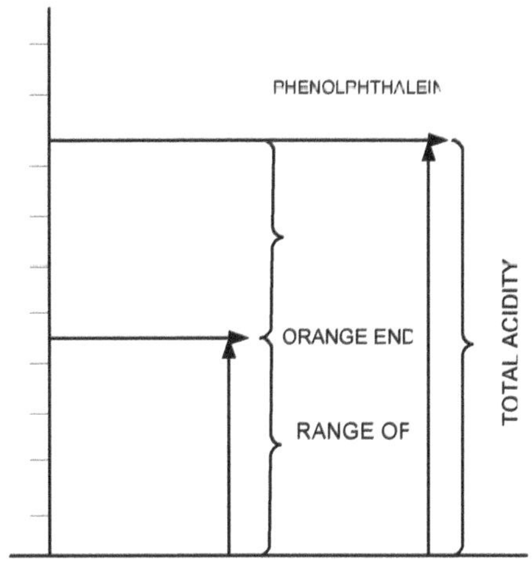

Fig 10.1 Types of Acidity and pH ranges in which they are significant

TITRIMETRIC METHOD

Interferences

Residual chlorine interferes by bleaching the color of the indicator. It can be removed by the addition of sodium thiosulfate solution. A fading and impermanent end point characterizes the phenolphthalein acidity.

Reagents

1. Methyl orange indicator solution: Dissolve 50 mg methyl orange powder in distilled water and dilute to 100 ml.

2. Phenolphthalein indicator solution: See Chaper – 11 "ALKALINITY".

3. Sodium hydroxide solution 0.02 N: Prepare 1N sodium hydroxide solution by dissolving 40g. NaOH* in distilled water and diluting to 1000ml. Standardize it against standard 1N.sulfuric acid (for preparation, see Chapter - 11 "ALKALINITY"). Dilute appropriate volume of 1N.sodium hydroxide solution (about 20 ml.) to 1000 ml. to give a 0.02N solution.

1.0 ml. of exactly 0.02N.NaOH solution ≡ 1.0 mg $CaCO_3$.

* Sodium hydroxide used must be of high purity and should not contain even traces of sodium carbonate.

Procedure

Methyl orange acidity(Mineral Acidity): Place 50 ml or suitable aliquot of the sample* (requiring not more than 25 ml of titrant) in a 250 ml conical flask. Add 2 drops of methyl orange indicator solution. The color will be now red. Titrate with 0.02 N.NaOH solution (to pH 4.5) to faint orange color.

Total acidity (Mineral Acidity + CO_2 Acidity): To a suitable aliquot of the sample in a 250 ml. conical flask add 3 drops phenolphthalein indicator. Titrate with 0.02N.NaOH solution to the appearance of faint pink color (to pH 8.3).

Note : If the sample is colored, titrations may conveniently be performed, using pH meter to the required pH values.

Calculation

If the normality of sodium hydroxide solution is exactly 0.02N,

$$Acidity\ (as\ CaCO_3),\ mg/l = \frac{ml.\ titrant\ (NaOH) \times 1 \times 1000}{ml.\ sample\ taken\ for\ titration}$$

While reporting results, state the indicator used and the temperature at which the titration has been performed.

As discussed earlier, the methyl orange acidity value indicates the mineral acidity, and the total acidity (using phenolphthalein indicator) is the sum of both mineral acidity and carbon dioxide acidity. However in the absence of mineral acidity, total acidity is only the CO_2 acidity and it is the case with most of the samples.

Problem 10 – 1

50 ml. of a water sample when titrated with 0.02N sodium hydroxide, gives the following titre values.

With methyl orange indicator = 1.0ml.

With phenolphthalein indicator = 2.5ml.

Calculate the mineral acidity and carbondioxide acidity.

---

* Dechlorinate the sample if it contains any chlorine, by the addition of 1 drop of 0.1 N sodium thiosulfate solution (25g. $Na_2S_2O_3.5H_2O$/1000 ml.)

Methyl orange acidity is mineral acidity.

$$\therefore \text{Mineral Acidity, mg/l (as } CaCO_3) = \frac{\text{ml. titrant} \times 1 \times 1000}{\text{ml. sample taken for titration}}$$

$$= \frac{1 \times 1 \times 1000}{50}$$

= **20**

Total acidity is phenolphthalein acidity.

$$\text{. Total acidity, mg/l (as } CaCO_3) = \frac{2.5 \times 1 \times 1000}{50}$$

= **50**

Total acidity is the sum of mineral acidity and carbon dioxide acidity.

. Carbon dioxide acidity, mg/l (as $CaCO_3$) } = Total acidity – Mineral acidity

= 50 – 20

= **30**

Problem 10 – 2

25 ml. of a sample of water when titrated with 0.02N sodium hydroxide, gives the following titre value.

With methyl orange indicator = 0.0ml.

With phenolphthalein indicator = 1.0ml.

Calculate the mineral acidity and carbon dioxide acidity.

Since titre value for methyl orange acidity is zero, there is mo methyl orange acidity (i.e) there is no mineral acidity. Then the total acidity is only due to carbon dioxide.

$$\therefore \text{Carbon dioxide acidity, mg/l (as } CaCO_3) = \frac{1.0 \times 1 \times 1000}{25}$$

= **40**

FREE MINERAL ACIDITY

Free Mineral Acidity (FMA) is the term used to denote the acidity present as mineral acid. Free mineral acid determination is important as acids are produced during cation exchange by ion exchange resins in hydrogen form. This parameter is useful to check the sodium slip that may occur in cation exchanger column. Free mineral acidity is found out by titrating a suitable aliquot of the cation exchanger effluent with 0.02 N. sodium hydroxide solution, which has been described in the previous section under the title "Methyl Orange Acidity" (Mineral Acidity). Normally this value should be equal to the value of Equivalent Mineral Acidity (EMA), which is calculated from the quantity of anions, namely, sulfate, chloride and nitrate present in water. If there is a significant decrease in free mineral acidity (FMA), it indicates sodium slip. See Chapter – 64, for further details on sodium slip.

CHAPTER - 11

ALKALINITY

The measurement of alkalinity is significant in many uses and treatment of natural and waste waters. Alkalinity is an important parameter involved in corrosion control. It must be known in order to calculate the Langelier Saturation Index. The alkalinity in excess of alkaline earth concentrations is significant in determining the suitability of a water for irrigation. Alkalinity measurements are used as the means of evaluating the buffering capacity of waste waters and sludges. Alkalinity and pH are the factors in determining the amenability of waste waters to biological treatment.

Alkalinity is the quantitative capacity of the water to react with hydrogen ions. The alkalinity of natural or treated waters is normally due to the presence of bicarbonate, carbonate and hydroxide compounds of calcium, magnesium, sodium and potassium. Borates and phosphates and silicates also contribute to alkalinity. Some other ions not ordinarily found in natural waters such as arsenate and aluminate could also increase the alkalinity. Because of the relative abundance of carbonate minerals and because of the ready availability of carbon dioxide that enters into equilibria with them in water solution, most waters contain bicarbonates and carbonates only. The occurrence of hydroxide ions in natural waters is very rare, unless artificial contamination has occurred. In surface waters where algae are flourishing the waters may contain appreciable amounts of carbonate and hydroxide alkalinities. Boiler waters usually contain carbonate and hydroxide alkalinities. Chemically treated waters, particularly softened waters (by lime or lime soda process), contain carbonates and hydroxides.

It is customary to calculate the hydroxide, carbonate and bicarbonate alkalinities from the titre values with a standard acid. Each form of the alkalinity exists separately or in combination.

Phenolphthalein and methyl orange are the indicators commonly used for alkalinity titrations. Phenolphthalein gives a pink color in the presence of hydroxide or normal carbonate. The change from pink to colorless occurs at a pH 8.3. Methyl orange gives orange yellow color in the presence of all the three types of alkalinity and is orange red in the presence of acid. The color change occurs at a pH value of about 4.5. (See Fig. 11 – 1)

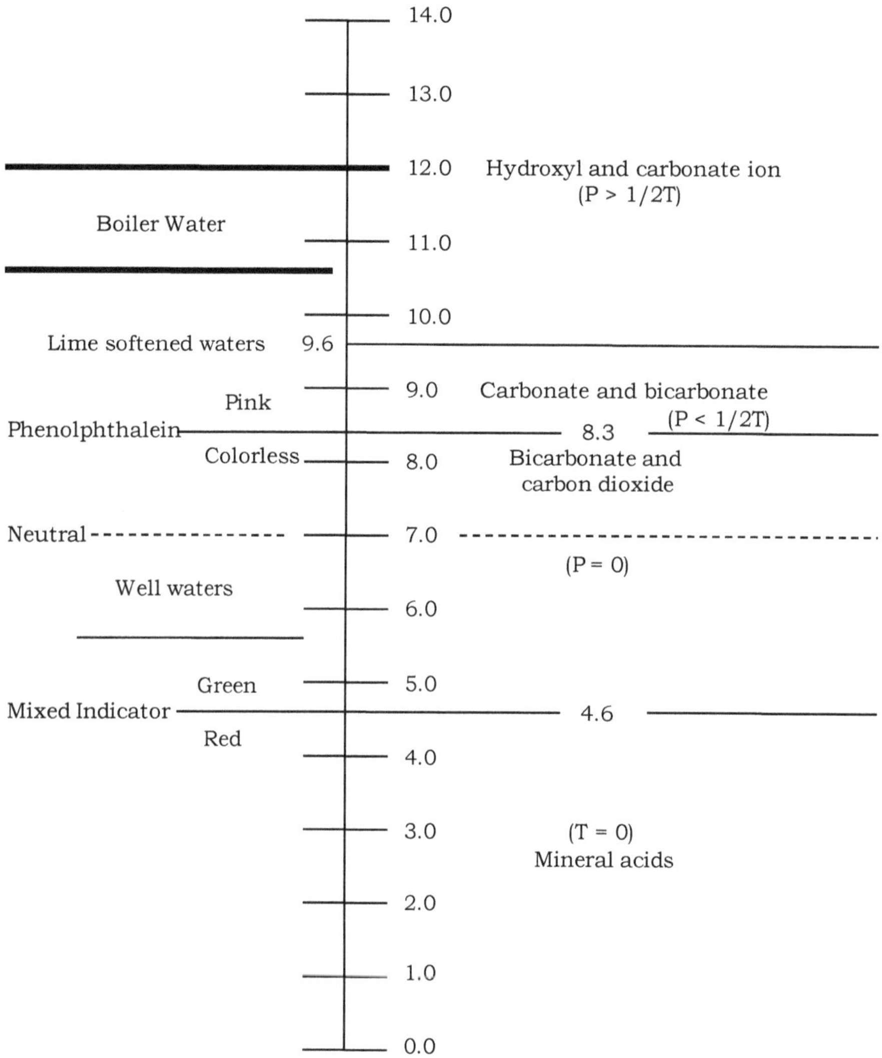

Fig 11 - 1. Zones of pH value where free carbondioxide, bicarbonates, carbonates and hydroxide alkalinities prevail*

P – Phenolphthalein alkalinity
T – Total Alkalinity

* Fig. 11 – 1 reproduced with the kind permission from Chemical Publishing Company, Newyork, from their publication, "The Chemical Treatment of Boiler Water" by James W.McCoy – 1981.

Since the color change of methyl orange is not sharp, the present day trend is to use mixed indicator the color change of which is very sharp. However the traditional term 'methyl orange alkalinity' is still used regardless of the indicator used. (See next paragraph).

If there is methyl orange alkalinity, it is due to any one of the three alkalinities or a combination of hydroxide and carbonate together, or carbonate and bicarbonate together. It is to be stated that hydroxide and bicarbonate alkalinities cannot co-exist. Since methyl orange is traditionally used for total alkalinity titration, the term 'total alkalinity' is also referred to as methyl orange alkalinity. The terms 'total alkalinity' and 'methyl orange alkalinity' are used interchangeably throughout this book. However, the following procedure for total alkalinity uses a mixed indicator solution containing bromocresol green and methyl red instead of methyl orange indicator since it gives a distinct end point. (The end point is from blue green to pink).

Among the methods available for the determination of alkalinity, the electrometric titration, and the color change titration procedure are commonly employed. The electrometric titration yields very accurate results. In the absence of color and other interfering matters, color change titration procedure is equally good and is easily adaptable.

TITRIMETRIC METHOD

Principle

Alkalinity is directly determined by titration with 0.02 N.H_2SO_4 using phenolphthalein and methyl orange indicators.

$$Ca(OH_2) + H_2SO_4 \rightarrow CaSO_4 + 2H_2O$$
$$2CaCO_3 + H_2SO_4 \rightarrow Ca(HCO_3)_2 + CaSO_4$$
$$Ca(HCO_3)_2 + H_2SO_4 \rightarrow CaSO_4 + 2CO_2 \uparrow + 2H_2O$$

Interferences

Natural color, or the formation of a precipitate during titration may mask the color change of the indicator. Waste materials present in some waters may interfere by destroying the indicator. Varied results may be obtained with waters containing oxidizing or reducing substances, depending on the equilibrium conditions and the manner in which the sample is handled.

Reagents

1. Sodium carbonate solution, 1N: Weigh accurately 13.25 g. anhydrous sodium carbonate, Na_2CO_3 (previously dried at 140°C for 2 hr). Dissolve it in little distilled water and make up to 250 ml. in a volumetric flask.

2. Sulfuric acid, 1N: Place 28 ml. conc. H_2SO_4 (Sp.gr 1.84) in a 1000 ml. volumetric flask and make up to the mark with carbon dioxide free distilled water. Standardize it against 1N sodium carbonate solution using methyl orange indicator. (End point is the appearance of faint orange color).

3. Sulfuric acid, 0.02N: Dilute appropriate volume of 1N. H_2SO_4 (about 20 ml.) to 1000 ml. in a volumetric flask with carbon dioxide-free distilled water, to give a 0.02N solution.

$$1.0 \text{ ml. exactly } 0.02 \text{ N } H_2SO_4 \equiv 1.0 \text{ mg } CaCO_3$$

4. Phenolphthalein indicator solution: Dissolve 500 mg phenolphthalein in 50 ml. ethyl or isopropyl alcohol and add 50 ml. distilled water. Add 0.02 N sodium hydroxide solution dropwise until a faint pink color appears.

5. Mixed indicator solution*: Dissolve 20 mg methyl red and 100 mg bromo cresol green in 100 ml. 95% ethyl or isopropyl alcohol (if the sodium salts of indicator are used, prepare the solution with distilled water instead of alcohol).

Procedure

Phenolphthalein alkalinity: Place 25 ml. or 50 ml. or appropriate volume of the sample† (requiring not more than 25 ml. of the titrant) in a 250 ml. conical flask. Add 5 drops of phenolphthalein indicator solution. If no pink coloration occurs, there is no phenolphthalein

* If the component indicator substances are not available, use 0.05% alcoholic solution of methyl orange. End point is from orange to red.

† If the sample contains any chlorine, dechlorinate by the addition of 1 drop of 0.1N.sodium thiosulfate solution (25g. $Na_2S_2O_3.5H_2O$/1000ml).

alkalinity. If a pink color appears then titrate with 0.02N sulfuric acid (to pH 8.3) until the solution becomes colorless.

Total Alkalinity: Add 3 drops of mixed indicator solution to the solution in which phenolphthalein alkalinity has been determined and titrate against 0.02N sulfuric acid (to pH 4.5) to light pink. (If methyl orange is used as indicator, the end point is from orange yellow to orange red).

Note: The pH to which the titration should be performed depends on the magnitude of the alkalinity and free carbon dioxide. pH 4.5 is sufficiently accurate for most purposes. However, if most accurate results are needed, choose the pH as follows:

Alkalinity as $CaCO_3$ mg/l	pH
30	5.1
150	4.8
500	4.5

If the sample contains suspended matter, remove by filtration or centrifugation and then determine the alkalinity. If suspended matter contains any alkalinity, add a known excess of standard acid to the sample, boil off carbon dioxide, cool and back-titrate with standard alkali.

Calculation

If the sulfuric acid used for the titration is exactly 0.02N,

Phenolphthalein alkalinity (as $CaCO_3$), mg/l

$$= \frac{ml.\ 0.02N.H_2SO_4 \text{ for phenolphtahlein end point} \times 1 \times 1000}{ml.\ \text{sample taken for titration}}$$

Total alkalinity (as $CaCO_3$), mg/l

$$= \frac{ml.\ 0.02N.H_2SO_4 \text{ for total alkalinity end point} \times 1 \times 1000}{ml.\ \text{sample taken for titration}}$$

Calculation of three forms of alkalinity

The titre values obtained from phenolphthalein and total alkalinity determinations are used to estimate the three forms of alkalinity as shown in the following table :

Table 11–1

CALCULATION OF THREE FORMS OF ALKALINITY

	Hydroxide alkalinity (as $CaCO_3$)	Carbonate alkalinity (as $CaCO_3$)	Bicarbonate alkalinity (as $CaCO_3$)
P = 0	0	0	T
P < ½ T	0	2P	T - 2P
P = ½ T	0	2P	0
P > ½ T	2P – T	2(T - P)	0
P = T	T	0	0

where P is phenolphthalein alkalinity, and T is total alkalinity.

(It is assumed that the entire alkalinity is due to the presence of bicarbonate, carbonate and hydroxide only and other ions contributing to alkalinity are absent).

EXAMPLES FOR THE CALCULATION OF ALKALINITY

Problem 11 - 1

The initial reading of a burette with 0.02N H_2SO_4 is 1.5 ml. After titration to phenolphthalein end point with 100ml. sample, the reading is 3.9 ml. The titration is continued after adding mixed indicator. When the end point is reached, burette reading is 6.0 ml. Calculate phenolphthalein and total alkalinities of the sample.

Volume of 0.02 N H_2SO_4 required for phenolphthalein alkalinity

$$= 3.9 - 1.5 = 2.4 \text{ ml.}$$

∴ Phenolphthalein alkalinity (as $CaCO_3$), mg/l

$$= \frac{\text{ml. } 0.02\text{ N.}H_2SO_4 \times 1 \times 1000}{\text{ml. sample taken for titration}}$$

$$= \frac{2.4 \times 1 \times 1000}{100}$$

$$= \mathbf{24}$$

Volume of 0.02 N.H_2SO_4 required for methyl orange alkalinity

$$= 6.0 - 1.5 = 4.5 \text{ ml.}$$

∴ Total alkalinity (as $CaCO_3$), mg/l

$$= \frac{4.5 \times 1 \times 1000}{100}$$

= **45**

Problem 11 - 2

A 100 ml. sample when titrated with 0.02N H_2SO_4, the following titre values are obtained:

To reach phenolphthalein end point = 19.5 ml.

To reach methyl orange end point = 24.5 ml.

Calculate hydroxide, carbonate and bicarbonate alkalinity of sample.

P > ½ T

∴ Hydroxide alkalinity = 2P – T
= 2 (19.5) – 24.5
= 39 – 24.5 = 14.5

∴ Hydroxide alkalinity (as $CaCO_3$), mg/l

$$= \frac{14.5 \times 1 \times 1000}{100}$$

= **145**

Since **P > ½ T**

Carbonate alkalinity = 2 (T – P) = 49 – 39 = 10

∴ Carbonate alkalinity (as $CaCO_3$), mg/l

$$= \frac{10 \times 1 \times 1000}{100}$$

= **100**

Since hydroxide and carbonate alkalinities are present, the bicarbonate alkalinity will be absent. The three forms of alkalinity cannot coexist together.

The determination of alkalinity provides an idea of the nature of salts present. If the alkalinity is equal to hardness, the contributors to alkalinity are only calcium and magnesium. If the alkalinity is greater than hardness, it indicates that sodium and potassium also contribute to alkalinity in addition to those of calcium and magnesium. If the alkalinity is lower than hardness, neutral salts of calcium or magnesium must be present that are not carbonates; usually these are sulfates.

DETERMINATION OF HYDROXIDE ALKALINITY - B VALUE

Many authors prefer to include 'B' value in alkalinity calculations. B–value is the concentration of free hydroxyl ions as distinguished from that formed by hydrolysis.

Determination

Place 50 ml or 100 ml or suitable volume of the sample in a conical flask. Add 5 drops of phenolphthalein indicator and 10 ml of neutralized 10% barium chloride solution. Mix well. Titrate with 0.02 N sulfuric acid until the solution becomes colorless. Do not mind even the pink color recurs.

Calculation

$$\text{B–value mg/l (as } CaCO_3) = \frac{\text{ml } 0.02 \text{ N } H_2SO_4}{\text{ml sample}} \times 1000$$

$$\text{B–value mg/l (as NaOH)} = \text{B–value mg/l, (as } CaCO_3) \times 0.8$$

SOLUBLE ALKALINITY

This is one of the important parameters used to assess the suitability of a water for silk reeling. This is the alkalinity of a water remaining after a prolonged boiling and is due to the presence of small quantities of calcium and magnesium carbonates which remain in solution and due to the presence of sodium and potassium carbonates. This is expressed in terms of mg. Na_2CO_3 per litre. This soluble alkalinity is determined by titrating 25 ml or 50 ml or suitable aliquot of water sample after boiling, with 0.02 N. H_2SO_4 using phenolphthalein and mixed indicators. (1 ml 0.02 N.$H_2SO_4 \equiv$ 1.06 mg Na_2CO_3).

$$\text{Soluble Alkalinity (as } Na_2CO_3\text{), mg/l} = \frac{\text{ml.}0.02\text{N.}H_2SO_4}{\text{ml. sample taken for determination}} \times 1.06 \times 1000$$

Water with a soluble alkalinity lower than 15 mg/l is regarded as having a lower soluble alkalinity; that between 15 to 40 mg/l as medium, and above 40 mg/l as strongly alkaline. Water having medium soluble alkalinity is desirable for silk reeling.

CHAPTER - 12

ALUMINIUM

Aluminum in water occurs as a result of coagulation of raw water. Its presence in natural waters may be due to the suspended clays or due to industrial wastes or more likely due to wash water from water treatment plants. Corrosion and erosion of aluminium tubing in condensers and feed water heaters can also be a source of aluminium in industrial waters. It is undesirable in industrial waters because of its sludge forming characteristics.

Aluminium is best determined by Eriochrome cyanine - R method which suffers little interference.

ERIOCHROME CYANINE - R METHOD

Principle

Aluminium reacts with Eriochrome cyanine – R dye at a pH 6.0 and produces a red to pink complex. The intensity of the developed color depends on the aluminium concentration, reaction time, temperature, pH, alkalinity and the concentration of other ions present in the sample.

Interferences

Fluoride interferes by giving negative errors. Known amounts of fluoride can be added to a set of standards to eliminate fluoride interference. Polyphosphates also interfere by giving negative type of errors. Sample should preliminarily be treated to remove polyphosphates and complex phosphates. Ortho phosphates do not interfere if present in concentration less than 10 mg/l. Alkalinity interferes if present in small quantities. However, it can be removed by acidifying the sample just beyond the neutralization point of methyl orange. Sulfate does not interfere up to a concentration of 2000 mg/l. Iron and manganese interfere and are eliminated by adding ascorbic acid. Color and turbidity are compensated by complexing aluminium in one portion of the sample with EDTA, the portion being used as a blank.

Reagents

1. Aluminium stock solution: Weigh 1000mg. aluminium metal (AR) accurately in a 100 ml. beaker and add 10 ml. conc. HCl. Dissolve by heating gently. Transfer to a 1000 ml. volumetric flask and make up to the mark with distilled water.

1.00 ml. = 1.0 mg Al

(or)

Dissolve 8.792g. aluminium potassium sulfate $AlK(SO_4).12H_2O$ in distilled water and make up to 1000 ml. in a volumetric flask.

1.00 ml. = 1.0 mg Al

2. Aluminium intermediate solution: Pipet 10.0 ml. stock solution and make up to 1000 ml. in a volumetric flask.

1.00 ml. = 10.00 μg Al

(Prepare freshly)

3. Aluminium working solution: Pipet 10.0 ml. aluminium intermediate solution and make up to 100 ml. in a volumetric flask.

1.00 ml. = 1.0 μg Al

4. Sulfuric acid, 0.02 N: See Chapter – 11, "ALKALINITY".

5. Ascorbic acid solution: Dissolve 100mg. ascorbic acid in distilled water and make up to 100 ml. in a volumetric flask. Prepare freshly.

6. Sodium acetate buffer solution: Dissolve 136g. sodium acetate $NaC_2H_3O_2$. $3H_2O$ in distilled water, add 40 ml. 1N acetic acid (prepared by mixing 28.7 ml. glacial acetic acid with 471.3 ml. distilled water) and dilute to 1000 ml.

7. Dye solution:

(a) Stock solution: Dissolve 150 mg Eriochrome cyanine - R if it is the product of Hartman – Leddon Co., or 300mg. Eriochrome cyanine – R if it is the product of Pfaltz and Bauer, Inc., in 50 ml. distilled water. Adjust the pH from about 9 to 2.9 with 1 + 1 acetic acid. (Approximately 2 ml. or 3 ml. will be required). Dilute to 100 ml. with distilled water.

(or)

Dissolve 100 mg Eriochrome cyanine - R or Solochrome cyanine – R200 in distilled water and dilute to 100 ml. in a volumetric flask. The solution should have a pH of about 2.9. Stock solution have excellent stability and can be kept for a year.

(b) Working solution: Pipet 10.0 ml. of any stock dye solution, to 100.0 ml. in a volumetric flask and make up to the mark with distilled water. This solution is stable for at least six months.

8. Mixed indicator (Bromocresol green + Methyl red): See Chapter – 11 "ALKALINITY" (Methyl orange indicator can also be used instead of mixed indicator).

9. EDTA solution 0.02 N: See Chapter – 31, "HARDNESS (TOTAL)".

Procedure

Standards

1. Pipet 0.0, 1.0, 2.0, 3.0, 4.0, 5.0, 6.0, 7.0 ml. of aluminium working solution in 50 ml. volumetric flasks or Nessler tubes. Dilute to 25 ml. with distilled water.
2. Add the following reagents in order with mixing after each addition.
   (i) 1 ml. 0.02 N sulfuric acid.
   (ii) 1 ml. ascorbic acid solution.
   (iii) 10 ml. buffer solution.
   (iv) 5.0 ml. of working dye solution.
3. Immediately make up to 50 ml. with distilled water. Mix and allow to stand for 5 to 15 minutes.
   Note : The color begins to fade after 15 minutes.
4. Use the blank to set the instrument to 100% transmittance or zero absorbance and measure either the transmittance or absorbance at 535 nm using 1 cm light path. Prepare a calibration curve.

Samples

Important Note: Do not use filter paper, absorbent cotton or glass wool for filtering any solution that is to be tested for aluminium, as aluminum is removed by them. If only soluble aluminium is to be determined, filter a portion of the sample through a 0.45 μm membrane filter. Discard the first 50 ml. filtrate and use the succeeding filtrate for estimation.

(a) Samples that do not contain fluoride and complex phosphates

(i) Pipet 25.0 ml. sample or an aliquot diluted to 25 ml. in a flask. Add a few drops of methyl orange indicator. Titrate with 0.02 N.H_2SO_4 to a faint pink color. Record the reading and discard it.

(ii) Take two 25ml. volumes of sample (or aliquots diluted to 25ml.) and add the same amount of 0.02 N.H_2SO_4 used for the above titration and 1 ml. in excess.

(iii) To one of the samples, add 1ml. EDTA solution. This will serve as the blank since it complexes the aluminium and compensate for color and turbidity.

(iv) Add 1ml. ascorbic acid, 10 ml. buffer reagent and 5.0ml. working dye reagent to both blank and the sample, mixing after each addition. Immediately make upto 50ml. with distilled water, allow to stand for 5 to 15 minutes.

(v) Set the instrument to 100% transmittance or zero absorbance using EDTA added sample as the blank. Read the transmittance or absorbance and find out the aluminium equivalent (as Al) per litre of the sample.

Note : For visual comparison no EDTA blank is needed. If the sample contains color or turbidity visual comparison results in considerable error.

(b) Samples containing phosphates

Place 100 ml. of sample in a 200 ml. conical flask. Add 1.7ml. 6N.H_2SO_4 (prepared by mixing 16.7 ml. conc.H_2SO_4 with 83.3 ml. distilled water) and keep it on a hot plate and maintain the temperature of the solution below its boiling point. At the end of the heating period the volume should be about 25 ml. Cool it well and neutralize to a pH 4.3 to 4.5 using 1N sodium hydroxide (40 g NaOH/l) at first and 0.1 N sodium hydroxide (4g NaOH/l) in the final stage with the aid of a pH meter. Make upto 100 ml. with distilled water. From this made up solution use 25.0 ml. and proceed as described for standards. Conduct a blank also in the same manner using 100 ml. distilled water. Use this blank to set the instrument at 100% transmittance or zero absorbance and measure the transmittance or absorbance of the sample and find out the mg equivalent of aluminium in the sample.

(c) Samples containing fluoride

Find out the fluoride concentration of the sample and add the calculated amount of fluoride to each aluminium standard and proceed as described above.

CHAPTER - 13
ARSENIC

Arsenic in water is highly undesirable because of its toxicity. It is a cumulative poison and possesses carcinogenic properties. The occurrence of arsenic in natural waters is rare. It may occur as a result of mineral dissolution, industrial discharges or the application of pesticides. It can be determined with greater precision by the following silver diethyl dithio carbamate method, using modified arsenic estimation apparatus [devised by NEERI (National Environmental Engineering Research Institute, Nagpur, India].

SILVER DIETHYL DITHIO CARBAMATE METHOD

Principle

Inorganic arsenic is reduced by zinc in acid solution to arsine which is absorbed in to silver diethyl dithiocarbamate solution to form a red complex and is measured colorimetrically at 535 nm.

$$AsH_3 + SDDC \rightarrow As\ [(C_2H_5)_2NCS_2]_3$$

Interferences

Copper, chromium, cobalt, nickel and silver ions interfere. But these metals are normally present in very small quantities and hence these interferences are insignificant. Antimony present in five times excess of arsenic concentration interferes by forming a red complex, stibine.

Apparatus

Modified Gutzeit apparatus (devised by NEERI) as shown in Figure 13-1.

Reagents

1. **Zinc:** 20 to 30 mesh, arsenic-free.
2. **Potassium iodide solution, 15%:** Dissolve 15g. potassium iodide KI in 100 ml. distilled water. Store in an amber bottle.

3. Lead acetate solution, 10%: Dissolve 10g. lead acetate $Pb(C_2H_3O_2)_2.3H_2O$ in 100 ml. distilled water.

4. Silver diethyl dithiocarbamate solution, 0.5%: Dissolve 0.5g. silver diethyl-dithiocarbamate in 100 ml. pyridine.

5.Hydrochloric acid: conc.

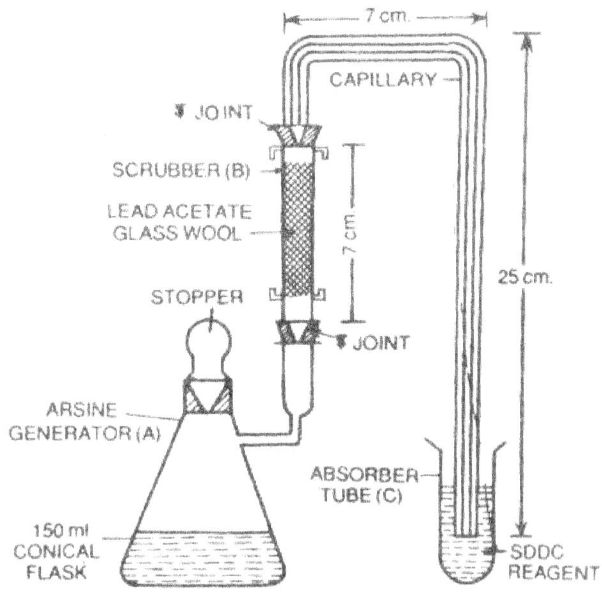

Fig. 13 –1. Modified Arsenic Estimation Apparatus

6. Stannous chloride reagent: Dissolve 40g. arsenic free $SnCl_2.2H_2O$ in 100 ml. conc. HCl.

7. Arsenic stock solution: Dissolve 1.320g. arsenic trioxide As_2O_3 in 10 ml. distilled water containing 4g. NaOH and makeup 1000 ml. with distilled water, or alternatively dissolve 1.734g. sodium meta arsenite $NaAsO_2$ in water and make up to 1000 ml. with distilled water

1.00 ml. = 1.0 mg As

8. Arsenic intermediate solution: Dilute 10.00 ml. arsenic stock solution to 1000 ml. with distilled water.

1.00 ml. = 10.0 μg As

9. Arsenic standard solution: Dilute 10.00 ml. intermediate solution to 100 ml. with distilled water.

$$1.00 \text{ ml.} = 1.00\ \mu g \text{ As}$$

Procedure

Standards

1. In to a series of arsenic generators, pipet, 0.0, 1.0, 2.0 .... 10.0 ml. of arsenic standard solution. Dilute to 50 ml. with distilled water.
2. Add 5 ml. conc. HCl, 2 ml. potassium iodide solution and 8 drops of stannous - chloride solution with thorough mixing after each addition. Allow 15 minutes for the complete reduction of As^{V} to As^{III}.
3. Meanwhile impregnate the glass wool in the scrubber with lead acetate solution. (Do not make too wet, since water will be carried over in to the reagent solution).
4. Take 4.0 ml. silver diethyl dithiocarbamate reagent in the absorber tube. Connect the generator scrubber-absorber assembly. Make sure that all the connections are tightly fitted.
5. After 15 minutes, add 3g. zinc into the generator by opening the stopper, and immediately replace the stopper. Allow 30 minutes for complete evolution of arsine.
6. Pour the solution from the absorber directly in to a 1 cm cell and measure the absorbance of the solution at 535 nm spectrophotometrically.
7. Prepare a calibration curve.

Samples

Natural waters excepting those from certain areas usually contain no arsenic or only trace quantities. Hence larger volumes of samples have to be taken and concentrated. Place 500 to 1000 ml. sample in a silica evaporating dish. Add 0.5 ml. conc. HCl and evaporate on a water bath. When the volume becomes 25 ml., transfer to arsine generator. Rinse the dish and pour washings to the generator.

Proceed from Step No. 2 to 6. Find out the microgram equivalent of arsenic to the net optical density.

Express the result as mg arsenic (as As) per litre of the sample.

CHAPTER - 14
BIO-CHEMICAL OXYGEN DEMAND (BOD)

The test Biochemical Oxygen Demand is usually not carried out on waters. Biochemical oxygen demand (BOD) is a test of great value in the analysis of sewage, industrial effluents and grossly polluted waters. It is an indirect measure of total amount of unstable organic matter contained in a waste. Also it provides a basis for assessing the effect of the discharge of the water on the oxygen balance of the receiving waters. BOD refers to the quantity of oxygen required by bacteria and other microorganisms in the biochemical degradation and transformation of organic matter under aerobic conditions. There are so many factors to be remembered to make the determination successful and are as follows.

Temperature and Period of Incubation

Temperature is an important factor in the determination of BOD. The standard temperature for the test is 20°C. The temperature should be kept constant throughout the 5 day period of incubation.

Presence of bacteriostatic and bactericidal substances

The presence of substances that inhibit or prevent the growth of bacteria, cause errors in the BOD determination.

Acids and alkalies have a marked effect upon bacterial activity and hence before proceeding the determination, the pH of the samples should be brought to the range of 6.5 to 8.3.

Free chlorine present in the sample affects the determination and the sample should be dechlorinated with sodium sulfite.

The presence of toxic metals are undesirable since they are bacteriostatic and only low BOD values are obtained. It is better to conduct 4 hours permanganate test instead of BOD test, on samples of river waters and industrial wastes containing both organic and metallic contaminants.

The BOD of tidal waters and estuaries is affected by the salinity and only low values are obtained. Separate determinations of organic carbon and nitrogeneous matter are preferable for these samples.

Presence of nutrient materials

Mineral nutrients are necessary for the growth and activity of bacteria. Carbon, nitrogen, phosphorus and sulfur are the most important elements. Traces of calcium, magnesium, potassium and iron are also needed. Low BOD values are therefore likely to be obtained with mineral deficient wastes unless they are compensated by the dilution water. Probably most stream waters contain sufficient minerals to give a satisfactory BOD results. Difficulties arise only when a stream consists largely of industrial effluents.

Composition of dilution water

A suitable dilution water should satisfy the following requirements: correct pH limits, proper salinity, reasonable buffer capacity, the presence of necessary mineral nutrients and the absence of bactericidal and bacteriostatic substances such as toxic metals and free chlorine. The presence of a balanced and viable seeding material is necessary. The suitability of a dilution water may be checked against a standard glucose-glutamic acid solution.

Dilution

The BOD of a sample generally depends on the particular dilution used. It is therefore advisable to make several dilutions and to base the value on the dilution showing about 50% depletion of the dissolved oxygen.

Seeding with bacteria

Seeding is not necessary for samples containing bacterial flora. Hence sewages, sewage effluents and most river waters need not be seeded. It is essential to seed the industrial effluents which usually do not contain any bacteria. The seeding material generally preferred is fresh settled sewage. Fresh settled final effluent of good quality from an aerobic biological process is also recommended. The seed should be kept at 20°C for 24 to 36 hrs. The recommended seed concentration is 1 - 2 ml. per litre of dilution water.

Precautions to be observed

Sample must be free from all added preservatives. The test should be performed as soon as possible after the sample is collected. When it is impracticable to conduct the test immediately, samples can be stored at 5°C. On any account, samples should not be stored at room temperature.

BOD DETERMINATION

Principle

The basic principle underlying the BOD determination is the measurement of dissolved oxygen content of the sample before and after 5 days incubation at 20°C. Samples devoid of oxygen, or containing less amount of oxygen are diluted several times with special type of dilution water saturated with oxygen in order to provide sufficient amount of oxygen for oxidation.

Reagents

Use good quality distilled water containing less than 10 μg/l copper and free from chlorine.

1. **Calcium chloride solution:** Dissolve 27.5g. anhydrous calcium chloride $CaCl_2$ (or equivalent amount, if hydrated calcium chloride is used) in distilled water and dilute to 1000 ml.
2. **Magnesium sulfate solution:** Dissolve 22.5g. magnesium sulfate hepta hydrate, $MgSO_4.7H_2O$ in distilled water and dilute to 1000 ml.
3. **Ferric chloride solution:** Dissolve 0.25g. ferric chloride hexa hydrate, $FeCl_3.6H_2O$ in distilled water and dilute to 1000 ml.
4. **Phosphate buffer solution:** Dissolve 8.5g. potassium dihydrogen phosphate, KH_2PO_4, 21.75g. dipotassium hydrogen phosphate, K_2HPO_4, 33.4g. disodium hydrogen phosphate hepta hydrate, $Na_2HPO_4.7H_2O$, and 1.7 g ammonium chloride, NH_4Cl in 500 ml. distilled water and make up to 1000 ml. The pH of this buffer solution should be 7.2.

Note: If biological growth is noticed in any of the above reagents, discard it and prepare freshly.

Standard dilution water: Aerate a suitable volume of good quality distilled water with a supply of clean compressed air, after bringing the temperature to 20°C. Add 1 ml. each of calcium - chloride solution, magnesium sulfate solution, ferric chloride solution, and phosphate buffer solution per litre of dilution water. Use the dilution water as soon as possible. Discard any dilution water remaining after use and clean the bottle preferably with a sterilizing agent.

Note: A satisfactory dilution water when incubated with or without seed under standard conditions should not absorb more than 200 µg/l of oxygen.

Seeded dilution water: Add 1-2 ml. of the standard seed material (fresh settled domestic sewage which has been incubated at 20°C for 24-36 hr) to the above dilution water. Use this dilution water only for samples containing no bacteria.

Procedure

[A] Pre-treatment

If the pH of the sample is not in the range of 6.5 to 8.5, add sufficient amount of 1N.H_2SO_4 or 1N.NaOH to bring the pH to that range (Determine the amount of acid or alkali needed on a separate aliquot of the sample).

If the sample contains any chlorine, dechlorinate using freshly prepared sodium - sulfite solution of 0.025N (1.575g. anhydrous Na_2SO_3/1000ml.). Determine the amount of sodium sulfite required on a separate aliquot of the sample by iodometric method.

If the sample contains sulfite, sulfide, thiosulfate and polythionate a separate procedure has to be followed. See the section "Immediate Dissolved Oxygen Demand" in this chapter.

Some samples may be supersaturated with dissolved oxygen, especially waters containing algae. The excess oxygen above saturation level in those samples has to be removed by partly filling them in a bottle and then agitating at 20°C or by aerating at 20°C.

Algae laden samples must be filtered to remove algae and BOD should be done on the filtered sample only, otherwise during incubation the algae may rot and escalate BOD value.

To suppress nitrification in some samples, it is necessary to add 1 ml. of 0.05% solution of allyl thiourea per litre of standard dilution water and the BOD determination can be proceeded as usual but the result should be reported as BOD, nitrification suppressed mg/l.

Samples that have been stored in refrigerator should be allowed to reach room temperature before dilutions are made. Samples must be thoroughly mixed before dilution.

Coarse suspended matter present in the sample usually causes difficulty. Those samples should be shaken vigorously just before dilutions are made.

[B] Preparation of Dilutions

Dilution

Usually, dilutions more than one have to be made for each sample. The degree of dilution has to be determined by the analyst and atleast one of the dilutions should contain about 50% of the DO after 5 days incubation. The permanganate value and COD can be taken as a guide to the dilution required.

A general scheme of dilution is as follows:

River waters: Samples as such may be incubated without dilution. If the river is polluted, dilutions have to be made in the ratio 1 : 1, 1 : 2, 1: 3, 1 : 4, where the first numeral indicates the volume of the sample and the second one denotes the volume of the dilution water.

Biologically purified sewage effluents: The ratio of sample and dilution water ranges from 1 : 1 to 1 : 9.

Oxidized effluents	:	1 : 3 to 1 : 19.
Settled sewage	:	1 : 19 to 1 : 99.
Raw sewage	:	1 :19 to 1 :199.

Industrial effluents: Since the nature of industrial effluents vary in character, no limits of dilution can be suggested. However, it may be necessary to make a series of dilutions, for example 1:4, 1:9, 1:24, 1:49, 1:99, 1:249 upto 1:999.

Dilution technic

(i) Carefully siphon the standard dilution water (or seeded dilution water when BOD of industrial wastes are to be determined), into a 1000 ml. graduated cylinder and half full it without entrainment of air.

(ii) Add the appropriate quantity* of the well mixed sample into the cylinder by the sides without producing any air bubble. Make upto the 1000 ml. mark with dilution water.

(iii) Mix well with a plunger–type mixing rod, (or any other suitable rod) without entrainment of air. (The rod as shown in the figure 14.1 serves the purpose).

(iv) Siphon the mixture into two BOD bottles of 300 ml. capacity. While siphoning allow about 50 ml. of mixture to waste. Fill the bottles and stopper carefully without allowing any air bubble to remain inside. Water seal both the bottles.

(v) Prepare succeeding dilutions of lower concentrations in the same manner as described above.

Note: Samples may also be measured directly into B.O.D. bottles of known capacity with a large tip volumetric pipet and diluted with dilution water.

(vi) Siphon out standard dilution water into two BOD bottles and water seal them after filling completely.

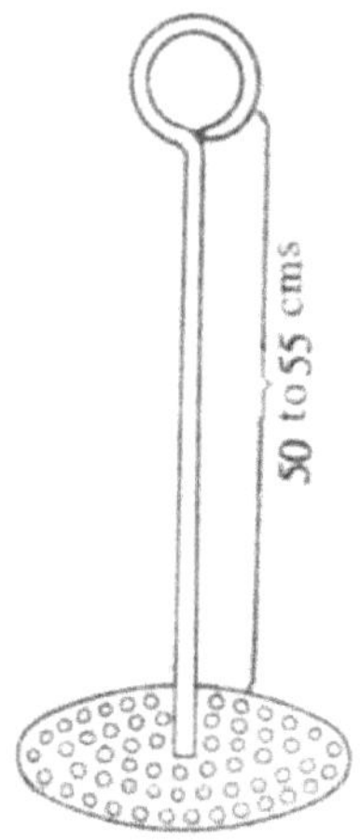

Fig. 14–1. Plunger-type mixing rod for BOD dilution.

[C] Determination of DO

Use one set of sample dilutions and dilution water to find out the initial dissolved oxygen content. Keep the other set of sample dilutions and dilution water in the BOD incubator at 20°C (in dark) for 5-days.

---

* If less than 5ml of sample is needed for dilution, dilute first a larger aliquot of the sample in a volumetric flask and use diluted sample for BOD dilutions.

Determine the DO content of dilution water and sample dilutions on the initial day and after 5 days. DO may be determined either by electrode method or by titrimetric method as given in Chapter – 49, ("OXYGEN DISSOLVED"), starting from Step No.2 of the procedure (i.e.,) from the addition of manganous sulfate (See Page No.222).

Calculation

If the sodium thiosulfate used is exactly 0.025N,

$$BOD\ (5\ days\ at\ 20^{o}C)\ mg/l = (V_1 - V_2 - BC) \times D$$

where

V_1 - is the volume in ml. of 0.025N thiosulfate for 200 ml. of the original dilution (1st day). (Initial day dissolved oxygen).

V_2 - is the volume in ml. of 0.025N thiosulfate for 200 ml. of the incubated dilution. (Dissolved oxygen after 5 days).

BC - is the difference between volumes of 0.025N thiosulfate required for 200 ml. of dilution water before and after incubation (blank correction).

D - Dilution factor

3 day BOD at 27°C

Taking into consideration the ambient temperature prevailing in the temperate and tropical countries, the river authorities and apex scientific bodies in those countries have recommended an incubation temperature of 27°C and an incubation period of 3 days. As regards the test procedure, it is the same as for 5 day BOD excepting the keeping of the second set of dilutions at 27°C for a period of 3 days. BOD calculations are also the same and the result is expressed as BOD (3 days at 27°C) mg/l.

Problem 14 – 1

The following results are obtained when conducting BOD test on a sewage sample.

Initial day dissolved oxygen	*= 7.2 mg/l*
Dissolved oxygen after 5 days	*= 4.5 mg/l*
Depletion of oxygen in blank	*= 0.2 mg/l*
Sample dilution	*= 50 times.*

Calculate BOD value of the sewage sample?

BOD(5 days at 20°C), mg/l	=	$(V_1 - V_2 - BC) \times D$
	=	$(7.2 - 4.5 - 0.2) \times 50$
	=	2.5×50
	=	**125**

It is usual to get two or more values for BOD for a sample as many dilutions are made. In that case, a question arises: which value is reliable and has to be taken? The BOD value for dilution that depletes more than 2 mg/l oxygen and has residual oxygen more than 0.5 mg/l is to be taken.

Problem 14 - 2

The following results are obtained when a waste water is tested for its BOD.

Sample Dilution	*1st day Titre value*	*5th day Titre value*
50	*7.4*	*0.3*
100	*7.4*	*3.9*
200	*7.4*	*5.8*
Blank	*7.4*	*7.2*

Calculate BOD values for the sample for all the dilutions?
Which BOD value is more reliable?

Blank correction = 7.4 – 7.2 = 0.2 mg/l
BOD = $(V_1 - V_2 - BC) \times D$
BOD value for 50 times dilution = $(7.4 - 0.3 - 0.2) \times 50 = 345$ mg/l
BOD value for 100 times dilution = $(7.4 - 3.9 - 0.2) \times 100 = 330$ mg/l
BOD value for 200 times dilution = $(7.4 - 5.8 - 0.2) \times 200 = 280$ mg/l

As regards 50 times dilution, almost all the oxygen is depleted. Hence BOD value obtained for this dilution is not reliable.

As regards 200 times dilution, only 1.6 mg/l oxygen is depleted. Hence BOD value for this dilution may not be reliable.

The 100 times dilution has depleted more than 2 mg/l oxygen leaving more than 0.5 mg/l oxygen. Hence the BOD value corresponding to 100 times dilution is appropriate and more reliable.

Immediate dissolved oxygen demand (IDOD)

Some effluents / highly polluted waters may contain large amounts of inorganic reducing substances such as sulfite, sulfide, ferrous iron etc., They consume the DO immediately and get oxidized. In these samples, the BOD value does not represent the oxygen demand due to organic substances only. To find out the real value of BOD, the oxygen depletion due to inorganic substances may be deducted from the observed BOD value. For this, each dilution has to be filled in three BOD bottles. The DO of one bottle has to be found out immediately. DO of another bottle has to be found out after 15 minutes. The third bottle has to be kept for incubation. Now calculate the real BOD value as follows:

DO of the dilution (titrated immediately) $= IDO_0$

DO of the dilution (titrated after 15 minutes) $= IDO_{15}$

$\therefore$ Immediate Dissolved Oxygen Demand (IDOD) $= (IDO_0 - IDO_{15}) \times$ Dilution factor

$\therefore$ Real BOD value = (observed BOD value) – IDOD

Checking the correctness of the technic

The technic can be checked by using a standard solution containing glucose and glutamic acid. Dissolve 150 mg glucose and 150 mg glutamic acid (both dried at 103°C for 1 hour) in distilled water and make upto 1000 ml. Prepare a 1 in 50 dilution using seeded dilution water and determine the BOD in the usual manner. The BOD should be approximately 220 mg/l. If the result is more than 240 mg/l or less than 200 mg/l, then seed, dilution water or experimental technic should be suspected.

Interpretation of the results

The BOD test may be considered as a wet oxidation procedure in which micro organisms act as the medium for the oxidation of organic matter into carbon dioxide, water and ammonia. Theoretically, an infinite time is required for the complete biological oxidation, but for practical purposes, the reaction may be considered to complete in 20 days. Even this 20 days period is too long and it has been found that a reasonably large percentage of the total

BOD is exerted in 5 days. Hence it should be remembered that the 5 days BOD value is only a portion of the total BOD and it amounts to about 70 – 80%. This is a large enough percentage and can be used for many considerations.

BOD test is found to be a more sensitive test for organic pollution, and it represents more or less the actual processes of deoxygenation that takes place in a stream and this is used as a laboratory model of deoxygenation process in the receiving waters. Generally, the higher the BOD, the greater is the amount of oxidizable organic matter present.

The Royal Commission on Sewage Disposal has classified rivers according to the 5–day BOD values as follows:

Approximate 5 - day B.O.D in mg/l	Classification
1	Very Clean
2	Clean
3	Fairly Clean
5	Doubtful
10	Bad

Toxic concentrations of bactericidal compounds would considerably reduce the BOD value owing to the suppression of bacterial activity. In such cases, the BOD test cannot be safely used as a measure of organic pollution and it is better to rely upon the permanganate test in those instances.

The BOD is determined by incubation in the dark and there are no chances for photosynthesis for samples containing algae, while algae in stream releases oxygen by photosynthesis which alters the polluting effect of an effluent. It is impossible to determine this effect quantitatively in 5 days BOD test and hence such cases should be considered on their merits.

The presence of suspended matter has also an important bearing on the BOD determination. Suspended matter in the discharged waste frequently will be deposited over a short distance immediately downstream of the out fall, where it may exert a very considerable effect on the local dissolved oxygen concentration. The effects of other bottom deposits, rooted plants and planktonic algae may also require consideration.

Another serious complication in the BOD determination is the nitrification. Though nitrification does not occur to any detectable extent during the 5 day BOD determination in raw and settled sewage and in industrial effluents, the BOD value of some of the treated effluents and river waters may be misleading in assessing plant performance or in calculating the effect of an effluent on a river.

In some cases such as the supernatant liquor from sewage digestion plants, abnormally low BOD values will be obtained, though the actual BOD is high. After about 5 days a sudden increase in BOD can be seen. This is because of the presence of anaerobic organisms and consequently the cause of lag period is related to a change from an anaerobic fermentation to aerobic. Hence, it can be concluded that any liquor which contains anaerobic organisms will behave in a similar manner and yield low BOD values.

Ratio of BOD and permanganate values

The 3 days BOD value of crude sewage is usually about 2 to 4 times the 4 hours permanganate value. It is to be mentioned that the permanganate value indicates the quantity of organic impurity to be treated while the BOD value indicates the relative ease with which biochemical oxidation takes place. The ratio of BOD : Permanganate value is less than three, the waste water will be less amenable to biochemical oxidation and if the ratio of three or more than three, it is easily treatable by aerobic biological processes.

Industrial wastes usually lower this ratio and ratios ranging from 0.2 to 1 are obtained. Well purified sewage effluents and other good effluents give ratios less than 1.0 (i.e) the BOD value is lesser than the permanganate value. With partially purified sewage and tank effluents the ratio is higher than 1. when the ratio exceeds 1 in river water, it is an indication of pollution by sewage, farm drainage or dairy wastes.

Application of BOD Data

There are many tests available for the determination of biologically oxidizable organic matter present in a waste water. But none of them are comparable to BOD test. This test almost reflects the natural conditions occurring in the stream and the BOD value is more or less represents the actual organic load of a waste water (despite many limitations). Therefore BOD test still remains the best available single test for assessing organic pollution and is the important criterion used in stream pollution control. This apart, for designing waste water treatment units such as trickling filter, activated sludge unit and oxidation ditches BOD value is the major criterion. It is the major factor in deciding the type of treatment and size of the units too. This test is also used to assess the efficacy of the above treatment units after commissioning.

CHAPTER - 15
BORON

Boron may enter into a water course by the discharge of industrial effluents containing boron compounds. The determination of boron is useful in assessing the suitability of a water for irrigation purpose. Boron is best determined by curcumin method.

CURCUMIN METHOD

Curcumin method is applicable to waters containing boron in the range of 0.10 – 1.0 mg/l.

Principle

When a water containing boron is acidified and evaporated with curcumin, a red colored product rosocyanin is formed which is dissolved in alcohol and the red color is measured colorimetrically.

Interferences

Nitrate nitrogen interferes at concentrations above 20 mg/l. If the total of calcium and magnesium hardness (as $CaCO_3$) exceeds 100 mg/l, opalescence is produced in the alcoholic solution of rosocyanin which affects the color measurement. However this interference can be eliminated by filtration of the final solution or by passing the original sample through a column of strongly acidic cation exchange resin, in the hydrogen form.

Reagents

1. Stock boron solution: Dissolve 571.6 mg anhydrous boric acid, H_3BO_3 in distilled water and make upto 1000 ml. in a volumetric flask.

$$1.0 \text{ ml} = 100 \ \mu g \text{ B}$$

2. Standard boron solution: Dilute 10.00 ml. of stock boron solution to 1000 ml. with distilled water in a volumetric flask.

$$1.0 \text{ ml} = 1 \ \mu g \text{ B}$$

3. Ethyl alcohol 95%

4. Curcumin reagent: Dissolve 40 mg finely ground curcumin and 5.0 g oxalic acid in 80 ml. of 95% ethyl alcohol in a beaker. Add 4.2 ml. conc. HCl and make up the solution to 100 ml. with ethyl alcohol in a volumetric flask. This reagent is stable for several days if kept in a refrigerator.

Procedure

1. Pipet 0.1, 0.20, 0.40, 0.60, 0.80, 1.00 ml. standard boron solution in to evaporating dishes (of 25 – 50 ml. capacity made up of vycor glass) and make up to 1 ml. with distilled water. Include a dish containing 1.0 ml. distilled water as the blank.
2. Add 4.0 ml. curcumin reagent to each dish and swirl gently.
3. Float the dishes on a water bath set at 55 ± 2°C until the contents are completely evaporated to dryness and the odor of HCl is gone.
4. Remove the dishes from the water bath and cool to room temperature, add 10.0 ml. 95% ethyl alcohol to each dish, stirring gently with a polyethylene rod to insure complete dissolution of the red colored product.
5. Transfer the contents of the dish to a 25 ml. volumetric flask or Nessler tube using 95% ethyl alcohol and make up to the mark with 95% ethyl alcohol and mix thoroughly.
6. Read the absorbance or transmittance of the standards at a wavelength of 540 nm after setting the blank at 100% transmittance using 20 mm cells.
7. Prepare a calibration curve relating optical density to mg boron.

   Note: Make the photometric reading within 1hour of drying.

Samples

Pipet 1.0 ml. or lower aliquot of the sample containing not more than 1 µg boron into an evaporating dish and dilute to 1.0 ml. with boron free distilled water.

(If the sample contains more than 1 mg/l boron dilute the sample with boron-free distilled water such that 1.0 ml. of the diluted sample contains approximately 0.5 µg boron).

Proceed from Step No. 2 to 6 as in the procedure for standards. Read the number of micrograms of boron equivalent to the optical density of the sample from the calibration graph.

If the sample contains a total quantity of calcium and magnesium greater than 100 mg/l (as $CaCO_3$) pass the sample through an ion exchange column which can be prepared as follows:

Soak the required quantity of ion exchange resin "Zeokarb 225" or equivalent grade in 1 + 5 HCl. Transfer the slurry to a column of approximately 50 cm. long and 1.2 cm. diameter fitted with a tap. Back wash the column with distilled water to remove air bubbles and to ensure even packing. Pass 50 ml. 1 + 5 HCl through the column and then wash with distilled water until free from chloride. Keep the resin remains covered with liquid at all times.

Pipet 25 ml sample (or a smaller aliquot if the sample is known to contain boron in high concentration) on to the resin column. Adjust the delivery tap so that the liquid flows out at a rate of 2 drops per second. Collect the eluate in a 50 ml. volumetric flask. Wash the column with several small portions of distilled water until the flask is full to the mark. Stopper the flask and mix well. Care should be taken to keep liquid remains over the resin at all times.

Pipet 1.0 ml. of the made up sample into an evaporating dish and proceed from step No.2 of the procedure for standards.

$$\text{mg/l Boron (as B)} = \frac{\mu\text{g.B}}{\text{ml.sample}}$$

CHAPTER - 16

CADMIUM

Cadmium salts are usually found in wastes from electroplating industries, textile printing, pigment works, lead mines and chemical industries. These effluents when discharged in to water courses, contribute cadmium. The following dithizone method is applicable to raw, potable and industrial waters and to some waste waters containing little interfering substances. When interferences are more, a different procedure has to be followed. (Refer "STANDARD METHODS", 20th edition).

DITHIZONE METHOD

Principle

Cadmium reacts with dithizone to form a red complex and is measured colorimetrically.

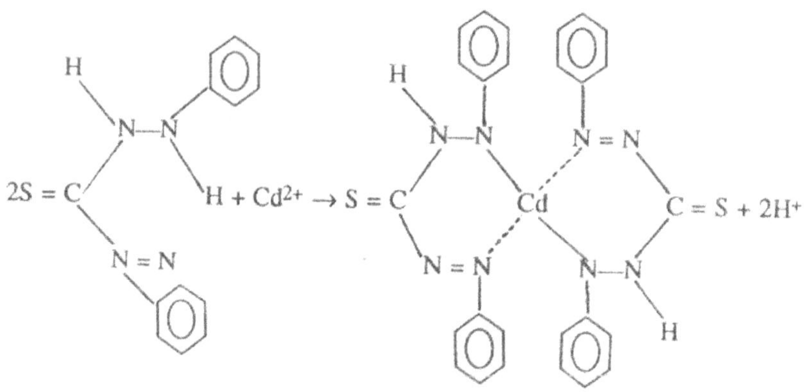

Interferences

The metal ions normally present in water do not interfere under the conditions of the test. Lead and zinc interfere above concentrations 6 mg/l and 3 mg/l respectively. The interference by copper and nickel are suppressed by complexing them with cyanide.

Reagents

1. Cadmium stock solution: Dissolve 100mg. pure cadmium metal in 50 ml. 10% nitric acid in 250 ml. beaker. Boil to expel the oxides of nitrogen and make up to 1000 ml. in a volumetric flask.

1.0 ml. = 100 μg Cd

2. Cadmium working solution: Pipet 10.0 ml. cadmium stock solution in to a 1000 ml. volumetric flask and make up to the mark with 1% nitric acid.

1.00 ml. = 1 μg Cd

3. Thymol blue indicator solution: Dissolve 400mg. thymol sulphonepthalein sodium salt in 100 ml. distilled water.

4. Sodium hydroxide solution: Dissolve 24g. NaOH in distilled water and dilute to 100 ml.

5. Tartaric acid solution: Dissolve 20g. tartaric acid $H_2C_4H_4O_6$ in distilled water and make up to 1000 ml. Keep it in a refrigerator. The solution must be cold when used.

6. Chloroform

7. Hydroxylamine hydrochloride solution: Dissolve 20g. hydroxylamine hydrochloride, $NH_2OH.HCl$ in distilled water and make up to 100 ml.

8. Dithizone stock solution: Dissolve 80mg. well powdered dithizone in 100 ml. chloroform. Place it in an amber bottle and keep it in refrigerator. Use the solution while still cold.

9. Dithizone working solution: Dilute 50 ml. dithizone stock solution to 500 ml. with chloroform.

10. Sodium hydroxide-potassium cyanide solution I: Dissolve 40g. NaOH and 1.0g. potassium cyanide, KCN in 100 ml. distilled water. Store it in a polythene bottle.

11. Sodium hydroxide-potassium cyanide solution II: Dissolve 40g. NaOH and 0.05g. potassium cyanide, KCN in 100 ml. distilled water. Store it in a polythene bottle.

12. Hydrochloric acid: conc. (Sp. gr 1.18)

13. Sodium potassium tartrate solution: Dissolve 25g. sodium potassium tartrate $C_4H_4O_3KNa.4H_2O$ in 100 ml. distilled water.

Procedure

Standards

1. In to a series of separating funnels measure appropriate volumes of cadmium working solution covering the range up to 20 μg Cd. Dilute to 25 ml. with distilled water.
2. Include a blank containing 25 ml. distilled water as the blank.
3. Add the following reagents in order with mixing after each addition.
   (i) 0.25 ml. hydroxylamine hydrochloride solution,
   (ii) 15.0 ml. dithizone working solution, and
   (iii) 5 ml. sodium hydroxide-potassium cyanide solution II.
4. Shake well, allow the phases to separate and filter the chloroform layer through a small plug of cotton wool in a funnel.

5. Measure the absorbance, using a spectrophotometer at 518 nm. or a filter photometer with a suitable green filter, taking water as the reference. Prepare a calibration curve.

Samples

a) Place appropriate volume of the sample containing 1 to 20 μg Cd in a separating funnel. If the sample contains less than 10 μg Cd, place 200 ml. of the sample in an evaporating dish. Add 0.5 ml. conc. HCl and evaporate to 20 ml. Add a few drops of thymol blue indicator solution and neutralize to the appearance of yellow color using sodium hydroxide solution. The pH of the sample now will be at about 2.8. Place it in a separating funnel and dilute to 25 ml. with distilled water.

(Digest appropriate volume of the sample if it contains any organic matter, as described in Chapter – 9 "PRELIMINARY ACID DIGESTION OF THE SAMPLE". Neutralise a suitable aliquot (containing not more than 20 μg Cd) of the digested sample to pH 2.8 using thymol blue indicator. Place it in a separating funnel and dilute to 25 ml. with distilled water).

b) To the sample in the separating funnel add the following reagents in order with mixing after each addition.
   (i) 1 ml. sodium potassium tartrate solution,
   (ii) 5 ml. sodium hydroxide-potassium cyanide solution I,
   (iii) 1 ml. hydroxylamine hydrochloride solution, and
   (iv) 15 ml. dithizone stock solution.

Shake well* and draw the chloroform layer in to a second separating funnel containing 25 ml. tartaric acid solution. Add 10 ml. chloroform to the first separating funnel and shake well. Drain the chloroform layer again in to the second separating funnel. Do not permit even a trace of the aqueous layer in to the second separating funnel.

Note 1: Complete the extraction operations as quickly as possible since cadmium dithizonate decomposes on prolonged contact with strong alkali saturated with chloroform.

Note 2: If lead and zinc are present in appreciable concentrations, the pink color of their dithizonates persists for several primary extractions. It should not be mistaken that cadmium has not been completely extracted.

---

* Release the vapor pressure by opening the stopper rather than opening the stop cock

Note 3: Absence of orange color of excess dithizone in the aqueous phase indicates that the volume of aliquot taken is too high. The determination has to be repeated then with smaller volumes of the sample.

c) Shake the separating funnel containing the combined chloroform extracts in tartaric acid solution for about two minutes. Allow the phases to separate and discard the lower chloroform layer. Wash the contents of the funnel by shaking with 5 ml. chloroform. Discard the chloroform layer making the separation as fine as possible. Gently blow air in to the funnel to evaporate the floating drops of chloroform. All the cadmium now will be in the tartaric acid solution and is free from interfering metallic ions.

Proceed from step No.3 of the procedure for standards and measure the absorbance. Find out the microgram Cd equivalent to the net optical density of the sample from the calibration graph. Visual comparison can also be made using nessler tubes and chloroform for dilution.

Express the result as mg cadmium (as Cd) per litre of the sample.

CHAPTER - 17

CALCIUM

Calcium is the common constituent of natural water. The presence of calcium in water is mainly due to its passage through or over deposits of lime stone, dolomite, gypsum and other gypsiferous materials. Calcium and magnesium are the two major scale-forming constituents in most raw water supplies. Calcium in industrial waters is undesirable mainly due to the formation of scales. It reacts with alkaline solutions to form precipitates and curds, especially during washing operations using soap. Calcium is usually determined on industrial waters and potable waters only. When it is necessary to determine calcium in an industrial effluent, or sewage the organic matter present in it should be destroyed first by evaporation and ignition.

Among the methods available for the determination of calcium, the EDTA-titrimetric method and gravimetric method are more accurate. The former can be conducted very quickly while the latter is a time consuming one but subject to less interference.

EDTA-TITRIMETRIC METHOD

Principle

The pH of the sample is made sufficiently high (12 – 13) to precipitate magnesium as hydroxide and calcium only is allowed to react with EDTA in the presence of a selective indicator.

$$Mg^{2+} + 2NaOH \rightarrow Mg(OH)_2 + 2Na^+$$

$$Ca^{2+} + 2EDTA \rightarrow Ca(EDTA)_2 + 2Na^+$$

Interferences

Alkalinity in excess of 300 mg/l may cause indistinct end-points with hard waters. Under the conditions of the test, the following concentrations of ions cause no interference.

Aluminium 5 mg/l	Manganese 10 mg/l	Lead 5 mg/l
Iron (Fe^{2+}) 20 mg/l	Zinc 5 mg/l	Tin 5 mg/l
Iron (Fe^{3+}) 20 mg/l	Copper 2 mg/l	

The gravimetric procedure may be employed when the concentration of the above interfering substances exceeds the limit and when orthophosphates are present which precipitate calcium at the pH of the test.

Reagents

1. Sodium hydroxide, 1N: Dissolve 40g. Sodium hydroxide, NaOH in about 100 ml. distilled water and make up to 1000 ml with distilled water.

2. Indicators: Any one of the following two indicators may be used.

- **(a) Murexide (Ammonium Purpurate) indicator:** It is available commercially as tablet. It can be also prepared by dissolving 150mg. of the dye in 100g. ethylene glycol or by grinding 200mg. murexide with 100g. solid NaCl to 40 to 50 mesh in a mortar.
- **(b) Patton and Reeder's indicator:** The indicator is 2-hydroxy-1-(2-hydroxy-4-sulfo-naphthyl azo)-3-naphthoic acid. Prepare the indicator by grinding in a mortar 1g. dye with 100g. sodium sulfate.

3. Standard EDTA titrant, 0.02N: See Chapter – 31 "HARDNESS (TOTAL)".

Procedure

1. Place in a conical flask 50.0 ml. of the sample or a suitable aliquot that consume not more than 15 ml. EDTA titrant. Dilute to 50 ml. with distilled water.
2. Add 2.0 ml. sodium hydroxide solution per 50 ml. volume to produce a pH of 12-13 and mix well by swirling the flask.
3. Add 0.1-0.2 g of the indicator powder (If murexide indicator tablet is used, add 1 tablet and grind well with a glass rod and wash the glass rod well into the conical flask with distilled water).
4. Titrate immediately with EDTA to the proper end point. With Patton-Reeder's Indicator, the end point is from wine red to pure blue; with murexide indicator, the end point is from pink to purple.

Note – 1: The titration should be performed within 5 minutes after the addition of alkali.

Note – 2: With hard waters of alkalinity exceeding 300 mg/l (as $CaCO_3$), neutralize the alkalinity with acid, boil for one minute. Cool and proceed from Step No.2.

Note – 3: For samples containing a high residual chlorine, dechlorinate with suitable amount of sodium thiosulfate solution, allow 15 minutes and then proceed.

Calculation

$$\text{mg/l Calcium (as } CaCO_3) = \frac{\text{ml.EDTA Titrant (0.02N) x 1 x 1000}}{\text{ml. sample taken for determination}}$$

If the result has to be expressed in terms of Ca,

$$\text{mg/l Calcium (as Ca)} = \text{mg/l Calcium (as } CaCO_3) \text{ x } 0.40$$

When very low harness waters are tested, colorimetric methods may be employed. For details refer, "Calmagite Colorimetric Method" described in "Hach Water Analysis Hand Book" – 4th Edition, (Revision – 2) – 2003, published by Hach Company, Colorado, U.S.A.,

GRAVIMETIC METHOD

(See Chapter - 37, "MAGNESIUM")

CHAPTER - 18

CALCIUM CARBONATE STABILITY TEST (OR) CORROSIVITY TEST (OR) MARBLE TEST

This test is used to determine the stability of a water with respect to calcium - carbonate and to determine the lime dose to be applied to the water to prevent its corrosive action.

Principle

The pH and alkalinity of the water is determined before and after treatment with calcium carbonate. Any increase in pH indicates corrosivity. The increased alkalinity is used to determine the lime dose.

Procedure

1. Fill a 250 ml. glass-stoppered bottle of pyrex or other resistant glass with the sample. Add about 0.25g. chemically pure precipitated calcium carbonate and stopper it without entrainment of air.
2. Shake well at frequent intervals for atleast three hours to make the calcium carbonate to saturate with the sample and to attain equilibrium.
3. Allow the sample to settle overnight. Carefully remove a portion of the supernatant and divide it in to two portions.
4. Filter one portion through a filter paper (Whatmann No. 42 or 50 or equivalent). Discard the first portion passing through the filter paper and determine the total alkalinity of the remainder.
5. Determine pH value of the unfiltered sample.

Any increase in pH value and alkalinity indicates that the water is corrosive. The observed increase in pH value and alkalinity should be obtained by the addition of lime, to make the water non-corrosive and for the formation of a calcium carbonate protective film on the surface of the piping. The needed lime dose can be calculated as follows:

1 ppm of hydrated lime will increase the alkalinity by 1.35 ppm. Suppose if the alkalinity of the sample has to be increased from 60 to 90 ppm, to make it non-corrosive then 30/1.35 or 22.2 ppm hydrated lime would be needed.

CHAPTER - 19

CARBON DIOXIDE (FREE)*

(See also, Chapter - 10, "ACIDITY")

Carbon dioxide is a normal component of all natural waters. It dissolves in water in varying amounts and the dissolution depends on the temperature, pressure and mineral content of the water. Surface waters normally contain less than 10 mg/l free carbon dioxide while some ground waters contain 30 to 50 mg/l. The presence of free carbon dioxide in water has an important consideration because of its corrosive properties. Hence free carbon dioxide in water is an important parameter and is usually determined by titration with a standard alkali to pH 8.3.

TITRIMETRIC METHOD

Principle

Sodium hydroxide reacts with carbon dioxide (present as carbonic acid) in water to form sodium bicarbonate and water. Phenolphthalein is used as the indicator which changes its color to pink at the end point.

$$CO_2 + H_2O \longrightarrow \underset{\text{Carbonic acid}}{H_2CO_3^-}$$

$$H_2CO_3^- \longrightarrow HCO_3^- + H^+$$

$$HCO_3^- + H^+ + NaOH \longrightarrow NaHCO_3 + H_2O$$

Reagents

1. Phenolphthalein indicator solution: See Chapter – 11, "ALKALINITY".

2. Standard sodium hydroxide solution, 0.0227N: Dilute 22.7 ml. of exactly 1N.NaOH solution (prepare as in Chapter – 10 "ACIDITY") to 1000 ml. with carbondioxide free distilled water in a volumetric flask. Place it in a pyrex bottle fitted with a carbon-dioxide trap. Prepare daily.

$$1.0\text{ml } 0.0227\text{N.NaOH} \equiv 1.0 \text{ mg } CO_2$$

* Free carbon dioxide is the carbon dioxide present in excess of that present as bicarbonate.

Procedure

When the Sample is known to contain only CO_2

1. Connect to the sample source a rubber tubing and discharge steadily to the bottom of a 100 ml. Nessler tube or graduated cylinder and allow the water to overflow for one to two minutes. Withdraw the cylinder while the water is still flowing. Draw off the surplus water above the 100 ml. mark with a pipet.

2. Add 5-10 drops of phenolphthalein indicator solution. If the sample turns pink, free carbon dioxide is absent. If no pink color appears, titrate with sodium hydroxide solution stirring gently using a plunger type stirrer until a pink color persists for 30 seconds when viewed through the depth of the sample.

$$\text{mg/l Free Carbon dioxide (as } CO_2) = \frac{\text{ml. NaOH used x 1 x 1000}}{\text{ml. sample taken for determination}}$$

When the sample is known to contain both free carbondioxide and other acids, follow the procedures given in the sections 'Methyl orange acidity' and 'Total Acidity' in Chapter – 10, "Acidity". The difference between total acidity and methyl orange acidity is the carbondioxide content of water. See next section.

FREE CARBON DIOXIDE IN THE DEGASSER EFFLUENT

Generally the pH value of most of the natural waters is more than 7 and consequently carbondioxide acidity is the only acidity present in them. However treated waters such as the effluent emerging from cation exchanger unit contains both mineral acidity and CO_2 acidity. Usually this effluent is passed on to degasser for the removal of free CO_2. The performance of the degasser can be assessed by determining the free CO_2 remaining after degasification. Free CO_2 in water can be determined by performing two separate titrations – one for the mineral acidity and the other for total acidity. The difference between total acidity and mineral acidity is the CO_2 acidity. The procedure for these titrations have been given in the previous Chapter – 10. An illustrative problem is given below.

Problem 19 – 1

An effluent from a degasser is titrated with 0.02N sodium hydroxide for methyl orange – acidity and total acidity. Volume of sample taken for both the titrations is 50 ml. The titre value for methyl orange acidity is 10 ml and total acidity is 12 ml. Findout the free CO_2 content of the degasser effluent? Also interpret the result?

Methyl Orange Acidity	=	$\frac{10 \times 1 \times 1000}{50}$
	=	200 mg/l (as $CaCO_3$)
Total Acidity	=	$\frac{12 \times 1 \times 1000}{50}$
	=	240 mg/l (as $CaCO_3$)
Free CO2	=	(Total Acidity – MO Acidity)
	=	(240 – 200)
	=	40 mg/l (as $CaCO_3$)

Here free CO_2 is expressed in terms of $CaCO_3$. It can be expressed in terms of CO_2 which is more convenient in interpreting the result.

1 mg $CaCO_3$	≡	0.44 mg CO_2
∴ The above sample contains	=	40 x 0.44
	=	17.6 mg/l CO_2

Interpretation of the result

Normally, the residual CO_2 content of degassed water will be 5 to 6 ppm (as CO_2). In the above case, 17.6 ppm CO_2 is present which indicates the poor performance of the degasser. Therefore the degasser has to be set right to leave only a residual fo 5 to 6 ppm CO_2.

FREE CARBONDIOXIDE IN STEAM

Free carbondioxide present in feed water is completely removed by deaeration. However steam contains carbondioxide in significant concentration. This carbondioxide is derived from the decomposition of bicarbonates and carbonates (present in feed water) under the influence of temperature and pressure in the boiler. Determination of the amount of such free carbondioxide is essential for subsequent determination of neutralizing and/or filming amines to protect the condensate system from corrosion. It can be calculated from the bicarbonate and carbonate values of the feed water. The following is an illustration of calculating the amount of carbondioxide that would present in the steam by the decomposition of bicarbonates and carbonates.

Problem 19 – 2

On analysis, a feed water sample shows a phenolphthanelin alkalinity (P) value of 2 ppm (as CaCO3) and total alkalinity (T) (Methyl orange alkalinity) value of 10 ppm. Find out the amount of carbondioxide that would present in steam.

Decomposition of carbonates and bicarbonates will not take place to the extent of 100 percent. Only 80% carbonates and 90% bicarbonates are decomposed and converted into carbondioxide. The following calculations are based on such percentage of decomposition taking place in the boiler.

CO_2 produced from carbonates and bicarbonates = [Carbonate (as $CaCO_3$) x 0.35]
+
[Bicarbonate (as $CaCO_3$) x 0.79]

Step–1 **CO_2 produced from Carbonates**

In this case, P alkalinity is lower than ½ T.

Therefore carbonate (as $CaCO_3$) = 2P (See Table 11-1)
= 2 x 2 = 4

∴ CO_2 produced = 4 x 0.35
= **1.45 ppm**

Step–2 **CO_2 produced from bicarbonates**

Since P < ½ T,

Bicarbonate (as CaCO3)	= T - 2P (See Table 11-1)
	= 10 – 4 = 6
∴ CO_2 produced	= 6 x 0.79
Total Carbondioxide	= **4.74 ppm**

Step–3

Total CO_2 produced	= (CO_2 produced from Carbonates) + (CO_2 produced from bicarbonates)
	= 1.42 + 4.74
	= 6.16 ppm

∴ The amount of free CO_2 that would present in steam = **6.16 ppm**

CHAPTER - 20

CHEMICAL OXYGEN DEMAND (COD)

This test is highly useful to find out the pollutional strength of industrial effluents and sewage. Chemical oxygen demand as the name implies is the oxygen requirement of a sample for oxidation of organic and inorganic matter. As the oxidizable inorganic matter is usually negligible in comparison with the quantity of organic matter, COD is generally considered as the oxygen equivalent of the amount of organic matter oxidisable by potassium dichromate.

Note that COD value is not the amount of organic matter but is the amount of oxygen required for oxidation. ASTM[*] defines chemical oxygen demand as the amount of oxygen (expressed in mg/l), consumed under specified conditions in the oxidation of organic and oxidisable inorganic matter corrected for the influence of chlorides.

Bio-chemical oxygen demand and chemical oxygen demand are two separate parameters. One should not confuse between these two.

BOD is the requirement of oxygen due to biodegradable organic matter only and is a biological phenomenon carried out at more or less natural conditions that exist in the receiving waters. In COD test, the sample is subjected to a chemical oxidation induced by chemical reagents and it is an artificial oxidation by which both biological oxidisable and biologically inert organic matter are oxidized. Hence the COD value for a sample is always higher than BOD value.

As the inert organic materials are also oxidized, this test (COD) is of great value particularly in evaluating the pollution potential of effluents such as pulp and paper industries, rayon industries and chemical industries. Though these effluents contain enough organic matter, the BOD of them will be much less as the materials are not biodegradable and lead to wrong conclusion that less organic matter is present and consequently to a less effective treatment system. In these cases COD reveals the real pollution potential.

---

[*] American Society for Testing and Materials.

The COD test is recommended as a supplement to BOD test. The major advantage of the COD test is the short time required to carryout rather than the longer 5 day (or 3 day) procedure for BOD. If waste water remains relatively uniform over a long period, a ratio between BOD and COD may be evolved and then BOD test substituted with COD test. In conjunction with BOD values, COD data are helpful in knowing toxic conditions and the presence of biologically resistant organic substances.

Though the COD test is advantageous, it is to be noted that it does not differentiate between biodegradable and bio-inert materials. Further, the rate at which the biologically oxidisable materials stabilize (under conditions that exist in nature) cannot be found out, whereas BOD test indicates that rate.

Sample collection and preservation

The sample collected for COD determination should be representative. It is best to conduct COD test as soon as possible after collection. If this is not possible, add conc.H_2SO_4 to the sample to bring the pH below 2. Analyze the sample within 7 days if the acidified sample is stored at room temperature or within 28 days if the sample is kept at 4^oC in a refrigerator.

Principle

The organic matter of the sample is oxidized to water, carbon-di-oxide and ammonia by refluxion with a known excess of potassium di chromate in a 50% sulfuric acid solution. The excess dichromate is titrated with a standard solution of ferrous ammonium sulfate solution.

$$6Fe^{2+} + Cr_2O_7^{2-} + 14H^+ \rightarrow 6Fe^{3+} + 2Cr^{3+} + 7H_2O$$

Usually silver sulfate is added as a catalyst to promote oxidation of certain classes of organic compounds such as straight chain aliphatic compounds like acetic acid, amino acids etc. Mercuric sulfate is added to eliminate the interference due to chlorides.

Note: Aromatic hydrocarbons such as benzene and its homologs, pyridine and other heterocyclic compounds containing nitrogen, [(e.g.) pyrrole, pyrolidine, proline, nicotinic acid] paraffins and naphthenes, are not oxidized by dichromate under the experimental conditions.

Interferences and Elimination

Chloride is the most common interfering ion. During COD test, it is oxidized to gaseous chlorine and for this oxidation some of the dichromate is utilized resulting in higher COD values.

$$6Cl^- + Cr_2O_7^{2-} + 14H^+ \rightarrow 3Cl_2 + 2Cr^{3+} + 7H_2O$$

Addition of mercuric sulfate suppresses chloride interference upto 2000 mg/l by forming poorly ionized soluble mercuric chloride complex.

At concentrations above 2000 mg/l, mercuric sulfate is less effective. In such cases the sample may be diluted appropriately such that the chloride concentration of the diluted sample is within 2000 mg/l. Nitrites interfere by oxidation to nitrate for which some dichromate would be utilized leading to high COD results (1 mg nitrite nitrogen exerts a COD of 1.1mg). Addition of sulfamic acid eliminates this interference. But nitrite seldom occurs in effluents excepting those from dye house and dye manufacture.

Ferrous iron and sulfides also consume dichromate for their oxidation leading to positive errors. If high concentration of these ions are present, a separate analysis and correction is necessary.

Apparatus

A round bottomed reflux flask of 500 ml. capacity with a ground glass 24/40 (B-24) neck and anyone of the condensers-Davies double surface, or Graham, or Liebig or West with 300 mm jacket. Use of reflux condenser prevents the loss of volatile matter during digestion. Do not apply any stop cock grease on the ground glass joint as this may lead to high COD values.

Reagents

1. Mercuric sulfate crystals

2. Sulfuric acid – Silver sulfate reagent: Dissolve 10.1 gm silver sulfate (Ag_2SO_4) in 1000 ml. conc. H_2SO_4. About 2 days will be required for dissolution.

3. Potassium dichromate, 0.25N: Dissolve 12.258g. potassium dichromate $K_2Cr_2O_7$ (previously dried at 103°C for 2 hrs) in distilled water and make upto 1000 ml. in a volumetric flask.

1.0ml. exactly 0.025 N potassium dichromate ≡ 2.0mg oxygen.

4. Potassium dichromate, 0.025N: Pipet 50 ml. of 0.25 N potassium dichromate solution into a 500 ml. volumetric flask and make upto the mark with distilled water.

5. Ferroin indicator: Dissolve 695 mg ferrous sulfate $FeSO_4.7H_2O$ and 1.485g. 1, 10-phenanathroline monohydrate together in distilled water and dilute to 100 ml.

6. Ferrous ammonium sulfate, 0.1N: Dissolve 39.29g. ferrous ammonium sulfate, $Fe(NH_4)_2(SO_4).6H_2O$ in distilled water. Add 20 ml. conc. sulfuric acid. Cool and make upto 1000 ml. in a volumetric flask. Standardize it daily against 0.25N potassium dichromate as follows:

Standardization of ferrous ammonium sulfate: Pipet 10.0 ml. 0.25 N potassium dichromate solution into a 250 ml. conical flask. Dilute to 100 ml. and add 30 ml. conc. H_2SO_4 and cool to room temperature. Add 2 to 5 drops of ferroin indicator and titrate with ferrous ammonium sulfate solution. The end point is the color change from blue green to reddish brown.

$$\text{Normality of ferrous ammonium sulfate solution} = \frac{(10 \times 0.25)}{\text{ml. Ferrous ammonium sulfate}}$$

7. Ferrous ammonium sulfate, 0.01N: Pipet 50 ml. of 0.1 N ferrous ammonium sulfate and make upto 500 ml. in a volumetric flask. Standardize daily against 0.025 N potassium dichromate as described above.

8. Sulfamic Acid

Procedure

Precaution: Do not use de-ionized water for dilution as this may lead to high results because of the organic matter content.

1. Shake the sample well, so that the contents are mixed thoroughly. (It is also recommended that if large amounts of settleable solids are present, the sample may be homogenized using a blender).
2. Place 50 ml. or an aliquot in the reflux flask. Add mercuric sulfate according to the chloride concentration of the aliquot (i.e.) a ratio of 1:10, chloride : mercuric sulfate has to be maintained. For example, if the aliquot contains 10 mg chloride, add 100 mg mercuric sulfate. Mix well
3. Measure suitable amount of sulfuric acid–silver sulfate reagent in a measuring cylinder. The amount should be equal to the sum of sample volume and dichromate volume. For example, if 50 ml. aliquot is taken and 25 ml. dichromate is added, then the amount of sulfuric acid–silver sulfate reagent to be added is 75 ml. From this amount, initially

transfer carefully and slowly 5 ml. only to the reflux flask and mix thoroughly to dissolve the mercuric sulfate. Cool while mixing to avoid possible loss of volatile matter in the sample.

4. Pipet a pre–determined quantity of 0.25 N (or 0.025 N) potassium dichromate solution into the flask and mix well. Use 0.25 N potassium dichromate solution for samples having a COD of 50 mg/l or more and use 0.025 N $K_2Cr_2O_7$ solution for samples having a COD of less than 50 mg/l.
5. Carefully add the remaining sulfuric acid–silver sulfate reagent in the measuring cylinder to the reflux flask, a little quantity at a time, cooling the flask simultaneously under tap.

   Note: Always maintain the ratio of volume of sulfuric acid to sample volume + dichromate volume as 1 : 1. If it is smaller, the oxidizing power of the solution will decrease and if it is larger, blank consumption of dichromate will be more.
6. Add a few antibumping granules (already cleaned with sulfuric acid and rinsed with distilled water) and place the flask in position. Attach the condenser and start the cooling water. Reflux for 2 hrs or for a suitable small period as required.
7. Remove the flame, allow the flask to cool and wash the condenser with distilled water into the flask.
8. Transfer the contents of the flask to a 500 ml. conical flask and dilute appropriately with distilled water and cool it well. The final volume after dilution should be about 2.5 times the volume of sample + dichromate + sulfuric acid.
9. Add 2 to 3 drops of ferroin indicator and titrate with 0.1 N ferrous ammonium sulfate solution. The end point is the sharp color change from blue green to reddish brown (Do not mind even if the blue-green color reappears again within a few minutes).
10. Conduct a blank using distilled water in place of sample. (The quantity of other reagents added are the same as that added for the sample). Note the titre value.
11. Standardize the ferrous ammonium sulfate solution, on each day of determination as described in the section "Reagents".

Calculation

$$mg/l\ COD = \frac{\text{(Blank titre value - sample titre value)} \times \text{Normality of ferrous ammonium sulfate} \times 8 \times 1000}{\text{ml. sample taken for determination}}$$

Note: The correctness of the analysis can be checked against a standard solution of glucose. Dissolve 486.6mg. glucose in distilled water and make upto 1000 ml. COD of this solution is 500 mg/l.

Procedure for samples of high nitrite content

Add 10 mg sulfamic acid per mg nitrite nitrogen to the reflux flask before the addition of dichromate solution and proceed as described above.

Procedure for polluted waters

The general procedure is applicable to polluted waters also. Use higher aliquots of sample and 0.025 N dichromate solution and titrate with 0.01 N Ferrous ammonium sulfate solution.

Problem 20 - 1

Following is a brief of COD test carried out on an untreated dairy waste. Calculate COD value of the waste.

ml. sample	*mg.$HgSO_4$*	*ml. 0.25N $K_2Cr_2O_7$*	*ml. H_2SO_4+Ag_2SO_4*	*Final volume (ml)*	*Titre value*
10	10	25	35	170	39.1
Blank (10ml distilled water)	10	25	35	170	53.2

Normality of ferrous ammonium sulfate used = 0.1037 N

$$mg/l\ COD = \frac{(\text{Blank titre value - sample titre value}) \times \text{Normality of ferrous ammonium sulfate} \times 8 \times 1000}{\text{ml. sample taken for determination}}$$

$$mg/l\ COD = \frac{(59.2 - 39.1) \times 0.1037 \times 8 \times 1000}{10}$$

= 1667.49

= Rounded to **1667**

CHAPTER - 21

CHLORIDE

Chloride is found practically in all natural waters. It may be derived from stratum or from contamination by sea water, sewage or industrial effluents. Analysis of chloride in industrial water is useful in many ways. Besides knowing the concentration of chlorides, it helps to determine the concentration of water that has occurred in boiler and cooling systems. Due to its high solubility, chloride remains in solution and gets concentrated in boiler water and evaporative cooling waters in an exact ratio to the amount of evaporation and thus useful in estimating how many times the water has concentrated.

Two titrimetric procedures are given for the determination of chloride. The first one involves the titration with standard silver nitrate solution and can be applied with reasonable accuracy to all types of water, particularly to waters of low purity and to waste waters. The mercuric nitrate method is applicable to relatively pure waters. Further this method is more accurate and can be applied to waters of low chloride concentrations. Phosphates do not interfere in this method.

Chloride ion selective electrodes are also available for direct measurement of chlorides. The chloride concentration range that could be measured with this electrode is stated to be 20 to 35000 mg/l.

SILVER NITRATE METHOD

Principle

Silver nitrate reacts with chloride ions to form silver chloride. The completion of reaction is indicated by the red color produced by the reaction of silver nitrate with potassium chromate solution which is added as an indicator.

$$AgNO_3 + Cl^- \rightarrow AgCl \downarrow + NO_3^-$$
$$2AgNO_3 + K_2CrO_4 \rightarrow Ag_2CrO_4 + 2KNO_3$$

Interferences

The substances normally found in potable waters do not interfere. Bromide, iodide and cyanide will titrate as chloride. Orthophosphate in excess of 25 mg/l interfere. Iron in excess of

10 mg/l masks the end point. Sulfide, sulfite, thiosulfate and thiocyanate ions also interfere. However, these interferences can be eliminated by the addition of hydrogen peroxide.

Reagents

Use chloride – free distilled water for the preparation of all reagents:

1. Standard silver nitrate titrant, 0.0282N: Dissolve 4.791g. silver nitrate, $AgNO_3$ in distilled water and make upto 1000 ml. in a volumetric flask. Standardize it against 0.0282N sodium - chloride solution as described in the 'procedure for clear samples'. Store it in an amber bottle.

$$1.0 \text{ ml. of } 0.0282N.AgNO_3 \equiv 1.0 \text{ mg Cl}$$

2. Standard sodium chloride titrant, 0.0282N: Dissolve 1.648g. sodium chloride, NaCl and make upto 1000 ml. in a volumetric flask.

$$1.00 \text{ ml.} = 1.0 \text{ mg Cl.}$$

3. Potassium chromate indicator solution: Dissolve 25g. potassium chromate, K_2CrO_4 in 100 ml. distilled water. Add silver nitrate solution dropwise until a slight red precipitate is formed. Allow to stand for 12 hours. Filter and dilute the filtrate to 500 ml. with distilled water.

4. Special reagents for removal of interferences:

(i) **Aluminium hydroxide suspension:** Dissolve 125g. potassium aluminium - sulfate, $K_2SO_4.Al_2(SO_4)_3.24H_2O$ or aluminium ammonium sulfate, $AlNH_4(SO_4)_2.12H_2O$ in 1000 ml. distilled water. Warm to 60°C and add slowly with stirring 55 ml. conc. ammonia solution. Allow the precipitate to settle, for about one hour. Wash by decantation with distilled water to make the precipitate free from chloride. Check it by treating a portion of the decantate every time with silver nitrate solution. After the precipitate is free from chloride, dilute it to 1000 ml. with distilled water.

(ii) **Sodium hydroxide, 1N:** Dissolve 40g. sodium hydroxide NaOH in distilled water and dilute to 1000 ml.

(iii) **Sulfuric acid, 1N:** Place 27.7 ml. conc. H_2SO_4 (Sp. gr 1.84) in 500 ml. distilled water and dilute to 1000 ml.

(iv) **Hydrogen peroxide:** 30%.

Procedure

For clear samples

Place 100 ml. of the sample or an aliquot containing not more than 10 mg chloride in a porcelain basin of about 300 ml. capacity. (If the pH of the sample is in the range of 7 to 9.5 it can directly be titrated. If the pH is not within this range, adjust the pH to be within this range using sulfuric acid or sodium hydroxide solution). Add 1 ml. potassium chromate indicator solution. Titrate against standard silver nitrate solution with constant stirring until a slightest perceptible reddish coloration persists. Conduct a blank by placing 100 ml. chloride-free distilled water instead of sample. A blank of 0.2 to 0.3 ml. is usual for the method.

For Colored and Turbid Samples

If the sample is highly colored or turbid, add 3 ml. aluminium hydroxide to a measured volume of the sample in a beaker. Stir well and allow to settle. Filter, and wash the precipitate with chloride-free distilled water. Combine the filtrate and washings, (bring the pH in the 7-9.5 range if necessary) and titrate as described above.

For Samples containing Cyanide, Sulfide, Thiosulfate and Thiocyanate

Measure a suitable aliquot of the sample in a 400 ml. beaker and dilute to 150 ml. with chloride-free distilled water. Add 2.5ml. 30% hydrogen peroxide and boil for five minutes. (see note). Decompose the excess peroxide by adding five drops of potassium chromate solution and by boiling for 2 minutes. Add 5 ml. of 20% sodium carbonate solution and boil for one minutes. Cool and filter. Make upto 250 ml. with distilled water. Place 100 ml. of the made up solution in a porcelain basin. Adjust the pH to be within the range 7.0 – 9.5 with sulfuric acid or sodium hydroxide solution and titrate with standard silver nitrate as described above.

Note: Test at this point for thiocyanate by spotting the solution with ferric chloride. If any red coloration or precipitate appears it indicates the presence of thiocyanate. Then repeat the heating with hydrogen peroxide until the solution is free from thiocyanate.

Calculation

$$\text{mg/l chloride (as } Cl^-) = \frac{(\text{ml. } AgNO_3 \text{ solution for sample - ml. } AgNO_3 \text{ for blank}) \times \text{Normality of } AgNO_3 \times 35.45 \times 1000}{\text{ml. sample taken for estimation}}$$

If the normality of $AgNO_3$ soltuion is exactly 0.0282 N, then

$$\text{mg/l chloride (as } Cl^-) = \frac{(\text{ml. } AgNO_3 \text{ solution for sample - ml. } AgNO_3 \text{ for blank}) \times \text{Normality of } AgNO_3 \times 1 \times 1000}{\text{ml. sample taken for estimation}}$$

MERCURIC NITRATE METHOD

Principle

Mercuric ions react with the chloride ions to form a stable and soluble complex. After all the chloride has been reacted, the excess mercuric ions react with the diphenyl carbazone (which is added as an indicator) to form a blue–violet complex and this is the end point of the titration.

$$Hg(NO_3)_2 + 2Cl^- \rightarrow HgCl_2 + 2NO_3^-$$

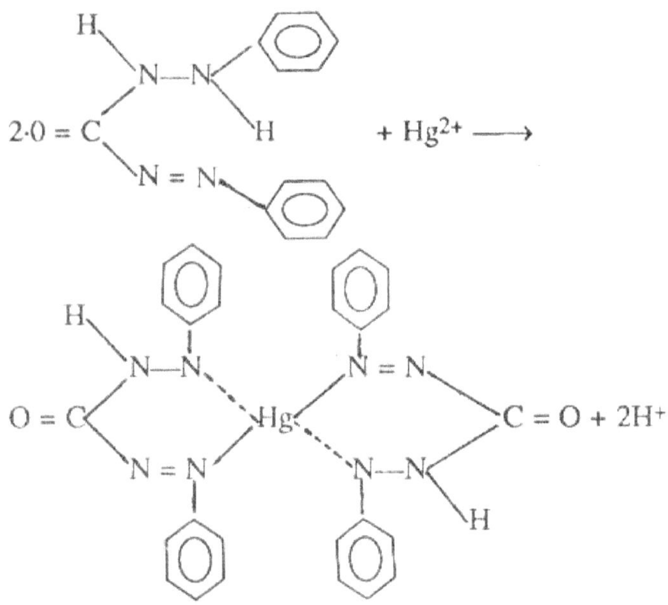

Interferences

Chromate and ferric ions interfere if present in concentration above 10 mg/l. However, they do not interfere if they are reduced to chromic and ferrous state. The titration is carried out in the pH range of 2.9 to 3.4.

Reagents

1. Mercuric nitrate solution, 0.0282N: Dissolve 5.04g. mercuric nitrate, $Hg(NO_3)_2 \cdot H_2O$ in 50 ml. distilled water containing 0.5 ml. conc. HNO_3 in a beaker. Transfer to a 1000 ml. volumetric flask and make upto the mark with distilled water. Standardize it against standard sodium chloride solution as described in the procedure. Store in a dark bottle and keep it away from light.

1.0 ml. of 0.0282 N mercuric nitrate soln ≡ 1.0 mg Cl^-

2. Standard sodium chloride solution, 0.0282N: See "Silver Nitrate Method".

3. Indicator solution: Dissolve 500 mg diphenyl carbazone and 50 mg bromophenol blue in 100 ml. 95% ethyl or isopropyl alcohol. Store it in an amber bottle.

4. Reagents for the removal of chromate and ferric ions interferences:

- **(i) Nitric acid, 0.05N (approximately):** Dilute 3.2 ml. conc.HNO_3 to 1000 ml. with distilled water.
- **(ii) Sodium hydroxide, 0.05N (approximately):** Dissolved 2g. NaOH in distilled water and dilute to 1000 ml.
- **(iii) Hydroquinone solution:** Dissolve 1.0g. hydroquinone in water and dilute to 100 ml. prepare daily.

Procedure

Place 50 ml. or an aliquot containing not more than 20 mg chloride in a conical flask. Dilute to 50 ml. with chloride-free distilled water. If the chloride content of the sample is much less, concentrate a suitable volume of the sample to 50 ml. Add 5 to 10 drops of the indicator solution and mix well. The color of the solution at this point should be blue, blue-violet or red. Add 0.05 N nitric acid solution drop wise until the color just turns yellow. Add exactly 1 ml. in excess. (If a yellow or orange color develops after the addition of indicator solution, add 0.05N sodium hydroxide dropwise until the color changes to blue violet and then dropwise 0.05N nitric acid until the color changes to yellow and then add exactly 1 ml. in excess). Titrate with standard mercuric nitrate solution to the first permanent dark purple.

Conduct a blank also using 50 ml. distilled water.

Calculation

$$mg/l\ chloride\ (as\ Cl^-) = \frac{(ml.\ HgNo_3\ solution\ for\ sample\ -\ ml.\ HgNO_3\ for\ blank) \times Normality\ of\ HgNO_3 \times 35.45 \times 1000}{ml.\ sample\ taken\ for\ estimation}$$

If the mercuric nitrate solution is exactly 0.0282 N, then

$$mg/l\ chloride\ (as\ Cl^-) = \frac{(ml.\ HgNo_3\ solution\ for\ sample - ml.\ HgNO_3\ for\ blank) \times Normality\ of\ HgNO_3 \times 1 \times 1000}{ml.\ sample\ taken\ for\ estimation}$$

The chlorides of calcium, magnesium, sodium and potassium are highly soluble in water and that is why they stay in solution when water is evaporated in boilers/when used for cooling. The concentration of chlorides before and after evaporation will reveal how much the water has concentrated. Thus the test for chloride constitutes a routine control method to determine how many times the dissolved solids in boiler water or cooling water gets concentrated. Take for example, the circulating cooling water: the number of concentration can be obtained from the following formula.

$$Number\ of\ Concentrations = \frac{Chlorides\ in\ circulating\ water}{Chlorides\ in\ makeup\ water}$$

When the total solids is allowed to build up in boilers, the concentration will reach a point at which foaming and /or priming will result. Both foaming and priming results in "carry-over" of salts and water which is most dangerous. Hence in every industry foaming and priming are avoided by keeping the total solids concentration below the dangerous level by boiler blow down. To keep the number of concentrations within safe range it is necessary to establish and follow a blow down schedule. The blow down schedule is based on total solids concentrations of boiler water. The total solids in the boiler water can be calculated by finding out chloride concentrations of boiler water and feed water.

$$T.S.\ in\ boiler\ water = \frac{Chlorides\ in\ boiler\ water}{Chlorides\ in\ feed\ water} \times Total\ Solids\ in\ feed\ water$$

Similarly, using chloride data, the amount of blow down can be expressed as a percentage of make up water.

$$\textit{Blow down, percent} = \frac{\textit{Chlorides in makeup water}}{\textit{Chlorides allowed in boiler water}} \times 100$$

For example, if the chloride content of make up water is 5 mg/l, and chloride allowed in boiler water is 50 mg/l, then the percent blow down is calculated as follows.

$$\textit{Blow down, percent} = \frac{5}{50} \times 100 = \mathbf{10\%}$$

CHAPTER - 22

CHLORINE DEMAND

Chlorine demand of a water is the amount of chlorine required to react with the organic matter and other substances present in water. It is determined as the difference between the amount of chlorine added and amount of residual remaining after a specific contact time. The smallest amount of residual chlorine considered significant is 0.1 mg/l chlorine.

Reagents

1. Standard chlorine solution: Prepare a bleaching powder solution such that it contains 100 mg/l to 1000 mg/l of chlorine. Weigh out 350 mg of 30% bleaching powder, grind it well in a mortar to a paste using little distilled water, transfer to a 1000 ml. volumetric flask and make up to the mark with distilled water. The solution contains approximately 100 mg/l chlorine. Standardize it against 0.0250 N sodium thiosulfate solution as follows.

Place 2 ml. glacial acetic acid and 25 ml. distilled water in a 250 ml. conical flask. Add about 1 g potassium iodide. Measure 25 ml. or a suitable aliquot of the prepared chlorine solution and titrate with 0.025 N sodium thiosulfate using starch as the indicator. (Disappearance of blue color is the end point).

$$mg.\ chlorine\ per\ ml.\ solution = \frac{ml.\ 0.025N\ sodium\ thiosulfate \times 0.886}{ml.\ sample\ taken\ for\ determination}$$

2. Sodium thiosulfate solution, 0.025 N: See Chapter – 14, "BIOCHEMICAL OXYGEN DEMAND".
3. Starch indicator solution: See Chapter – 51, "PERMANGANATE VALUE".

Procedure

1. Measure 500 ml. portions of the well mixed sample into each of ten amber bottles.
2. Add increasing amounts of chlorine solution in increments of 0.1 mg/l.
3. Allow to stand for 10 minutes or for a specific contact time.
4. At the end of the contact period, determine the free* available chlorine and combined† available chlorine either by starch iodide method or by orthotolidine-method.

---

* Free chlorine – chlorine in the form of hypochlorous acid (HOCl) or hypochlorite ion (OCl^-) or both.
† Combined chlorine – chloramines and other chloro derivatives.

Calculation

$$mg/l \text{ chlorine demand} = mg/l \text{ chlorine added} - mg/l \text{ residual chlorine}$$

If contact time has to be determined, conduct the above experiment with three or more series of the samples at various contact periods.

For a particular water, the chlorine demand will vary with the amount of chlorine applied, the length of contact time and the temperature. Determination of chlorine demand must take all these factors into consideration.

CHAPTER - 23

CHROMIUM

Chromium is not a natural constituent of water. It enters into a water course through the discharge of industrial effluents containing chromium. Electroplating wastes contain a high concentration of chromium. Chromium compounds are extensively used in cooling water systems as corrosion inhibitors. When such waste waters are discharged, chromium gains access to water.

Chromium exists in hexavalent and trivalent forms, the latter being rare in occurrence. (Chromates and dichromates are hexavalent chromium compounds). Since chromium exists in two forms, two separate determinations are necessary. Hexavalent chromium is best determined by the following s-diphenyl carbazide method. If trivalent chromium is to be determined it is oxidized to hexavalent from and then the determination is carried out.

S-DIPHENYL CARBAZIDE COLORIMETRIC METHOD

[For hexavalent chromium (chromium present as chromate and dichromate)]

Principle

Under acidic conditions hexavalent chromium reacts with s-diphenyl carbazide to form a reddish violet colored complex which can be determined either visually or photometrically.

Interferences

Iron above 1 mg/l interferes but can be eliminated by preliminary treatment with cupeferron. (See procedure). Other common ions generally present in water do not interfere.

Reagents

The distilled water used for the preparation of reagents and dilutions should be free from chromium and organic matter.

1. Chromium stock solution: Dissolve 283mg. dried potassium dichromate $K_2Cr_2O_7$ in distilled water and make up to 1000 ml. in a volumetric flask.

$$1.0 \text{ ml.} = 100 \ \mu g \text{ Cr}$$

2. Chromium working solution: Pipet 10.0 ml. chromium stock solution into a 1000 ml. volumetric flask and make up to the mark with distilled water. Prepare freshly.

1.0 ml. = 1.0 μg Cr

3. s-Diphenyl carbazide reagent: Dissolve 200mg. s-diphenyl carbazide in 100 ml. 95% ethyl alcohol. Add 400 ml. 1+ 9 H_2SO_4 into it. Place the reagent in an amber bottle and keep it in a refrigerator. Discard the solution when it becomes appreciably brownish.

Procedure

1. Place appropriate volumes of chromium working solution in 100 ml. volumetric flasks (or Nessler tubes) covering the range upto 20 μg Cr and make upto the mark with distilled water. Include a flask containing 100 ml. distilled water as the blank.
2. Place 100 ml. or an aliquot of the neutralized sample in a 100 ml. volumetric flask (or Nessler tube). Dilute to 100 ml. if the aliquot taken is lower than 100 ml. (If the sample is turbid, centrifuge or filter through a sintered glass crucible and then place in the volumetric flask or Nessler tube. Samples should not be filtered through filter paper as it may absorb some chromium).
3. To the blank, standards and sample add 5.0 ml. s-diphenyl carbazide reagent and mix well and set aside for 5 minutes.
4. Measure the optical densities of the blank, standards and sample using a spectrophotometer at a wavelength of 540 nm (or filter photometer equipped with a suitable green filter) taking water as the reference. Prepare a calibration curve. Find out the mg Cr equivalent to the observed optical density.
   (If nessler tubes are used, visual comparison can be made).

   Express the result as mg chromium (as Cr) per litre of the sample.

If the sample contains iron in excess quantities, acidify a suitable aliquot of the sample with HCl or H_2SO_4 such that it contains 5 to 10% acid. Precipitate iron with 6% aqueous solution of cupferron under cold condition. Filter and use the filtrate for chromium determination.

TOTAL CHROMIUM

(Hexavalent chromium + Trivalent chromium)

This method is applicable to potable waters, treated waters and waste waters.

Principle

Chromium is oxidized by permanganate to chromate and then treated with diphenyl - carbazide to obtain the violet color complex which is measured colorimetrically.

Interferences

Mercury and vanadium interfere but they occur only in negligible concentrations. Iron interferes when its concentration exceeds 1 mg/l but can be eliminated by preliminary treatment with cupferron.

Reagents

In addition to the reagents required for the determination of hexavalent chromium prepare the following reagents.

1. **Potassium permanganate solution, approximately 0.1N:** Dissolve 316mg. potassium permanganate, $KMnO_4$ in little distilled water and dilute to 100 ml.
2. **Sodium azide solution:** Dissolve 500 mg sodium azide, NaN_3 in 100 ml. chromium free distilled water.
3. **Reagents for the destruction of organic matter**
   - (i) Sodium sulfate
   - (ii) Conc. sulfuric acid
   - (iii) Conc. nitric acid
   - (iv) Ammonium oxalate solution saturated
   - (v) Conc. ammonia solution

Procedure

Samples containing no organic matter

Place appropriate volume of the sample containing not more than 20 μg Cr in a 100 ml. beaker. Add 10 ml. 10% sulfuric acid. Add potassium permanganate solution dropwise until the solution becomes pink in color. Add about 30 ml. water and place on a water bath for 20 minutes. Add sodium azide solution dropwise until the solution becomes colorless and continue heating for one minute. Cool and filter the solution through a sintered crucible if necessary. Transfer to a 100 ml. volumetric flask (or to 100 ml. Nessler tube) and add 5.0 ml. diphenyl carbazide solution and make up to the mark with chromium-free distilled water. Mix well and compare the color visually or photometrically. Find out the chromium content per litre of the sample.

Samples containing organic matter

Destroy the organic matter first as described below. Place a suitable volume of the sample containing not more than 20 μg Cr in a 250 ml. Kjeldahl flask. Dissolve 100 mg. sodium sulfate in it and add 2 ml. conc.H_2SO_4. Evaporate until white fumes are evolved. If the organic matter is hard to destroy, add 1 to 2 ml. of conc. HNO_3 dropwise. Add 10 ml. saturated ammonium oxalate solution and evaporate to fuming again until a minimum of residual acid liquor is obtained. Cool and dilute with 10 ml. distilled water. Transfer the contents of the Kjeldahl flask to a 25 ml. volumetric flask and make up to the mark with distilled water. Place the whole of the made up solution or an aliquot in a 100 ml. beaker. Neutralize with conc. ammonia solution. Then proceed as described for samples containing no organic matter.

Calculate the chromium content of the sample.

The value obtained indicates the total chromium content of the sample. The trivalent chromium can be calculated as follows:

mg/l trivalent chromium = mg/l Total chromium – mg/l Hexavalent chromium

VOLUMETRIC DETERMINATION OF CHROMATE IN COOLING WATER

This method is applicable to waters containing 100 to 1500 mg/l chromium.

Principle

This method is based upon the oxidation of iodide to iodine by dichromate in acidic solution. The liberated iodine is titrated with standard thiosulfate solution using starch as the indicator.

Interferences

Iron, copper and nitrite interfere. Iron is made innocuous by the addition of ammonium bifluoride and nitrite by the addition of sulfamic acid. If copper is present, the solution has to be made alkaline and filtered. However copper does not occur in such concentrations to interfere with the test.

Procedure

a) Available chromate (Hexavalent chromium): Place 100 ml. of the cooling water sample in a 500 ml. conical flask. Add 2 ml. 10N.H_2SO_4, 1g. ammonium bifluoride and 1g. sulfamic acid. Swirl to dissolve and add 2g. potassium iodide. Swirl again and allow to stand in dark for five minutes. Titrate with standard 0.1N sodium thiosulfate solution using starch as indicator.

Calculation

$$\text{Available Chromium (as Cr)} = \frac{\text{ml. Thio} \times \text{Normality of Thio} \times 17.332 \times 1000}{\text{ml. sample taken}}$$

$$\text{Available Chromium as Chromate(as } CrO_4^{2-}) = \frac{\text{ml. Thio} \times \text{Normality of Thio} \times 38.66 \times 1000}{\text{ml. sample taken}}$$

b) Total chromium as chromate: Filter the sample of cooling water and pipet 100 ml. of the filtrate in to a 500 ml. conical flask. Add the following reagents in order with mixing after each addition.

(i) 5 ml. of 10 N.H_2SO_4,

(ii) 2 ml. of 1 N silvernitrate solution,

(iii) 2g. potassium persulfate.

Add a few glass beads and boil on a hot plate for about 15 minutes. Cool the solution and dilute to 100 ml. with distilled water. Add 1g. ammonium bifluoride and swirl the flask. Add 2g. potassium iodide and keep the flask in dark for five minutes. Titrate with standard 0.1N sodium thiosulfate solution using starch as the indicator.

Calculation

$$\text{Total Chromium (as Cr)} = \frac{\text{ml. Thio} \times \text{Normality of Thio} \times 17.332 \times 1000}{\text{ml. sample taken}}$$

$$\text{Total Chromium as Chromate(as } CrO_4^{2-}) = \frac{\text{ml. Thio} \times \text{Normality of Thio} \times 38.66 \times 1000}{\text{ml. sample taken}}$$

CHAPTER - 24

COAGULANT DEMAND (OR) JAR TEST

The purpose of this test is to determine the most effective doses of alum and lime for a specific water in the control of coagulation and flocculation at a filter plant, especially when water quality fluctuates rapidly. The coagulation and flocculation of water is influenced by a number of factors such as pH, alkalinity, temperature, color, turbidity and the degree of agitation during flocculation and the characteristics of the coagulant in use.

Collection of sample: Collect atleast 25 litres of the sample, to permit atleast four series of tests of six portions each.

Apparatus

Multiple stirrer: A multi-position stirrer with continuous speed variation from about 20 to 150 rpm. The stirring paddles should be of light-gage corrosion-resistant material all of the same configuration and size. The area of the paddles should not be more than one-fourth the vertical cross section of the water in each beaker. Use only beakers of 1.5 litres capacity.

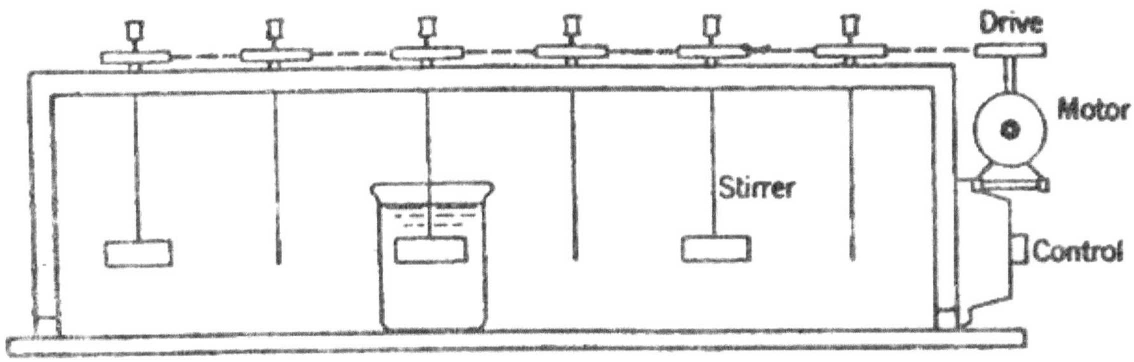

Fig. 24 - 1. Laboratory apparatus for the determination of the optimum dosage of coagulating chemicals. A variable-speed motor is provided.

Reagents

1. Alum stock solution: Dissolve 1g. of filter alum (aluminium-sulfate, non-ferric) in water and make up to 1 litre.

2. Lime stock solution: Mix well 1g. powdered hydrated lime with 1 litre distilled water. Shake well every time before use. 1 ml. of this suspension when added to 1 litre of water will give a dose of 1 mg/l.

Procedure

1. Determine the color, turbidity, pH and alkalinity of the water.
2. Measure 1000 ml. of the sample in to each of the four beakers (or six beakers) and place them in the multiple stirrer and start the motor of the stirrer.
3. Add graded doses of alum and lime* as quickly as possible to the beakers. (Conduct preliminary trials to find out whether alum only is enough or both alum and lime are necessary).
4. Agitate for about 30 minutes or for a prefixed time.
5. Observe the time needed for the first appearance of the visible floc in each of the four portions (or six portions) of the sample. Observe the appearance, size and quantity of the floc after the end of the flocculation period. Allow to settle for 1 hr.
6. Using a large pipet withdraw a portion of the supernatant from each portion without disturbing the floc and determine the color, turbidity, pH and alkalinity of each of the supernatant.
7. If none of the test portions has a color below 10 to 20 units or turbidity 5 to 10 units, discard the flocculated samples in the beakers and use a fresh set of samples and continue the test with another range of chemical doses. Repeat, if necessary, until several portions are shown to be effectively coagulated.
8. When effective coagulation is secured in one or two portions, determine the color, turbidity, alkalinity and pH of the supernatant settled water.
9. The test portion giving the best flocculation and reduction of color and turbidity will indicate the optimum doses of alum and lime. After determining the dosage, the optimum degree of agitation can be determined using different degrees of agitation with constant selected dose.

---

* An arbitrary ratio of alum : lime = 3:1

CHAPTER - 25

COLOR

The determination of color is rapid one and is useful in detecting a change in the character of the water. When waters from the source as from river, are being regularly examined, the variations in color often serve as indices of quality. Generally yellow color indicates the presence of appreciable amounts of organic matter and sometimes chromium. The presence of peaty matter is indicated by the characteristic red brown color. The yellowish–red indicates the presence of iron. Any unusual color would indicate the necessity for a special examination to ascertain the cause. Color determination is also useful to check the functioning of filter beds and to determine the rate of filtration.

The term 'color' means the 'true color' i.e., the color due to substances in solution and due to the substances present as fine colloids. For true color determination, the sample has to be centrifuged to remove turbidity. 'Apparent color' is the color due to both suspended and dissolved matters. Apparent color determination hence needs no centrifugation.

The standard method for color comparison involves the use of 'standard color solution' prepared by using potassium chloroplatinate and cobaltous chloride. Colors can be compared visually or photoelectrically.

Note: Determine the color on the day of collection itself, otherwise ferric iron will precipitate as ferrichydroxide which may contribute color to the sample. Carbonates may also precipitate and exhibit color.

Interferences

Turbidity in excess of 5 mg/l seriously affects the true color determination. Hence it should be removed prior to determination by centrifugation. (Do not filter the sample to remove turbidity or suspended matter, since filtration has a decolorizing effect).

Reagents

Standard color solution: Dissolve 1.246g. potassium chloroplatinate, K_2PtCl_6 and 1.0g. cobaltous chloride hexa hydrate, $CoCl_2.6H_2O$ in distilled water. Add 100 ml conc. HCl and make up to 1000 ml. in a volumetric flask.

$$1.0 \text{ ml} \equiv 500 \text{ Hazen units}$$

Prepare a series of working color standards to cover the range from 5 to 70 units using the following table:

Color in Hazen units	ml. stock solution to be diluted to 50 ml.
5	0.5
10	1.0
15	1.5
20	2.0
25	2.5
30	3.0
35	3.5
40	4.0
45	4.5
50	5.0
60	6.0
70	7.0

Procedure

Place 50 ml. of the sample (centrifuged) in a 50 ml. Nessler tube. Compare the color of the sample with that of the working color standards by looking vertically downwards against a pure white surface placed at such an angle that light is reflected upwards through the columns of liquid. If the color exceeds 70 units dilute the centrifuged sample with distilled water such that the color is within the range.

Note: With samples containing much clay or other fine suspensoids, normally it is impossible to get a bright centrifugate. In these circumstances, the measurement of 'apparent color' is useful, particularly in water works control.

For 'apparent color' determination view the sample vertically downwards contained in a jar (of 50 cm height and 25 mm internal dia.) and express the color hue as observed.

Expression of results

Report the color results in Hazen units in whole numbers and record as follows:

Color in Hazen units	Record to the nearest
1-50	1
51-100	5
101-250	10
251-500	20

Since the color of a water is extremely pH dependent, report the pH value also at which the color has been determined.

CHAPTER - 26
COPPER

Copper is rarely found in natural waters and when the amount exceeds 0.05 mg/l, it is attributable to the corrosive action of the water on copper and brass tubing, to industrial effluents, or frequently to the use of copper compounds in controlling the growth of algal and other undesirable planktons in reservoirs.

Ammonia when present in concentrations more than 10 mg/l along with dissolved oxygen usually results in fairly significant copper dissolution. Carbon dioxide and sodium - bicarbonate also cause corrosion of copper and copper alloys. Determination of copper in industrial water is necessary to monitor corrosion of copper, and copper alloy fittings and tubes.

NEOCUPROINE METHOD

Principle

This method is applicable to natural waters and all types of industrial waters and industrial waste waters containing 0.05 mg/l Cu or more. Copper is reduced with hydroxylamine hydrochloride and the pH is adjusted to 4.0 to 6.0. The cuprous ion on reaction with neocuproine forms yellow colored complex which is extracted by chloroform and the color is measured.

Interferences

None are likely. Possible interferences from organic matter, sulfide, cyanide and chromium are eliminated. (See procedure).

Reagents

Use copper free distilled water (glass distilled water or deionised distilled water) for the preparation of solutions and dilutions.

1. **Hydroxylamine hydrochloride solution:** Dissolve 50g. hydroxylamine hydrochloride, $NH_2OH.HCl$ in redistilled water and dilute to 250 ml.

2. Neocuproine solution: Dissolve 0.2g. neocuproine (2, 9-dimethyl-1, 10-phenanthroline hemihydrate) in 200 ml methyl alcohol. Keep it in a refrigerator. This reagent is stable under ordinary storage conditions for a month or more.

3. Chloroform

4. Sodium citrate solution: Dissolve 125g. sodium citrate dihydrate $Na_3C_6H_5O_7.2H_2O$, in redistilled water and dilute to 500 ml. Transfer to a 1000 ml. separating funnel, add 10 ml. hydroxyl amine-hydrochloride solution and 10 ml. neo-cuproin solution and shake well. Extract with 50 ml. chloroform to remove copper impurities. Discard the choloroform layer.

5. Copper stock solution: Place 100.0mg polished electrolytic copper in a conical flask and add 10 ml. redistilled water and 3 ml. conc. HNO_3. After the reaction has subsided, warm gently to complete dissolution and then boil to expel oxides of nitrogen. Take care to avoid loss of copper. Cool and add about 50 ml. redistilled water. Transfer quantitatively to a 1000 ml. volumetric flask and make up to the mark with redistilled water.

1.0 ml. = 100 μg Cu.

6. Copper standard solution: Pipet 50.0 ml. of stock solution and make up to 1000 ml. in a volumetric flask.

1.0 ml. = 5 μg Cu.

7. Iso-propyl alcohol

8. Reagents for the removal of interferences

- (i) Conc. nitric acid
- (ii) Conc. sulfuric acid
- (iii) Conc. perchloric acid
- (iv) Conc. hydrochloric acid
- (v) Conc. ammonia solution, (sp. Gr 0.90).

Procedure

Samples containing no interferences

1. Place 50 ml. of the well shaken sample or an aliquot diluted to 50 ml. in a beaker. Add 0.1 ml. of conc. HCl. (If the sample is already acidified during collection, omit the addition of acid). Boil gently for about 10 minutes and cool to room temperature. Transfer to a 125 ml. separating funnel.
2. To a series of separating funnels, pipet copper standard solution covering the range from 5 μg to 100 μg. Dilute the contents of each separating funnel to 50 ml. with

distilled water. Include a separating funnel containing 50 ml distilled water as blank. Add 0.1 ml. conc. HCl to each separating funnel.

3. To the sample, standards and blank add 5 ml. hydroxyl amine-hydrochloride solution, and 10 ml. sodium citrate solution, and 10 ml. neocuproin solution shaking the funnels for 30 seconds after each addition.
4. Add 20 ml. chloroform and shake the funnel vigorously. Allow the phases to separate and drain the chloroform layer into 50 ml. Nessler tubes if visual comparison is to be made. (If photometric comparison is made, drain the chloroform layer in to a 50 ml. volumetric flask). Extract with another 20 ml. portion of chloroform and combine with the first chloroform extract. Make upto 50 ml. with isopropyl alcohol.
5. Compare the color of the sample with that of the standards visually or measure the absorbance of standards and samples using the blank as the reference at 457 nm using a light path of 1 cm.
6. Prepare a calibration curve using a semilog graph paper. Find out the mg equivalent of the sample to the observed optical density.

Express the result as mg copper (as Cu) per litre of the sample.

Samples containing interferences

If the sample contains organic matter, chromium*, sulfide or cyanide, the sample has to be subjected to acid digestion with sulfuric and nitric acids. (See Chapter – 9 "PRELIMINARY ACID DIGESTION OF THE SAMPLES"). Adjust the pH of the acid digested sample to 4-6, using conc. ammonia solution and suitable pH paper. Add 0.2 ml. conc. HCl and make up the solution to 100 ml. in a volumetric flask. Transfer appropriate volume of the made up solution to a separating funnel. If the volume is less than 50 ml, dilute to 50 ml with distilled water. Proceed from Step No.3 of the procedure for sample containing no interferences.

SODIUM DIETHYL DITHIOCARBAMATE METHOD

Though the sodium diethyl dithiocarbamate method lacks sensivity and selectivity, it is still used as a widespread method for feed waters and cooling waters containing chromium. It

* When the ratio of chromium to copper is greater than 5 to 1.

can also be applied to natural waters, since extraction with carbon tetra chloride removes all the possible interferences.

Feed waters and cooling waters

(i) Place 50 ml. of the filtered sample in a 125 ml. separating funnel. (If lower aliquots are chosen, dilute to 50 ml. with distilled water).

(ii) Place 1.0, 2.0, 3.0, 4.0, 6.0, 8.0, 10.0 ml. of copper standard solution in a series of separating funnels, and dilute to 50 ml. with distilled water. Include a separating funnel containing 50 ml distilled water as blank.

(iii) To the blank, standards and sample add 2 ml. conc. ammonia solution, 5 ml. 20% ammonium citrate solution and 1.0 ml. 0.1% aqueous solution of sodium diethyl dithiocarbamate with mixing after each addition.

(iv) Add to each funnel 15.0 ml. carbon tetra chloride and shake well for 2 minutes. Allow the layers to separate and drain the CCl_4 layer through a plug of glass wool in to the absorption cell. Measure the optical density of samples and standards at 440 nm using blank as the reference solution.

(v) Prepare a calibration curve using a semilogarithmic paper. From the graph find out the copper content of the sample.

CHAPTER - 27

CYANIDES (TOTAL)

Cyanide is best determined by bis-pyrazolone method. The major problem in the determination of cyanide is the removal of interfering ions. Most of the interferences however, are eliminated by preliminary distillation.

DISTILLATION – PHOTOMETRIC METHOD
(USING BIS-PYRAZOLONE)

The method is applicable to all types of water, waste water, and industrial effluents.

Principle

Simple cyanides such as potassium cyanide, reacts readily and exhibit the characters of CN^- whereas complex cyanides such as potassium ferricyanide though contain CN group do not exhibit properties of CN^-, since the anion in them is not the CN group but radical $M(CN)_x$ where M is the heavy metal. In order to convert the CN present in all the cyanides (particularly in alkali iron cyanide complexes) to HCN, distillation with acid cuprous chloride solution is necessary. The distillate containing hydrogen cyanide is then taken for colorimetric determination. The CN^- ion in the distillate is then converted to cyanogen bromide. (CNBr), by the addition of bromine water. After the reaction is complete, CNBr forms a blue dye upon the addition of pyridine–pyrazolone reagent - the intensity of which is measured by visual comparison or by using spectrophotometer.

Apparatus

Distillation apparatus containing a distillation flask of 1000 ml. capacity, fitted with a splash head and water cooled condenser. (For an efficient and special type of distillation apparatus, refer 'STANDARD METHODS FOR THE EXAMINATION OF WATER & WASTE WATER' 20th edition, 1998).

Reagents

For Distillation

1. Acid cuprous chloride solution: Dissolve 2g. cuprous chloride Cu_2Cl_2 in 100 ml. 5N. HCl. Prepare freshly. (Prepare 5N.HCl by diluting 43 ml.conc. HCl to100 ml. with distilled water).

2. Sodium hydroxide solution, approximately 0.5N: Dissolve 10g. NaOH in distilled water and make upto 500 ml.

For photometric Measurement

3. Sodium hydroxide, approximately 1N: Dissolve 20g. NaOH in distilled water and dilute to 500 ml.

4. Cyanide stock solution: Dissolve 2.5g. potassium cyanide KCN in distilled water, add 100 ml. 1N sodium hydroxide solution and make upto 1000 ml. in a volumetric flask. Standardize the solution against 0.1N silver nitrate solution, as follows:

(The strength of the solution decreases on standing. Hence it should be checked before use).

Standardization*: Place 25.0 ml. cyanide stock solution in to a 250 ml. conical flask. Dilute to 100 ml. Titrate with 0.1N silver nitrate solution (16.9893 g. $AgNO_3$/1000 ml.) using rhodanine indicator solution (0.02% in acetone). The color change is from canary yellow to salmon pink.

Calculate the cyanide in the whole of the solution as follows:

$$\text{Total Cyanides (as CN), mg/l} = \frac{5.200 \times \text{ml. of } AgNO_3 \times 1000}{\text{ml. solution taken}}$$

5. Cyanide intermediate stock solution: Dilute appropriate volume of cyanide stock solution to 100 times. During dilution add 1 ml. of 1 N sodium hydroxide solution for each 10 ml. of the final volume. Prepare freshly.

1.0 ml. = 10 μg CN

6. Cyanide working solution: Dilute appropriate volume of cyanide intermediate stock solution 10 times. During dilution add 1 ml. of 1N sodium hydroxide solution for each 10 ml. of the final volume.

1.0 ml. = 1 μg CN

Prepare this solution freshly.

7. Hydrochloric acid, conc.

8. Bromine water, saturated solution.

9. Arsenious acid solution: Dissolve 2g. arsenious oxide As_2O_3 in 100 ml. distilled water by boiling under reflux.

* Standardization itself constitutes one method for the determination of cyanide. Higher concentrations of cyanide can be determined accordingly, after distillation.

10. 1-phenyl-3 methyl-5 pyrazolone reagent: Prepare a saturated aqueous solution (approximately 0.75 gm in 150 ml.) by adding pyrazolone to water at 75°C. Agitate occasionally as the solution cools to room temperature. Filter (prepare freshly).

11. Pyridine, AR.

12. Bis-Pyrazolone: Dissolve 25 mg bis pyrazolone in 25 ml. pyridine. (prepare freshly).

13. Mixed pyridine-pyrazolone reagent: Mix 125 ml. of the filtered aqueous solution of 1-phenyl-3-methyl-5-pyrazolone with 25 ml. of bis-pyrazolone in pyridine.

Procedure

Distillation

If the sample contains sulfide, oxidizing agents and fatty acids, remove them first and then proceed with the distillation.

Removal of sulfide

Treat the alkaline sample at pH 11.0 with small increments of lead carbonate powder until no more precipitate is formed. Filter and wash with distilled water.

Removal of oxidizing agents

If the sample contains any oxidizable matter shown by starch-iodide paper test, titrate the sample with sodium sulfite solution (12.6g. Na_2SO_3/l) until a negative result is obtained.

Removal of fatty acids

Acidify the sample to pH about 6.5 with acetic acid and extract with chloroform or preferably by hexane.

Place into the distillation flask appropriate volume of the sample containing not more than 50 μg cyanide.

Caution: Do not use mouth pipets for dispensing volumes of cyanide containing samples and solutions

Dilute to about 200 ml. with distilled water if necessary. Add 5 ml. acid cuprous - chloride solution and a few antibumping granules. Assemble the apparatus immediately. Place 20 ml. 0.5 N sodium hydroxide solution in the receiving flask. Make the tip of the condenser dipped in sodium hydroxide solution. Start the distillation and collect a total volume of about 40 ml. in the receiving flask. Transfer the contents of the receiving flask to a 50 ml. volumetric flask and make upto the mark with distilled water.

Photometric Measurement

1. Place a suitable aliquot of the distillate not exceeding 25 ml. in a 50 ml. volumetric flask. Place appropriate volumes of cyanide working solution covering the range upto 20 μg cyanide, into a series of 50 ml. volumetric flasks and dilute to 25 ml. with distilled water. Include a 50 ml. volumetric flask containing 25 ml. distilled water as the blank.
2. To each flask add 0.5 ml. conc.HCl and 0.5 ml. bromine water. (Dispense bromine water by means of an automatic pipet) stopper the flask immediately, mix well and allow to stand for about two minutes.
3. Add 0.5 ml. arsenious acid solution to each flask to remove excess bromine. Mix well and add 5 ml. mixed pyridine-pyrazolone reagent. Stopper and mix well and allow to stand for 20 minutes for the development of characteristic blue color. Make upto the mark with distilled water.
4. Measure the optical densities of the blank, standards and sample using a spectrophotometer at 620 nm. Prepare a calibration curve and find out the mg cyanide equivalent to the optical density. Visual comparison can also be made.

Express the result as mg total cyanide (as CN^-) per litre of the sample.

CHAPTER - 28

ELECTRICAL CONDUCTIVITY

Electrical conductivity is a measure of a water's capacity to convey electric current. Electrical conductivity of water is the sum of the several conductivities resulting from the electrolytes present. Hence electrical conductivity of a water is directly proportional to its dissolved mineral matter content. The unit of conductivity is micromhos/cm (reciprocal megohm/cm). Since electrical conductivity varies directly with the temperature of the sample, the result is usually reported at 25°C. If the conductivity is measured at temperatures other than 25°C, the conductivity value has to be multiplied with appropriate factors. Commercial instruments for the determination of electrical conductivity are available in various forms. While using these instruments manufacturers' instructions should be followed carefully. The instrument has to be checked up constantly by using standard potassium chloride solutions the electrical conductivities of which are given below:

Concentration		Specific conductance
g/l	M	μ mhos/cm*
0.0074	0.0001	14.94
0.0373	0.0005	73.90
0.0745	0.001	147
0.3728	0.005	717
0.7456	0.01	1,413
1.4912	0.02	2,767
3.728	0.05	6,668
7.456	0.1	12,900
14.192	0.2	24,820
37.280	0.5	58,640
74.56	1	1,11,900

Dissolved gases also contribute to electrical conductivity but silica and organic matter do not contribute.

* Also termed as micro Siemens/cm.

Applications of Electrical Conductivity Data

1. Electrical conductivity determination can be used as a test to determine the amount of total dissolved solids in water. Due to the rapidity of electrical conductivity measurement, the quantity of dissolved salts in a water can be ascertained at once.
2. Usual analytical methods for the determination of dissolved solids of boiler water or circulating cooling water are time consuming, whereas conductivity measurements are instantaneous and the quantity of solids can be known immediately there by the blow down rate could be adjusted immediately.
3. Continuous monitoring of the quality of process water is made by measurement of conductivity of flowing samples using online conductivity meters. This is especially so in monitoring quality of high purity waters (such as demineralized waters)
4. Conductivity is a useful guide to the purity of distilled water and is used for evaluation of steam purity in power industry. The purity of steam leaving a boiler may be determined by measuring the conductivity of a condensed steam sample. Freshly distilled water has a specific conductance of 0.5 to 2 micromhos/cm.
5. Where waters from the same source are periodically examined, the variation in the dissolved solids content is indicated by conductivity measurements.
6. Based upon the conductivity of the water, the quantity of the sample needed for the other determinations can be decided.

NEUTRALIZED CONDUCTIVITY

In case of boiler water, the neutralized conductivity is measured. Boiler water sample contains significant amounts of hydroxide. The hydroxyl ion (OH^-)contribution to conductivity is disproportionately high compared to other ion species. Normally conductivity of boiler water is measured to find out the dissolved solids content. Hence the hydroxide ions are neutralized with organic acids such as gallic acid. (3, 4, 5 – trihydroxy benzoic acid). Since organic acids do not contribute to conductivity, they are used for neutralization. The conductivity value obtained after neutralization is termed as 'neutralized conductivity'.

To find out neutralized conductivity, place 50 ml or 100 ml sample in a conical flask. Add 5 drops phenolphthalein indicator. Add gallic acid solution dropwise until the pink color disappears completely. Then determine the conductivity.

CHAPTER - 29

EQUIVALENT MINERAL ACIDITY (EMA)

Equivalent mineral acidity (EMA) is an important parameter used to design demineralization plants etc., The sum of the anions which are capable of forming strong acids with H^+ ions is called the Equivalent Mineral Acidity (EMA). The ions contributing to equivalent mineral acidity are sulfate, chloride and nitrate. If a water containing these ions is passed over a cation exchanger (in the H^+ form), then this latent acidity (EMA) becomes Free Mineral Acidity (FMA) EMA is usually expressed in terms of calcium carbonate. Following is an illustration of calculating the equivalent mineral acidity.

Problem 29 - 1

The results of analysis of a sample of water is given below. Calculate the equivalent mineral acidity of the water.

Sulfate (as SO_4), mg/l = 119

Chloride (as Cl), mg/l = 204

Nitrate (as NO_3), mg/l = 35

$$Sulfate\ in\ terms\ of\ CaCO_3 = \frac{CaCO_3\ Equivalent.\ Wt}{Sulfate\ Equivalent.\ Wt} \times Quantity\ of\ sulfate$$

$$= \frac{50}{48} \times 119 = 123.96$$

$$Chloride\ in\ terms\ of\ CaCO_3 = \frac{CaCO_3\ Equivalent.\ Wt}{Chloride\ Equivalent.\ Wt} \times Quantity\ of\ chloride$$

$$= \frac{50}{35.45} \times 204 = 287.73$$

$$Nitrate\ in\ terms\ of\ CaCO_3 = \frac{CaCO_3\ Equivalent.\ Wt}{Nitrate\ Equivalent.\ Wt} \times Quantity\ of\ nitrate$$

$$= \frac{50}{62} \times 35 = 28.23$$

. Equivalent Mineral Acidity (as $CaCO_3$) = 123.96+ 287.73+ 28.23

= **439.92**

CHAPTER - 30

FLUORIDE

Fluoride occurs naturally in all ground waters. The concentration of fluoride to an extent of 1.0 mg/l is desirable to prevent dental caries. However higher concentrations are objectionable as such high concentrations are responsible for the discoloration of tooth enamel called 'mottling'. Hence the concentration of fluoride has to be kept within 1.0 mg/l both in drinking water and industrial waters intended for food processing, soft drinks manufacture, carbonated beverages etc.

Two colorimetric methods are given for the estimation of fluoride. The SPADNS method tolerates many interferences and determines fluoride in the range from 0.00 to 1.40 mg/l. The reaction is immediate, color is stable and no waiting period is necessary. Regarding alizarin method, a period of 20 minutes is necessary for color comparison. Also, it suffers from interferences. However it is equally a good method and can be applied, when interferences are known to be absent or present in low concentrations. When the sample is highly colored or turbid, distillation may be adopted to remove the interferences or electrode method using fluoride ion sensitive electrode may be employed.

SPADNS METHOD

Principle

Fluoride ion reacts with the Zirconium – SPADNS reagent to form zirconium fluoride with the result the color of the zirconium-SPADNS complex is bleached, and the color change is proportional to the fluoride ion concentration. (Reaction – see next page).

Interferences

None is likely. Highly colored and turbid samples should however be subjected to a preliminary distillation.

Reagents

1. **Fluoride stock solution:** Dissolve 221.0 mg anhydrous sodium fluoride, NaF, in distilled water and make upto 1000 ml. in a volumetric flask.

$$1.00 \text{ ml.} = 0.1 \text{ mg } F^-$$

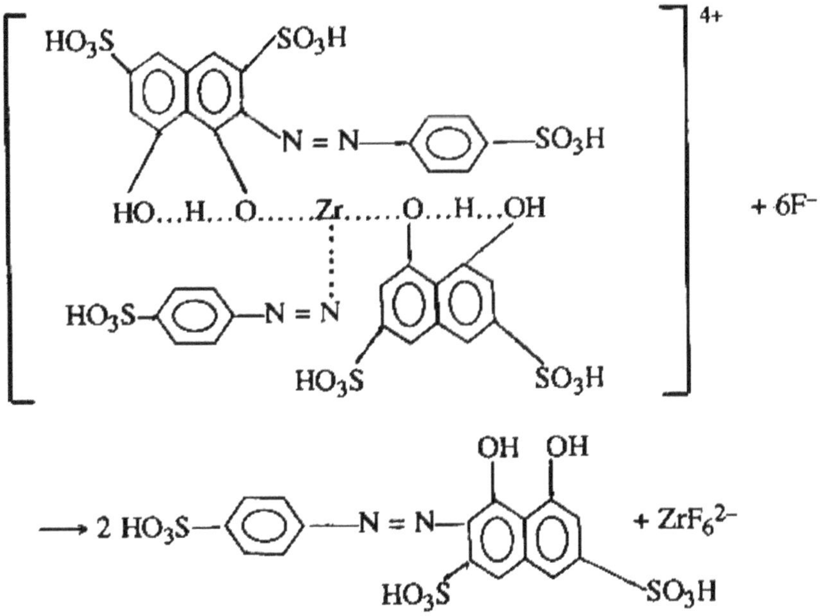

2. Fluoride standard solution: Dilute 100 ml. stock fluoride solution to 1000 ml. in a volumetric flask with distilled water.

$$1.00 \text{ ml.} = 0.01 \text{ mg } F^-$$

3. Acid Zirconyl – SPADNS reagent:

(a) Zirconyl-acid reagent: Dissolve 133 mg zirconyl chloride octa hydrate, $ZrOCl_2.8H_2O$ in 25 ml. distilled water. Add 350 ml. conc. HCl and dilute to 500 ml. with distilled water.

(b) SPADNS solution: Dissolve 958mg. SPADNS, sodium 2-(parasulfophenyl azo)–1, 8–dihydroxy– 3, 6-naphthalene disulfonate*, in distilled water and dilute to 500 ml. This solution is highly stable if protected from sunlight.

Mix equal volumes of Zirconyl-acid reagent and SPADNS solution. This is acid Zirconyl – SPADNS reagent and is very stable and can be kept for 2 years.

4. Reference solution: To 100 ml. distilled water in a beaker add 10 ml. SPADNS solution. Dilute 7 ml. conc. HCl to 10 ml. with distilled water and add to the above solution. Mix well. Use this solution as reference to set the spectrophotometer. This solution is stable and can be reused.

* Also known as 4,5-dihydroxy-3-(para sulfophenyl azo)-2,7-naphthalene disulfonic acid tri sodium salt.

5. Sodium arsenite solution (app. N/70): Dissolve 1.0g. sodium arsenite, $NaAsO_2$ in distilled water and dilute to 1000 ml.

$$1.0 \text{ ml} \equiv 0.5 \text{ mg } Cl_2$$

Procedure

Standards

1. Pipet 0.0, .05, 1.0, 2.0, 3.0, 4.0, 5.0, 6.0 and 7.0 ml. fluoride standard solution in to 100 ml. Nessler tubes and make up to 50 ml mark with distilled water.
2. Add exactly 10.0 ml. acid-zirconyl-SPADNS reagent to each tube and mix well.
3. Set the spectrophotometer or filter photometer to zero absorbance using reference solution at 570 mμ and measure the absorbance of the standards immediately.
4. Plot a calibration curve.

Samples

Place 50.0 ml. of the sample or an aliquot diluted to 50.0 ml. in a 100 ml. Nessler tube. If the sample contains any residual chlorine add 1 drop of sodium arsenite solution for each 0.1 mg Cl_2 and mix well. Add exactly 10.0 ml. of the acid-zirconyl-SPADNS reagent and mix well. After setting the spectrophotometer to zero-absorbance with the reference solution, measure the absorbance of the sample. Find out the mg fluoride of the sample equivalent to the observed optical density from the calibration graph.

Express the result as mg Fluoride (as F^-) per litre of the sample.

SCOTT-SANCHIS METHOD

Principle

Fluoride decolorizes the zirconium alizarin complex and the de-colorization is proportional to fluoride concentration.

$$\underset{\text{(Reddish color)}}{(Zr\text{ - alizarin lake})} + 6F^- \text{ - } Alizarin + \underset{\text{(Yellow)}}{(ZrF_6)^{2-}}$$

Interferences

Aluminium in excess of 0.1 mg/l interferes. Phosphate as P_2O_5 in excess of 0.5 mg/l interferes. Residual chlorine must be completely removed with arsenite reagent. Color and turbidity must be removed. Highly colored or turbid samples should be subjected to preliminary distillation.

Reagents

1. Fluoride stock solution: See "SPADNS Method".

2. Fluoride standard: See "SPADNS Method".

3. Acid zirconlum – alizarin solution:

(a) **Alizarin solution:** Dissolve 700mg. alizarin red S. (sodium alizarin sulfonate) in 100 ml. distilled water.

(b) **Zirconyl chloride solution:** Dissolve 450mg. zirconyl chloride $ZrOCl_2.8H_2O$, in 100 ml. distilled water.

(c) **Sulfuric acid solution:** Add carefully, with stirring 70 ml. conc. H_2SO_4 to 700 ml. water in a beaker and cool.

Pour solution (a) slowly into solution (b) and then add solution (c) and dilute to 1000 ml. Use the reagent only after 4 hrs. Keep it in dark.

4. Sodium arsenite solution: See "SPADNS Method".

Procedure

1. Pipet 0.0, 1.0, 2.0, 4.0, 6.0, 8.0, 10.0, 12.0 ml. standard fluoride solution into 50 ml. Nessler tubes and make up to the mark with distilled water.
2. Place 50 ml. of the sample in a 50 ml. Nessler tube. If necessary suitable aliquots may be taken and diluted to 50 ml.

   (If the sample contains any suspended matter filter through whatman No. 42 or equivalent filter paper. If the sample contains residual chlorine, dechlorinate it by adding sodium arsenite solution to the aliquot taken in the Nessler tube. 1.0 ml. of the above sodium arsenite solution will dechlorinate 0.5mg. Cl_2. Calculate the chlorine content in the aliquot and add suitable amount of sodium arsenite solution and mix well.)
3. Add 1.0 ml. acid zirconium-alizarin solution to the standards and sample. Mix well immediately and compare the color after 20 minutes.

Express the result as mg Fluoride (as F^-) per litre of the sample.

PRELIMINARY DISTILLATION

If the sample is highly colored or turbid, or known to contain excess concentration of interfering substances, distillation of the sample is necessary. Fluoride is separated from other constituents by distillation as hydrofluoric acid.

Apparatus

The apparatus for distillation is as shown in the figure.

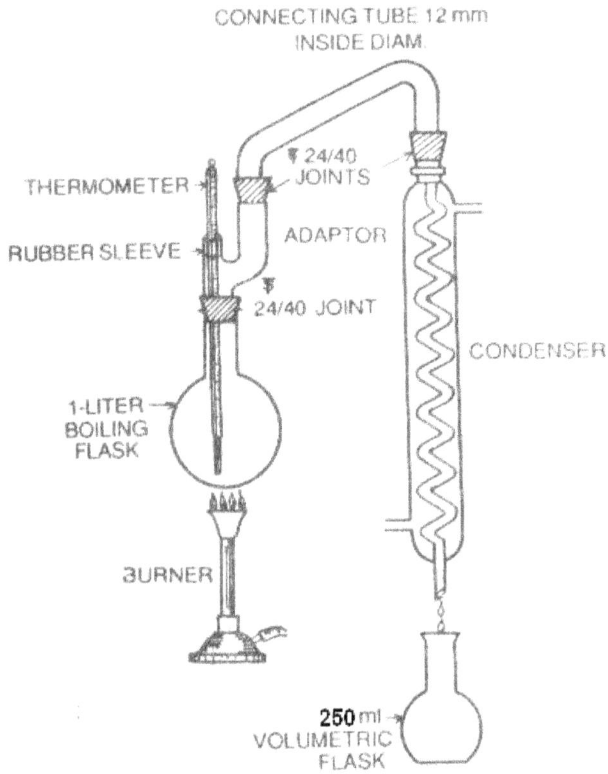

Fig 31.1 Direct Distillation apparatus for fluoride*

Procedure

Place 400 ml. distilled water in a 1 litre boiling flask. Carefully add 200ml. conc. H_2SO_4 and swirl the flask to make the contents homogeneous. Add glass beads and connect the flask as shown in the figure. Make sure that all joints are tightly fitted. Heat the flask slowly at first and then rapidly. (The distillate must be cool, whatever rapid the heating may be). As the temperature reaches 180°C stop the heating and allow it to cool. Discard the distillate. This

* Reproduced with the kind permission of American Public Health Association from Standard Methods for the Examination of Water and Waste water" – 1976.

process serves to remove fluorides if any in the glassware or in the reagents and to adjust the acid water ratio for subsequent distillation. After cooling to 120°C add 250 ml. sample, mix thoroughly and distill until the temperature reaches 180°C. (Do not allow the temperature to exceed 180°C as it may cause sulfate carry over). Collect the distillate in a 250 ml. volumetric flask.

If the sample contains high amounts of fluoride, continue the distillation with 250 ml. distilled water after the sample has been distilled and collect the second distillate in a 250ml volumetric flask. Combine both the distillates for fluoride determination.

If the chloride content of the sample exceeds 2000 mg/l, add silver sulfate crystals to the distilling flask at a rate of 5 mg per milligram of chloride.

Note: The sulfuric acid solution in the distilling flask can repeatedly be used until the accumulation of contaminants from water samples affect the fluoride recovery.

ELECTRODE METHOD

Fluoride ion–selective electrode is very useful for the rapid determination of fluoride in water samples, particularly, when contaminated with color, turbidity and other interfering substances. The fluoride-sensitive electrode is of the solid state type, consists of lanthanum - fluoride crystal. In combination with a reference electrode (normally calomel electrode) it forms a cell. An expanded scale-pH meter or specific ion meter may be employed for measuring the potential. For use of the electrode, the supplier's instructions should be followed. Generally the solution under test in which electrodes are dipped should be thoroughly agitated for about 3 minutes to attain a steady reading. To eliminate the interferences, a total ionic-strength adjustment buffer (TISAB) should be added to the test solution. Addition of TISAB is also made to adjust the pH to an optimum value* and introduce sufficient chloride to mask variations in ionic strength. It may be purchased from the suppliers and added to the test solution in the recommended proportion. The preparation of one of such buffers is as follows:

To 500 ml distilled water in a 1 litre beaker add 57 ml conc. (glacial) acetic acid, 58 g sodium chloride and 12g. sodium - citrate dihydrate, $Na_3C_6H_5O_7.2H_2O$. Stir well to dissolve the contents and cool to room temperature. Adjust the pH of the solution

* The hydroxide ion is measured directly by the electrode. The addition of TISAB minimises the effect of hydroxide ion by buffering the solution.

between 5.0 and 5.5 using 6N sodium hydroxide solution. (About 125 ml is required). Cool to room temperature. Transfer to a 1 litre volumetric flask and make up to the mark with distilled water.

To 50.0 ml of the sample or an aliquot diluted to 50 ml, add 50 ml of this ionic strength adjustment buffer. Fluoride standards diluted to 50 ml should also be added with 50 ml of this buffer. After dipping the electrodes in the sample, fluoride concentration can be read directly from the meter.

CHAPTER - 31

HARDNESS (TOTAL)

Hardness is deemed to be the capacity of water for reducing and destroying the lather of soap. Hardness in water is due to the natural accumulation of salts from contact with soil and geological formations or it may enter from direct pollution by industrial effluents. Calcium and magnesium are the principal cations causing hardness. Though calcium and magnesium are the principal cations causing hardness, iron, alumininum, manganese, strontium and zinc also cause hardness but to a relatively little extent or to a negligible amount. The term "Total Hardness" thus practically represents the concentration of calcium and magnesium ions only. However if present in significant amounts, the other metallic ions should also be included. **The total hardness is expressed in terms of calcium carbonate.**

The hardness of water varies considerably from place to place depending on the geological formations. Generally surface waters are softer (containing much less quantities of calcium and magnesium) than ground waters.

Among the methods available for the determination of hardness, the EDTA titrimetric method is the precise one and can be performed rapidly. It can be applied to any kind of water with appropriate modifications. This method, however, is not applicable to sewage and industrial effluents because of the possible presence of interfering cations.

EDTA TITRIMETRIC METHOD

Principle

Calcium and magnesium ions react with EDTA to form soluble complexes and the completion of reaction is indicated by the color change of a suitable indicator such as Eriochrome black T.

$$Ca^{2+} + \underset{\text{(EDTA)}}{H_2Y^{2-}} \rightarrow CaY^{2-} + 2H^+$$

$$Mg^{2+} + H_2Y^{2-} \rightarrow MgY^{2-} + 2H^+$$

$$\underset{\text{(wine red)}}{MgD^-} + H_2Y^{2-} \rightarrow MgY^{2-} + \underset{\text{(blue)}}{HD^{2-}} + H^+$$

In this titration, calcium ions do not react with the indicator dye. Magnesium ions only will react and change the color of the dye. Therefore a small amount of complexometrically neutral magnesium salt of EDTA is introduced to the titre through the addition of buffer to obtain end point by color change of the indicator.

Interferences

Several metal ions interfere in this determination. They may cause fade or indistinct end points. These interferences can be eliminated by the addition of sodium sulfide inhibitor before titration. Suspended and colloidal organic matter may also interfere. A separate procedure has to be followed for those samples.

Reagents

1. Calcium standard solution: Weigh accurately 1.000 gm of pure calcium carbonate (previously dried at 105°C) and place in a 250 ml. conical flask using 50 ml. distilled water. Place a funnel in the neck of the flask. Add 20.5ml. 1N.HCl. Warm until the solution is complete. Cool and transfer to a 1,000 ml. volumetric flask and make up to the mark with distilled water.

$$1.0 \text{ ml.} = 1.0 \text{ mg } CaCO_3$$

2. Standard EDTA titrant, 0.02N (0.01M): Weigh accurately 3.723g. AR grade disodium - ethylene diamine tetra acetate dihydrate, also called (ethylene dinitrilo) – tetra acetic acid disodium salt EDTA $Na_2H_2C_{10}H_{12}O_8N_2.2H_2O$, and dissolve in distilled water and make up to 1000 ml. Check the titer by standardizing against standard calcium solution as in the 'Procedure for clear samples'. Store it preferably in polyethylene bottle or in a pyrex bottle.

$$1.0 \text{ ml. of exactly } 0.02 \text{ N EDTA} \equiv 1.0 \text{ mg } CaCO_3$$

3. Ammonia – Ammonium chloride buffer

(a) Dissolve 16.9g. ammonium chloride, NH_4Cl in 143ml. conc. ammonia solution (0.90 Sp.gr*).

(b) Dissolve 1.179g. EDTA di-sodium salt and 780mg. $MgSO_4.7H_2O$ (or 644 mg $MgCl_2.6H_2O$) in 50 ml. distilled water.

(c) Add solution (b) into solution(a) and mix well. Then make up to 250 ml. with distilled water.

* If 0.91 Sp.gr. is used, take 165 ml

4. Sodium sulfide inhibitor: Dissolve 5.0g. $Na_2S.9H_2O$ or 3.7g $Na_2S.5H_2O$ in 100 ml. distilled water. This inhibitor deteriorates on oxidation and air must be excluded. Keep the bottle tightly fitted with a rubber stopper. When appreciable concentration of heavy metals are present, they give sulfide precipitates which obscure the end point.

5. Eriochrome Black T indicator: Alcoholic solutions of this dye are not so stable and one of the following methods may be adopted:

(i) Mix 0.5 to 1.0 g of the dye in 100g. of triethanolamine (also termed as 2, 2', 2" nitrilotriethanol).

(ii) Mix together 0.5 g of the dye and 100 g sodium chloride to prepare a dry powder mixture [this is more stable than (i)].

Procedure

A. Clear samples

1. Place a suitable volume of sample (filtered if necessary) in a conical flask and dilute to 50 ml.
   For a total hardness below 100 mg/l, use 100 ml sample.
   For a total hardness greater than 100 mg/l, use 50 ml sample.
   For other samples use 25 ml, or smaller aliquots diluted to 50 ml.
   Whatever be the volume of sample taken, it should not consume more than 15 ml EDTA titrant.
2. Add 1 ml of buffer solution per 50 ml volume. The pH of the titre should be 10.0 ± 0.1. Add 1 ml sodium sulfide inhibitor. Add 1 drop of indicator solution or an appropriate amount of dry powder.
3. Titrate with standard EDTA solution slowly, until a reddish tinge appears and add the last few drops within 3-5 seconds. At the end point, the solution will be blue. The whole titration procedure should be completed within 5 minutes after the addition of buffer.

B. Samples containing Suspended Matter and Colloidal Organic Matter

Evaporate an aliquot of the sample to dryness on a steam bath. Heat it in a muffle furnace to 550°C until the organic matter is oxidized. Dissolve the residue in 20 ml 1N.HCl, neutralize to pH 7 with 1N.NaOH and dilute to 50 ml with distilled water. Cool to room temperature and proceed from step No.2 of the procedure described for clear samples.

C. Low-hardness samples

Ion exchanger effluents or softened waters and some natural waters will have only low hardness (less than 5 mg/l). For those samples take a larger volume for titration ranging from

100 to 1,000 ml. Add proportionately higher amounts of buffer, inhibitor and indicator. Titrate with EDTA slowly, using a micro burette. Carry out a blank using the same volume of deionised or redistilled water.

Alternatively, colorimetric methods may be employed. For detailed procedure, refer "Water Analysis Handbook" – 4th edition (Revision – 2) – 2003, published by M/s. Hach Company, Colorado, U.S.A.,

D. Samples having a pH of 10 and more

Neutralize the aliquot of the sample with hydrochloric acid to bring down the pH to 7.0 to 8.0. Then proceed from Step No.2 of the procedure described for clear samples.

Calculation

$$\text{Total Hardness (as } CaCO_3\text{), mg/l} = \frac{\text{ml. EDTA titrant} \times E \times 1000}{\text{ml. sample taken for titration}}$$

where E is mg $CaCO_3$ equivalent to 1.0ml EDTA

If the normality of the EDTA is exactly 0.02N, then

$$\text{Total Hardness (as } CaCO_3\text{), mg/l} = \frac{\text{ml. EDTA titrant} \times 1 \times 1000}{\text{ml. sample taken for titration}}$$

Classification of Hardness

Hardness has traditionally been classified into "Temporary Hardness" and "Permanent Hardness". The portion of hardness that disappears on prolonged boiling is referred as temporary hardness and is mainly due to the bicarbonates of calcium and magnesium which are precipitated as normal carbonates by the loss of CO_2 on heating. The hardness that remains after boiling is known as permanent hardness and is due to the sulfates, chlorides and nitrates of calcium and magnesium. However these terms are obsolete and the more precise terms are "Carbonate Hardness" and "Non-carbonates Hardness". The carbonate hardness is due to bicarbonates and carbonates of calcium and magnesium while non-carbonate hardness is due to sulfates, chlorides and nitrates of calcium and magnesium. When the hardness is numerically greater than the sum of the bicarbonate and carbonate alkalinities, the amount equal to the alkalinity is carbonate hardness, and the excess amount is non-carbonate hardness. If the hardness is equal to or less than the sum of the bicarbonate and carbonate alkalinities, all of the hardness is carbonate hardness and there is no non-carbonate hardness.

CALCULATION OF CARBONATE AND NON-CARBONATE HARDNESS-EXAMPLES

Problem 31 – 1

Calculate "carbonate hardness" and "non-carbonate hardness" of a sample having an alkalinity (carbonate + bicarbonate) of 232 mg/l (as $CaCO_3$) and hardness 264 mg/l (as $CaCO_3$).

Carbonate hardness = Alkalinity
= **232 mg/l (as $CaCO_3$)**

In this sample, hardness is greater than alkalinity

. Non-carbonate hardness = Total hardness – alkalinity
= 264 – 232
= **32 mg/l (as $CaCO_3$)**

Problem 31 – 2

The hardness and alkalinity (carbonate + bicarbonate) of a water sample are 244 mg/l (as $CaCO_3$) and 268 mg/l (as $CaCO_3$) respectively. Calculate carbonate and non-carbonate hardness

.

The value of hardness is lower than alkalinity.

∴ Carbonate hardness = Total hardness
= **244 mg/l (as $CaCO_3$)**

Non-carbonate hardness = **0**

Problem 31 – 3

50 ml of a sample when titrated with 0.02 N. EDTA consumes 8.3 ml of EDTA. If the sum of carbonate and bicarbonate alkalinities is 74 mg/l (as $CaCO_3$), calculate the total hardness, carbonate and non-carbonate hardness.

$$Total\ Hardness\ (as\ CaCO_3),\ mg/l = \frac{ml.\ EDTA\ titrant \times E \times 1000}{ml.\ sample\ taken\ for\ titration}$$

ml EDTA = 8.3
E (mg $CaCO_3$) equivalent to 1.0 ml EDTA = 1.0
(1.0 ml exactly 0.02 N EDTA ≡ 1.0 mg $CaCO_3$)

ml sample for titration = 50

$$\therefore Total\ Hardness\ (as\ CaCO_3),\ mg/l = \frac{8.3 \times 1 \times 1000}{50} = \mathbf{166}$$

Since the value of hardness is more than alkalinity,
Carbonate hardness = Alkalinity (carbonate + bicarbonate)

= **74 mg/l (as $CaCO_3$)**

Non-carbonate hardness = Total hardness – alkalinity
= 166 – 74
= **92 mg/l (as $CaCO_3$)**

The metals mentioned in Table - 31.1 also titrate as hardness along with calcium and magnesium. If the hardness due to calcium and magnesium alone is to be found out, the other metal ion concentration (found out by some other method) has to be multiplied by the conversion factor given in the table to get its calcium carbonate equivalent. Then it has to be subtracted from the total hardness value obtained.

Table 31–1
IONS OTHER THAN CALCIUM AND MAGNESIUM CAUSING HARDNESS AND EQUIVALENCE FACTORS

Metal	Hardness equivalence factor (mg/l as $CaCO_3$)
Aluminium	5.56
Barium	0.729
Cobalt	1.698
Copper	1.574
Iron (Ferric)	1.791
Manganese	1.820
Nickel	1.704
Strontium	1.141
Zinc	1.529

Units of Hardness

The normal expression of hardness is mg/l calcium carbonate. Whenever needed, they can be expressed in terms of calcium and magnesium. The conversion factors are given in Table 31 – 2.

Table 31 – 2
CALCIUM CARBONATE – CONVERSION FACTORS

To convert from	To	Multiply by
mg/l $CaCO_3$	mg/l Ca	0.4
mg/l $CaCO_3$	mg/l Mg	0.24
mg/l Ca	mg/l $CaCO_3$	2.5
mg/l Mg	mg/l $CaCO_3$	4.166

Hardness is also expressed in various other units. The interrelation and equivalents of these units are tabulated (Table 31–3). It is to be noted that while all other units are expressed in terms of calcium carbonate, German degree of hardness is based on calcium oxide.

Table 31–3
HARDNESS CONVERSION FACTORS

Units of Measure	mg/l $CaCO_3$	British gr/gal (Imperial) $CaCO_3$	American gr/gal (US) $CaCO_3$	French Degree Parts/100000 $CaCO_3$	German Degree Parts/100000 CaO	meq/l	g/l CaO
mg/l $CaCO_3$	1.0	0.07	0.058	0.1	0.056	0.02	5.6×10^{-4}
British gr/gal (Imperial) $CaCO_3$	14.3	1.0	0.83	1.43	0.83	0.286	8.0×10^{-3}
American gr/gal (US) $CaCO_3$	17.1	1.2	1.0	1.72	0.96	0.343	9.66×10^{-3}
French Degree Parts/100000 $CaCO_3$	10.0	0.7	0.58	1.0	0.56	0.2	5.6×10^{-3}
German Degree Parts/100000 CaO	17.9	1.25	1.04	1.79	1.0	0.358	1.0×10^{-2}
meq/l	50.0	3.5	2.9	5.0	2.8	1.0	2.8×10^{-2}
g/l CaO	1790.0	125.0	104.2	179.0	100.0	35.8	1.0

CHAPTER - 32

HYDRAZINE

Because of the increased use of hydrazine as an oxygen scavenger* in boiler feed waters (especially in power plants), the determination of hydrazine has become increasingly important. It is best determined by reaction with para dimethyl amino benzaldehyde.

PARADIMETHYL AMINO BENZALDEHYDE METHOD

Principle

Hydrazine reacts with para-dimethyl amino benzaldehyde and produces a specific yellow colour which is measured photometrically.

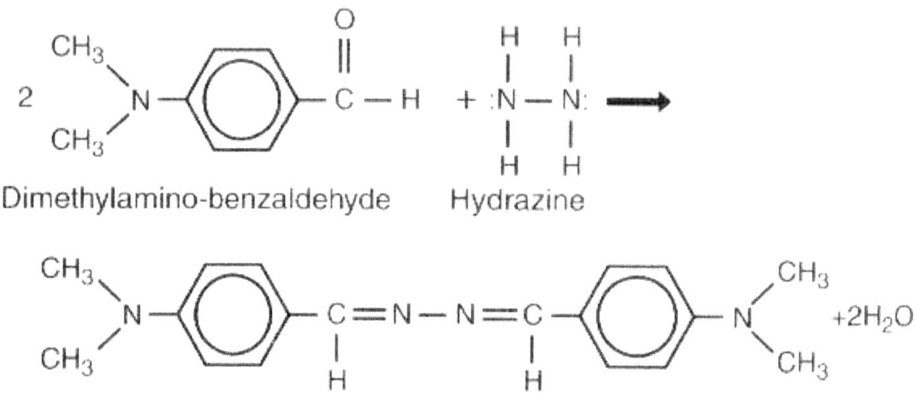

Azine Complex (Yellow in dilute solution)

Interferences

The substances normally found in industrial water do not interfere.

Reagents

1. Hydrochloric acid, 0.5N (approximately): Dilute 130 ml. conc. HCl to 3000 ml. with distilled water.

---

* Any residual oxygen remaining in the boiler feed water is eliminated by the addition of hydrazine. Oxygen is a well known corrosive agent and attack metallic parts, pipes and fittings and corrode severely. Hence hydrazine is added to boiler water to remove even the traces of oxygen.

2. Hydrazine stock solution: Dissolve 328mg. hydrazine hydrochloride $N_2H_4.2HCl$ in 50 ml. 0.5 N.HCl and make up to 1000 ml. in a volumetric flask with 0.5N HCl.

1 ml. = 0.1 mg Hydrazine (N_2H_4)

3. Hydrazine standard solution: Pipet 10.0 ml. hydrazine stock solution into a 1000 ml. volumetric flask and make up to the mark with 0.5N HCl.

1 ml. = 0.001 mg Hydrazine (N_2H_4)

4. Para-di-methyl amino benzaldehyde solution: Dissolve 4.0g. para dimethyl amino - benzaldehyde in 200 ml. of methyl alcohol and 20 ml. conc.HCl. Place in an amber bottle and keep in dark.

Procedure

1. Place 15 ml. of 0.5 N HCl into a series of 50 ml. Nessler tubes. Pipet 0.5, 1.0, 1.5, 2.0, 2.5, 3.0, 3.5, 4.0, 4.5 and 5.0 ml. hydrazine working solution in to them. Dilute to 40 ml. with distilled water. Include a Nessler tube containing 15 ml. 0.5 N HCl and 25 ml. distilled water as the blank.
2. In to another Nessler tube, place 15 ml. 0.5 N HCl and pipet a suitable aliquot (not exceeding 25 ml) of the sample containing not more than 5 μg hydrazine and dilute to 40 ml.with distilled water.
3. Add 10 ml. para-dimethyl amino benzaldehde solution to the blank, standards and sample and mix thoroughly.
4. After 10 minutes, measure the absorbances of standards and sample at 458 nm after setting the blank at 100% transmittance. Prepare a calibration curve and find out the mg. equivalent of hydrazine of the sample to the observed optical density. When the photometer is not available, visual comparison can also be made.

Express the result as mg Hydrazine (N_2H_4) per litre of the sample.

CHAPTER - 33

IODIDE

Iodide is present in natural waters only in ppb levels. Higher concentrations may be found in sea waters, brines and waters treated with iodine. Iodide is best determined by the following photometric method.

PHOTOMETRIC METHOD

Principle

Iodide can be determined by using its ability to catalyse the reduction of ceric ions by arsenious acid, the effect being proportional to the amount of iodide present. The reaction is stopped at specific time interval by the addition of ferrous ammonium sulfate, the resulting ferric ions which are directly proportional to the remaining ceric ions, develop a color complex with potassium thio cyanate which is relatively stable (as the concentration of the iodide increases, the intensity of the color of ferric thiocyanate decreases. The color of the sample is compared with that of the standards prepared simultaneously).

$$2Ce^{4+} + H_3AsO_3 + H_2O \xrightarrow{I^- \text{ Catalyst}} 2Ce^{3+} + H_3AsO_4 + 2H^+$$

$$Ce^{4+} + Fe^{2+} \longrightarrow Ce^{3+} + Fe^{3+}$$

$$Fe^{3+} + 3CNs^- \longrightarrow Fe(CNS)_3$$

Interferences

The interference by chloride and the non-catalytic forms of iodine are eliminated by the addition of sodium chloride.

Apparatus

Spectrophotometer or Filterphotometer equipped with a filter having maximum transmittance near 525 mμ.

Reagents

1. Distilled Water: Redistill the distilled water in an all glass distillation apparatus.

2. Sodium Chloride Solution: Dissolve 200g. NaCl in 100ml. distilled water.

Note: The sodium chloride used must be iodine free.

3. Arsenious acid solution, 0.1N: Dissolve 4.946g. arsenious oxide(As_2O_3) in distilled water. Add 0.20 ml. conc.H_2SO_4 and dilute to 1000ml.

4. Sulfuric acid: Conc.

5. Ceric ammonium sulfate solution, 0.02N:

(i) Dissolve 13.38g. ceric ammonium sulfate [$Ce(NH_4)_4(SO_4)_4.4H_2O$] in distilled water.

(ii) Add 44 ml. conc.sulfuric acid

(iii) Make upto 1000ml with distilled water.

6. Ferrous ammonium sulfate solution: Dissolve 1.50g. of ferrous ammonium-sulfate ($FeSO_4.6H_2O$) in 100ml. distilled water to which 0.6 ml conc. sulfuric acid has already been added. (Prepare daily).

7. Potassium thio cyanate solution: 4% Dissolve 4g. potassium thio-cyanate KCNS in 100ml. distilled water.

8. Stock Potassium Iodide Solution: Dissolve 0.2616g. anhydrous potassium iodide, KI in distilled water and dilute to 1000ml.

1.0 ml = 0.2 mg Iodide.

9. Working Standard Iodide solution: Dilute 5.0ml of the above stock solution to 1000ml. with distilled water.

1.0ml = 0.001mg. Iodide.

Procedure

1. Take 100ml of sample in a beaker and concentrate to 10ml. on a water bath.
2. Transfer to a test tube of capacity 20ml.
3. Place 10 ml distilled water in a test tube which is blank.
4. Into a series of test tubes place 1.0ml, 2.0ml, 3.0ml, 4.0ml, 5.0ml and 6.0ml working standard iodide solution and dilute to 10 ml with distilled water.
5. To the blank, standards and sample add the following reagents with mixing after each addition.
   (i) 1.0 ml. sodium chloride solution.
   (ii) 0.50ml. arsenious acid solution.
   (iii) 0.50ml. conc. sulfuric acid.
   (iv) 1ml. ceric ammonium sulfate solution.

6. Start a stop clock.
7. After exactly 15 minutes, add immediately 1 ml. ferrous ammonium sulfate solution and mix well.
8. When the yellow ceric ion color disappears add 1ml potassium thiocyanate solution.
9. Compare the color with that of the standards produced in the same manner within one hour.

Express the result as mg. Iodide (as I^-), per litre of the sample.

CHAPTER - 34
IRON

Iron usually exists in natural water both in ferric and ferrous forms. The form of iron however may be altered as a result of oxidation or reduction or due to the growth of bacteria in the sample during storage. Usually the ferric form is predominant in most of the natural waters. Iron in water may be either in true-solution, or in colloidal state or in the form of relatively coarse suspended particles.

The analysis of iron in water is usually done colorimetrically or photometrically. Higher concentrations of iron can be determined by titricmetric methods. A colorimetric method (using thiocyanate) and a titrimetric method are given here for total iron determination. A separate determination for ferrous iron using α-α bipyridyl as the complexing agent is also given.

TOTAL IRON
(BY THIOCYANATE COLORIMETRIC METHOD)

Total iron is the sum of both ferric and ferrous iron concentrations.

Principle

Ferric iron combines with thiocyanate ions to form red colored ferric thiocyanate which is measured colorimetrically.

$$Fe^{3+} + 3CNS^{-} \rightarrow Fe(CNS)_3$$

Interferences

Chromium (in higher concentrations), zinc, copper, cobalt, mercury, phosphates and molybdate ions interfere.

Reagents

1. Conc. Hydrochloric acid.

2. Dil.Hydrochloric acid, 1 + 1.

3. Potassium permanganate solution approximately 0.2 N: Dissolve 600 mg potassium permanganate, $KMnO_4$ in 80 ml distilled water and dilute to 100 ml. This is about 0.2 N.

4. Iron stock solution

Using electrolytic iron wire: Clean the electrolytic iron wire with fine sand paper to remove any oxide coating. Weigh 100.0mg. cleaned wire and place it in a 1000 ml volumetric flask. Dissolve in 20 ml 6N.H_2SO_4 (3.3 ml conc. H_2SO_4 + 16.7 ml distilled water) and make up to the mark with distilled water.

1.0 ml = 0.1 mg Fe

Using ferrous ammonium sulfate: Add 20 ml conc. H_2SO_4 to 50 ml distilled water. Dissolve 0.7022 g of ferrous ammonium sulfate crystals, $FeSO_4.(NH_4)_2SO_4.6H_2O$ in the above acid solution. Warm it gently. Add potassium permanganate solution dropwise until a slight pink color persists. Make upto 1000 ml in a volumetric flask with distilled water.

1.0 ml = 0.1 mg Fe

5. Iron standard solution: Dilute 50.0 ml stock solution to 500 ml with distilled water.

1.0 ml = 0.01 mg Fe

6. Thiocyanate solution: Dissolve 10g. potassium thiocyanate, KCNS or ammonium - thiocyanate, NH_4CNS in 80 ml distilled water and dilute to 100 ml.

Procedure

A. For clear samples

1. Place 100 ml of the well shaken sample in a 250 ml beaker, and add 5 ml 1 + 1 HCl. Reduce the volume to 40 ml by placing on a hot plate. Cool and add potassium permanganate solution drop by drop until a pink color persists for atleast 5 minutes. Transfer to a 50 ml Nessler tube and make up to the mark.
2. Pipet 1.0 ml., 2.0 ml., 3.0 ml., 4.0 ml., 5.0 ml. and 6.0 ml. iron standard solution in to 50 ml Nessler tubes. Add 1 ml dil. HCl and 2 drops of potassium permanganate solution. Mix well and make up to the mark with distilled water.
3. Add to the standards and sample, 1 ml thiocyanate solution and mix well. Compare the color of the sample with that of the standards and find out the mg equivalent of iron present in the sample as follows.

Calculation

$$mg/l\ Iron\ (as\ Fe) = \frac{matching.\ std \times 0.01 \times 1000}{ml.\ sample\ taken\ for\ determination}$$

If the quantity of ferric iron only is needed, subtract the ferrous iron value from this total iron value.

B. For samples containing appreciable amounts of organic matter

Place 50 ml or 25 ml or a suitable aliquot of the well shaken sample in to a 250 ml beaker and evaporate to dryness on a hot plate. Cool the residue and add dil. HCl and warm it. Wash the inner surface of the beaker with hot acid. Add about 20 ml distilled water and warm it for several minutes. Cool and add potassium permanganate solution dropwise until a faint pink color persists. Transfer to a 50 ml Nessler tube and make up to the mark with distilled water. Add 1 ml thiocyanate solution and mix well. Compare the color with that of the standards prepared simultaneously as in Procedure – A for clear samples.

C. For samples having higher concentrations of Iron, and/or silt and organic matter

Measure 20 ml or 10 ml or a suitable aliquot of the sample in an evaporating dish and add 10 ml (iron-free) conc. HCl. Evaporate to dryness on a hot plate and ignite by heating until a greyish white ash is obtained.

Note: Do not heat to a red glow

Cool the ash and add 5 ml dil. HCl and transfer to a 100 ml or 250 ml volumetric flask and make up to the mark with distilled water. Pipet aliquots from this made up solution in to 250 ml beakers and proceed as for clear samples.

TOTAL IRON
(TITRIMETRIC METHOD)

Some of the water such as derived from oil fields contain high amounts of iron. The concentration of iron in these waters can be determined by the following method.

Principle

All the ferric iron present in water is reduced to ferrous by stannous chloride and this ferrous iron is titrated with standard potassium dichromate solution.

Procedure

Prepare the sample as given in the preceding section. *'For samples having higher concentration of iron and / or organic matter'*.

Place 25 ml of the treated sample in a 500 ml conical flask. Add 20 ml conc. HCl. Bring the solution to near boiling (70° to 90°C). Reduce the Fe^{3+} to Fe^{2+} by adding stannous chloride solution, (prepared by dissolving 30 g. $SnCl_2$. $2H_2O$ in 100 ml conc. HCl and diluting to 200 ml. with distilled water, drop by drop until the yellow color disappears. Care must be taken not to add excess stannous chloride. Cool the solution rapidly under tap to about 20°C (with protection from air). Then rapidly add 10 ml of 5% mercuric chloride solution and mix thoroughly to remove the slight excess of stannous chloride. A slight silky white precipitate of mercurous chloride should be obtained at this stage (if a heavy or grey or black precipitate is formed it indicates the presence of excess stannous chloride. In this case the reduction should be repeated).

Allow to stand for 5 minutes. Add 200 ml. 2N. sulfuric acid and 0.5 ml. N-phenyl - anthranilic acid indicator (prepared by dissolving 100 mg. in 4.8 ml. of 0.1N. NaOH and diluting to 100 ml with distilled water). Titrate with standard potassium dichromate solution of appropriate normality. The color change is from green to violet red. From the titre value calculate the iron content.

1 ml. of exactly 1.0 N. $K_2Cr_2O_7$ = 55.85 mg. Fe

[Preparation of potassium dichromate solution (0.25N) is given in Chapter – 20 'CHEMICAL OXYGEN DEMAND']

FERROUS IRON
(α-α BIPYRIDYL COLORIMETRIC METHOD)

Principle

Ferrous iron has the tendency to oxidize readily to ferric state. A separate sample for ferrous iron determination is collected in a 100 ml or 60 ml bottle containing α-α bipyridyl which reacts immediately with the ferrous iron to form to a red colored compound. This compound is stable. It is to be noted that ferric iron does not form such complexes.

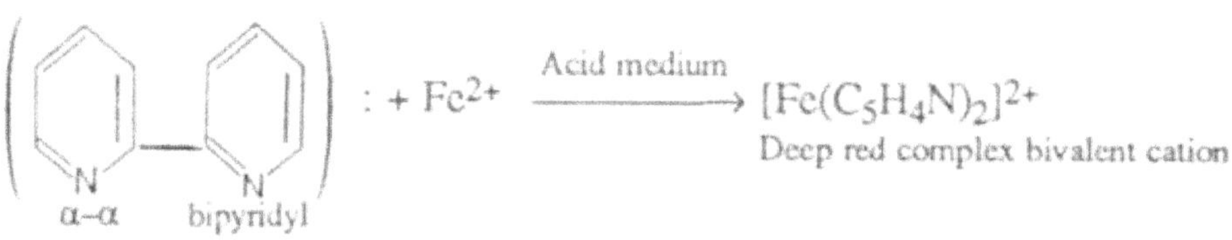

Reagents

1. Ferrous stock solution: Dissolve 3.511g. ferrous ammonium sulfate $Fe(SO_4).(NH_4)_2SO_4.6H_2O$ in distilled water and make up to 1000 ml in a volumetric flask.

$$1.0 \text{ ml} = 0.5 \text{ mg } Fe^{2+}$$

2. Ferrous working solution: Pipet 10.0 ml ferrous stock solution in to a 1000 ml volumetric flask and make up to the mark with distilled water.

$$1.0 \text{ ml} = 0.005 \text{ mg } Fe^{2+}$$

3. Hydroxylamine hydrochloride solution, 10%: Dissolve 10g. hydroxylamine- hydrochloride, $NH_2OH.HCl$ in 90 ml distilled water. Prepare freshly.

4. α-α bipyridyl solution: Carefully add 17 ml conc. HCl to about 200 ml distilled water in a 2000 ml standard flask. Dilute to the mark with distilled water. This is approximately 0.1 N HCl. Dissolve in it 1 g. α-α bipyridyl.

5. Sodium acetate buffer solution: Dissolve 68g. sodium acetate trihydrate, $NaC_2H_3O_2.3H_2O$ in 100 ml distilled water. Add 60 ml glacial acetic acid and dilute to 500 ml.

6. Ethylene diamine.

Procedure

1. Prepare the permanent color standards as follows:
   - (i) Into a series of 100 ml volumetric flasks (or 100 ml Nessler tubes), pipet 1.0, 2.0, 3.0, 4.0.... 20.0 ml standard ferrous solution.
   - (ii) Add to each flask 1 ml hydroxyl amine hydrochloride solution.
   - (iii) Mix well and add 10 ml α-α bipyridyl solution.
   - (iv) After 10 minutes, add 2 ml sodium acetate buffer solution.
   - (v) Make up to the mark with distilled water.
   - (vi) Add 2 drops of ethylene diamine reagent and mix thoroughly.

   Fill the prepared ferrous iron standards in 100 ml or 60 ml bottles of same size and shape and stopper tightly.
2. Collect the sample in a 100 ml or 60 ml bottle in which 5 ml of α-α bipyridyl has been previously added. Add 5 ml acetate buffer solution and shake well.
3. Compare the color of the sample with that of the prepared standards.

Express the result in mg per litre of the sample as Fe^{2+}.

Though iron ranks next to aluminium in abundance of metals in the earth's crust, natural waters contain only minor amounts of iron. The iron content of some deep wells and industrially contaminated streams may be very high. Iron in natural waters usually occurs in the ferrous form; however it is readily oxidized to ferric form and ferric salts are precipitated as rust-colored deposits.

Corrosion in boiler and preboiler systems releases iron in ionic form and iron oxides which are not removed by blow down. These oxides form adherent coatings and lead to tube failures. Further, ferric iron complexes with sodium silicate (if present) to form a very hard adherent deposit.

Iron in cooling waters may cause serious fouling problems by the deposition of iron oxide.

Apart from boiler waters and cooling waters, other process waters used for brewing, fabric dyeing, paper manufacture, ice making and food processing should also be free from iron. Otherwise, the iron present may cause discoloration or turbidity to the finished products.

Suspended and/or dissolved ferrous and ferric iron can also foul ion exchange resins used for softening of water.

CHAPTER - 35

LANGELIER SATURATION INDEX, RYZNAR STABILITY INDEX AND PUCKORIUS SCALING INDEX

Langelier Saturation Index is a measure of the waters' ability to dissolve or deposit calcium carbonate and is often used as an indicator to find out whether the water is corrosive or scale forming in nature. This index is based on the difference between actual pH value and theoretical pH value which is calculated from the total dissolved solids, temperature, calcium content and total alkalinity. Langelier Index though not quantitative, is more useful in estimating the water treatment requirements for low pressure boilers and cooling towers. There is another index termed as 'Ryznar Stability Index' which is also based on the actual pH value and theoretical pH value and is also widely employed to assess the particular water to be corrosive or to form scale. A somewhat more accurate index than Langelier and Ryznar indices is developed by P.R.Puckorius and J.M.Brooke recently. All these three indices are described in this chapter.

LANGELIER AND RYZNAR INDICES

As a general rule, Langelier Index is most useful in bulk systems, with a very low velocity of flow such as reservoirs, filters and lime softeners etc., Ryznar Index is applicable to flowing systems where the velocity is greater than about 0.6m/s (2ft/s) or sufficient to prevent sedimentation.

Langelier Saturation Index, $I_{sat} = pH - pH_s$

Ryznar Stability Index, $I_{stab} = 2pH_s - pH$

where pH is the actual pH observed and pH_s is the theoretical pH value calculated as follows:

$$pH_s = (9.3 + A + B) - (C + D)$$

where A is the factor for total dissolved solids,

B is the factor for temperature ^oC,

C is the factor for calcium hardness (as $CaCO_3$), and

D is the factor for total alkalinity.

Calculations of Langelier and Ryznar Indices

Obtain values of A, B, C and D from Table 35–1 and calculate pH_s and find out the Langelier Saturation Index (I_{sat}) and Ryznar Stability Index (I_{stab}). If I_{sat} is zero, the water is in chemical balance; if it is a plus quantity, scale forming tendencies are indicated; if it is a negative quantity, corrosive tendencies are indicated.

Water with Ryznar stability index 6 to 7 will be neither scale forming nor corrosive. Water with stability index below 6 will be scale forming and above 7 will be corrosive.

TABLE 35 - 1

DATA FOR CALCLUATION OF LANGELIER AND RYZNAR INDICES (CALCIUM CARBONATE SATURATION INDEX)

A

TOTAL SOLIDS, mg/l	A
50 to 300	0.1
400 to 1000	0.2

B

TEMPERATURE°C	B
0 to 1.1	2.6
2.2 to 5.5	2.5
6.7 to 8.9	2.4
10.0 to 13.3	2.3
14.4 to 16.7	2.2
17.8 to 21.1	2.1
22.2 to 26.7	2.0
27.8 to 31.1	1.9
32.2 to 36.7	1.8
37.8 to 43.3	1.7
44.4 to 50.0	1.6
51.1 to 55.6	1.5
56.7 to 63.3	1.4
64.4 to 71.1	1.3
72.2 to 81.1	1.2

C

CALCIUM HARDNESS (as $CaCO_3$), mg/l	C
10 to 11	0.6
12 to 13	0.7
14 to 17	0.8
18 to 22	0.9
23 to 27	1.0
28 to 34	1.1
35 to 43	1.2
44 to 55	1.3
56 to 69	1.4
70 to 87	1.5
88 to 110	1.6
111 to 138	1.7
139 to 174	1.8
175 to 220	1.9
230 to 270	2.0
280 to 340	2.1
350 to 430	2.2
440 to 550	2.3
560 to 690	2.4
700 to 870	2.5
800 to 1000	2.6

D

METHYL ORANGE ALKALINITY (as $CaCO_3$), mg/l	D
10 to 11	1.0
12 to 13	1.1
14 to 17	1.2
18 to 22	1.3
23 to 27	1.4
28 to 35	1.5
36 to 44	1.6
45 to 55	1.7
56 to 69	1.8
70 to 88	1.9
89 to 110	2.0
111 to 139	2.1
140 to 176	2.2
177 to 220	2.3
230 to 270	2.4
280 to 350	2.5
360 to 440	2.6
450 to 550	2.7
560 to 690	2.8
700 to 880	2.9
890 to 1000	3.0

Problem 35-1

On analysis, a water sample is found to have total solids 810 mg/l, calcium hardness 184 mg/l (as $CaCO_3$) methyl orange alkalinity 420 mg/l (as $CaCO_3$). The temperature of the sample is 30^0C and pH is 7.9. Calculate the Langelier Saturation Index and Ryznar Stability Index.

Langelier Saturation Index $I_{sat} = pH - pH_s$

$pH_s = (9.3 + A + B) - (C + D)$

From the Table 35-1, the values for A, B, C and D are,

A = 0.2

B = 1.9

C = 1.9

D =2.6

$\therefore pH_s = (9.3 + 0.2 + 1.9) - (1.9 + 2.6)$

$= 11.4 - 4.5 = 6.9$

Langelier Saturation Index,

$I_{sat} = pH - pH_s$

$= 7.9 - 6.9 = +1.0$

Since it is a plus quantity, it indicates the scale forming tendencies of water.

Ryznar Stability Index,

$I_{stab} = 2pH_s - pH$

$= 2(6.9) - 7.9 = 13.8 - 7.9 = 5.9$

Since the value is below 6.0, the water will have the tendency to deposit scale.

Problem 35-2

The values for total solids, temperature, calcium hardness, methyl orange alkalinity and pH for a sample of water are 170 mg/l, 32°C, 95 mg/l (as $CaCO_3$) 112 mg/l (as $CaCO_3$) and 7.1 respectively. Calculate the Langelier and Ryznar indices.

Langelier Saturation Index

$I_{sat} = pH - pH_s$

$pH_s = (9.3 + A + B) - (C + D)$

From the Table 35 – 1, the values of A, B, C and D are as follows:

$$A = 0.1$$
$$B = 1.8$$
$$C = 1.6$$
$$D = 2.1$$

$$\therefore pH_s = (9.3 + 0.1 + 1.8) - (1.6 + 2.1)$$
$$= 11.2 - 3.7 = 7.5$$

Langelier Saturation Index,

$$I_{sat} = pH - pH_s$$
$$= 7.1 - 7.5 = -0.4$$

Since it is a negative quantity, the water will have corrosive tencencies.

Ryznar stability index = $2pH_s - pH$

$$= 2(7.5) - 7.1 = 7.9$$

Since the value is above 7, the water will be corrosive.

Although information obtained from Langelier Saturation Index and Ryznar Stability Index are not quantitative*, they serve to know immediately about the corrosivity of a water just by calculations. They are useful in estimating water treatment requirements for low pressure boilers, cooling towers and water treatment plants. For example, in the treatment of water by Reverse Osmosis (for the removal of dissolved solids) the concentrate should always be negative or else would form precipitate at the interface of the membrane. To obtain a negative value to avoid such scale formation, the feed water is added with acid supplemented by a low dosage of polyphosphates. Thus LI is used as a control parameter. However care must be exercised while interpreting the results of Langelier Index. Sometime the Langelier Index may show a positive index (i.e) scale forming tendency but actually the water would be corrosive due to the high concentrations of sulfate, chloride and other ions that may interfere with calcium carbonate film formation.

---

* It is useful only for determining the corrosivity of waters containing more than 40 mg/l alkalinity, sufficient calcium ion concentration and pH value ranging from 6.5 to 9.5. The quantitative method for the determination of corrosivity of a water is described under Chapter - 18 "Calcium- Carbonate Stability Test"

PUCKORIUS SCALING INDEX

Puckorius Scaling Index provides a better and more consistent indications of scaling conditions of a water. It is based on using pH of equilibrium (pH_{eq}) rather than the actual pH. The practical scaling index is calculated as follows.

$$\text{Practical Scaling Index (PSI)} = 2\ pH_s - pH_{eq}$$

A value less than 6.0 in natural water indicates a scale-forming tendency. A value greater than 6.0 in natural water indicates a corrosive tendency.

Problem 35–3

The values for total solids, temperature, calcium hardness and total alkalinity for a sample of water are 535 mg/l, 28°C, 68 mg/l (as $CaCO_3$) and 408 mg/l (as $CaCO_3$). Calculate the Practical Scaling Index of that water.

$$pH_s = (9.3 + A + B) - (C + D)$$

From the Table 35 - 1, the values of A, B, C and D are as follows:

$$A = 0.2$$
$$B = 1.9$$
$$C = 1.4$$
$$D = 2.6$$
$$\therefore pH_s = (9.3 + 0.2 + 1.9) - (1.4 + 2.6)$$
$$= 11.4 - 4.0 = 7.4$$
$$pH_{eq} = 1.465 \times \log \text{Total Alkalinity} + 4.54$$
$$= 1.465 \times \log 408 + 4.54$$
$$= 1.465 \times 2.6106 + 4.54$$
$$= 3.82 + 4.54$$
$$= 8.36$$
$$\text{Practical Scaling Index (PSI)} = 2\ pH_s - pH_{eq}$$
$$= 2(7.4) - 8.36$$
$$= 14.8 - 8.36$$
$$= 6.44$$

Since the value is above 6, the water can be regarded to have corrosive tendencies.

Some other indices of less importance are also used in water analysis. They are, Stiff-Davis Index (Langelier Index modified for oil-brines), Aggressive Index, Driving Force Index and Momentary Excess. They are used in specific circumstances when Langelier index and Ryznar Index fail to give correct picture.

CHAPTER - 36
LEAD

DITHIZONE METHOD

Principle

When an acidified sample containing microgram quantities of lead is mixed with an ammoniacal-citrate-cyanide reducing solution and allowed to react with dithizone, a cherry red lead dithizonate is formed, which is measured colorimetrically.

Interferences

Bismuth, stannous tin and thallium (monovalent) interfere. But due to the high pH of the reaction, interference by these elements are suppressed. However when present in higher concentrations, bismuth and tin interfere. In such circumstances a modified procedure has to be followed.

Preliminary Treatment of Sample

As soon as the sample is collected, bring the pH of the sample to below 2 using nitric acid. Avoid excess nitric acid. Add 5 ml 0.1N iodine solution to avoid loss of volatile organo-lead compounds. Prepare a blank with lead-free distilled water in the same way.

Reagents

Use lead-free distilled water for the preparation of solutions and dilutions.

1. **Lead Stock solution:** Dissolve 0.1599g. lead nitrate, $Pb(NO_3)_2$ in about 200ml distilled water in a 1000ml volumetric flask. Add 10ml conc. nitric acid and makeup to the mark with distilled water.

$$1.0 \text{ ml} = 100\ \mu g \text{ Pb}$$

2. **Lead Intermediate solution:** Pipet 50.0ml lead stock solution to a 500ml volumetric flask and makeup to the mark with distilled water.

$$1.0 \text{ ml} = 10\ \mu g \text{ Pb}$$

Prepare daily.

3. Lead Working solution: Dilute 50.0ml lead intermediate solution to 500ml with distilled water in a volumetric flask.

$$1 \text{ ml} = 1.0\ \mu g \text{ Pb}$$

4. Nitric Acid, 1+4: Dilute 100ml conc. HNO_3 to 500ml with distilled water.

5. Citrate-Cyanide Reducing solution: Dissolve 400g. dibasic ammonium citrate $(NH_4)_2HC_6H_5O_7$, 20g. anhydrous sodium sulfite, Na_2SO_3, 10g. hydroxyl amine- hydrochloride, $NH_2OH.HCl$ and 40g. potassium cyanide, KCN in distilled water and dilute to 1000ml. Add the solution to 2000ml conc. ammonia solution and mix thoroughly.

Caution: Poison. Do not use mouth pipets for dispensing volumes of this solution.

5. Stock Dithizone solution: Weigh 100mg. dithizone (finely ground) in a beaker. Dissolve it in about 50ml chloroform and transfer to a 1000ml volumetric flask. Then rinse the beaker several times with 50ml portions of chloroform and transfer these rinsings to the volumetric flask. Makeup to the mark with chloroform. Place it in an amber bottle (lead-free) and keep in a refrigerator.

6. Working Dithizone solution: Dilute 100ml stock dithizone solution to 250ml with chloroform.

Procedure

Clean all glassware (including sample bottle) with 1+1 nitric acid thoroughly followed by rinsing with distilled or deionized water.

A. Preparation of Calibration Curve

1. Place 100ml lead-free distilled water into a series of 250ml separating funnels.
2. Pipet 0.0, 1.0, 2.0, 3.0, 4.0, 5.0 ml of lead working standard solution into them.
3. Include a separating funnel containing 100ml lead-free distilled water as the blank.
4. To each separating funnel add the following reagents with mixing after each addition.
   i) 20ml 1+4 nitric acid
   ii) 50ml citrate-cyanide reducing solution
   iii) 10ml dithizone working solution.
5. Shake the funnels vigorously for about 30 seconds.
6. Allow the layers to separate.
7. Draw off the lower chloroform layer [after drying the stem of the funnel with cotton swab (lead-free)] into the absorption cell.
8. Measure the absorbance of the extracts at 510nm, after zeroing the spectrophotometer with dithizone working solution.

B. Procedure for Sample

If the sample is colorless and clear lead may be determined without digestion. Carryout the determination as follows:

Place 100ml sample (already acidified with HNO_3 to pH 2) in a 250ml separating funnel and proceed from step number 4 of the Procedure A 'Preparation of Calibration Curve'. Similarly carryout the test for blank using distilled water (this blank water is already prepared as detailed in the section Preliminary Treatment of Sample – Page No. 181). Detect the absorbance of this blank (if any) from that of the sample. Find out the microgram lead equivalent to the net absorbance.

Express the result as mg Lead (as Pb) per litre of sample.

If the sample is colored and turbid and contains suspended impurities it has to be digested using nitric acid-sulfuric acid or nitric acid-perchloric acid. (for digestion procedure see Chapter 9 - 'PRELIMINARY ACID DIGESTION OF SAMPLE')

To a suitable aliquot of the digested sample (containing not more than 1ml conc. acid) add 20ml 1+4 HNO_3 and filter directly to a separating funnel through a lead-free filter paper (Whatman No. 41 or 541). Rinse digestion beaker with 50ml distilled water and add to the separating funnel. Add 50ml ammoiniacal-citrate-cyanide solution and mix well. Allow to cool to room temperature. Add 10ml dithizone working solution and shake well for about 30 seconds. Then proceed from Step no. 6 of the Procedure A 'Preparation of Calibration Curve'.

Modified Procedure for samples containing high amounts of interferences

If the sample is known to contain higher concentration of bismuth, stannous tin and monovalent thallium, following procedure has to be adopted.

To 100ml sample (undigested) add 5ml of sodium sulfite solution (5% aqueous solution) to reduce the iodine preservative. Adjust the pH of the sample to 2.5 using a pH meter and 1+4 HNO_3 or 1+9 NH_4OH. Transfer to a 250ml separating funnel and extract with 10ml portions of special dithizone solution (100mg. dithizone in 100ml chloroform), 3 times or until the chloroform layer is distinctly green. Remove the traces of dithizone by extraction with chloroform until the chloroform layer becomes colorless. Proceed from Step no. 4 of the Procedure - A 'Preparation of Calibration Curve'.

CHAPTER - 37

MAGNESIUM

Magnesium salts occur in significant concentrations in natural waters. Sea waters and estuary waters contain high amounts of magnesium. Next to calcium, magnesium is the major scale forming ion in water and it is also undesirable in some process waters. Hence determination of magnesium in industrial waters is essential.

Magnesium in industrial water is undesirable since it is the major scale forming cation. Compounds of magnesium such as magnesium silicate and magnesium hydroxide generally form soft deposits and sludges. These deposits interfere seriously with heat transfer and reduce efficiency. If such deposits are not removed periodically, failures occur frequently.

Magnesium may be calculated as the difference between total hardness and calcium hardness or it may be determined directly by gravimetric or photometric method.

MAGNESIUM BY CALCULATION

Magnesium (as $CaCO_3$), mg/l = mg/l Total Hardness (as $CaCO_3$) – mg/l Calcium Hardness (as $CaCO_3$)

Magnesium (as Mg), mg/l = [mg/l Total Hardness (as $CaCO_3$) – mg/l Calcium Hardness (as $CaCO_3$)] x 0.244

GRAVIMETRIC METHOD

Principle

After the removal of calcium, magnesium is precipitated as magnesium ammonium phosphate and the precipitate is ignited and weighed as magnesium pyrophosphate.

$$Mg^{2+} + NH_4OH + Na_2HPO_4 \rightarrow \underset{\text{(Magnesium ammonium phosphate)}}{MgNH_4PO_4} + 2Na^+ + H_2O$$

$$2MgNH_4PO_4 \rightarrow \underset{\text{(Magnesium pyrophosphate)}}{Mg_2P_2O_7} + 2NH_3 + H_2O$$

Interferences

Iron, aluminium, manganese and silica interfere but can be eliminated. (See the procedure).

Reagents

1. Hydrochloric acid , conc.

2. Hydrochloric acid 1+1

3. Nitric acid, conc.

4. Methyl red indicator solution: Dissolve 100 mg methyl red sodium salt in distilled water and dilute 100 ml.

5. Ammonia solution, conc: (sp.Gr. 0.90)

6. Ammonia solution, 1+19: Dilute 5ml conc. ammonia solution to 100ml with distilled water.

7. Ammonium oxalate solution, saturated: Prepare a solution, saturated at room temperature, of ammonium oxalate, $(NH_4)_2C_2O_4.H_2O$ in water.

8. Diammonium hydrogen phosphate solution, 30%: Dissolve 30g $(NH_4)_2HPO_4$ in distilled water and make up to 100 ml.

Procedure

Place a suitable volume of the sample containing 10 to 60 mg. Mg in a dish. Acidify with conc. HCl and add 2 ml in excess. Evaporate to dryness. Add 2 ml 1+ 1 HCl to the dried residue and dilute to about 75 ml with distilled water. Heat to boiling and filter. Wash the filter with two successive additions of hot 1 + 1 HCl and then thrice with hot distilled water.

Removal of iron and aluminium*: Heat the combined filtrate and washings to boiling and add a few drops of conc.HNO_3, cool and add conc. ammonia solution slowly until the solution turns yellow. Boil for about 2 minutes and filter the precipitated iron and aluminium hydroxides. Wash the filter thrice with hot water. The filtrate has to be used for subsequent determination of calcium and magnesium.

* Usually iron and aluminium only will be present in significant concentrations. Manganese will not be present in such concentrations to affect the determination of magnesium. If it is present in excess concentration, it has to be eliminated using ammonium thiosulfate and 1 + 19 ammonia solution.

Precipitation of calcium as calcium oxalate: Add 10 ml conc. ammonia solution to the filtrate and heat it gently (Do not allow the solution to boil). Add drop by drop ammonium oxalate solution to hot filtrate in excess. Keep the solution on water bath for about 2 hrs. Filter the precipitate through whatman No. 40 filter paper. Wash the filter thrice with 1 + 19 ammonia solution. Use the filtrate for the determination of magnesium. [If calcium has to be determined ignite the precipitate in a crucible (preferably platinum crucible) covering with the lid, at 1100° – 1200°C until constant weight is obtained. Weigh the crucible and find the weight of CaO.

$$Calcium\ (as\ Ca),\ mg/l = \frac{mg\ CaO \times 714.7}{ml.\ sample}$$

Precipitation of magnesium as magnesium ammonium phosphate: Acidify the filtrate obtained above with conc.HCl. Add 10 ml diammonium hydrogen phosphate solution and warm the solution. Add 2-3 drops of methyl red indicator solution and conc. ammonia solution dropwise until the solution becomes yellow and add 5 ml in excess. Stir the solution for several minutes until the precipitation is complete. Allow to stand overnight (a minimum of 4 hrs) and then filter through Whatman No. 42 filter paper. Wash with 1 + 19 ammonia solution. Discard the washings.

Determination as magnesium pyrophosphate: Heat the precipitate after placing it in a previously weighed crucible to 500°C until the residue is white. Then ignite for 30 minutes period at 1100°C to constant weight. Find out the weight of magnesium pyrophosphate. Then calculate the magnesium content as follows:

$$Magnesium\ (as\ Mg),\ mg/l = \frac{mg.\ Magnesium\ pyrophosphate \times 218.5}{ml.\ sample\ taken\ for\ determination}$$

$$Magnesium\ (as\ CaCO_3),\ mg/l = \frac{mg.\ Magnesium\ pyrophosphate \times 218.5}{ml.\ sample\ taken\ for\ determination} \times 4.166$$

PHOTOMETRIC METHOD

This method is applicable to waters containing magnesium in lower concentration.

Principle

When magnesium hydroxide is precipitated in the presence of the dye Brilliant yellow, the dye is adsorbed on the colloidal precipitate and its color changes from orange to red.

Interferences

Calcium and aluminium are the major interfering ions. There is an indirect method of elimination of these ions. The concentrations of these ions are increased sufficiently to a level where their influence is constant and predictable. To the magnesium standards also these ions are added and thus the effect of these ions are nullified. Iron also interferes above 2.5 mg/l. Residual chlorine above 0.5 mg/l also interferes. Manganese and zinc interfere, and must be absent.

Reagents

1. **Sulfuric acid, approximately N/50:** Add 0.6 ml conc. H_2SO_4 to 1000 ml distilled water and stir well.

2. **Calcium sulfate solution, saturated:** Add a large excess of calcium sulfate, $CaSO_4$ to 1000 ml distilled water and allow the solid to stand in contact with water for atleast 24 hours with occasional shaking. Filter the supernatant liquid.

3. **Aluminium sulphate solution:** Dissolve 310 mg aluminium sulphate, $Al_2(SO_4)_3.18H_2O$ in water, add 0.3 ml conc.H_2SO_4 and dilute to 1000 ml.

4. **Stabilizing solution:** Dissolve 1.0 g methyl cellulose in 100 ml water. Keep it in a refrigerator. If any mould growth or sediment is observed, reject the solution and prepare freshly. Before use, bring the solution to room temperature. This solution is added in order to maintain the $Mg(OH)_2$ in colloidal suspension.

5. **Brilliant yellow solution:** Dissolve 250 mg solid dye in 500 ml distilled water. Prepare freshly.

6. **Sodium hydroxide, 6N:** Dissolve 120g. sodium hydroxide, NaOH in 500 ml distilled water.

7. **Magnesium stock solution:** Place 1.000 g pure magnesium metal turnings, (99.9% Mg) in 500 ml conical flask. Add about 150 ml distilled water. Place a funnel in the neck of the flask. Add 5 ml of 1 + 1 H_2SO_4 in 1 ml portions. Mix well after each addition and allow the reaction to subside before adding the further 1 ml portion of acid. Transfer to a 1000 ml volumetric flask and make upto the mark with distilled water.

1.00 ml = 1.0 mg. Mg.

If Mg metal turnings are not available, weigh 10.27 g. $MgSO_4.7H_2O$ (AR Grade), dissolve in little water and make up to 1000 ml with distilled water.

1.00 ml = 1.0 mg. Mg.

8. **Magnesium working solution:** Pipet 100.00 ml magnesium stock solution into a 1000 ml volumetric flask and make up to the mark with distilled water.

1.00 ml = 100 μg.Mg

9. Dechlorinating solution: Dissolve 1.0g. anhydrous sodium sulfite, Na_2SO_3 in 100 ml water. Prepare freshly.

Procedure

1. Place in 100ml volumetric flasks, appropriate volume of magnesium working solution covering the range up to 600 μg Mg and dilute to 50 ml with distilled water. Include a 100 ml volumetric flask containing 50 ml distilled water as the blank.
2. Place a suitable aliquot of the sample (not exceeding 50 ml and containing not more than 600 μg Mg) in a volumetric flask. Dilute to 50 ml with distilled water. If necessary dechlorinate the sample by adding 1ml dechlorinating solution.
3. To the blank, standards and sample, add 1 ml 0.02 N sulfuric acid, 20 ml calcium sulfate solution, and 5 ml aluminium sulfate solution. (If any precipitate is formed in the sample, repeat using an increased amount of sulfuric acid).
4. Bring the volume to about 80 ml with distilled water. Add with mixing 5.0 ml stabilizing solution, 2.0 ml brilliant yellow solution and 3.5 ml sodium hydroxide solution. Make up to the mark with distilled water and mix well. Set aside for five minutes for maximum color development.

Measure the optical densities of blank, standards and sample using a spectrophotometer at a wave length of 550 nm or a filter photometer equipped with a suitable blue-green filter. Use water as the reference. Prepare a calibration curve and find out mg. magnesium equivalent to the observed optical density.

Express the result as mg magnesium per litre of the sample. If the magnesium content has to be expressed in terms of $CaCO_3$, then multiply the result by 4.166.

CHAPTER - 38
MANGANESE

Manganese occurs in soils and rocks as manganese dioxide and can be dissolved in natural waters by the action of anaerobic bacteria. Under reducing conditions, manganese can be leached from the soil and occur in considerable concentrations in ground water.

Since manganese exhibits objectionable characteristics similar to those of iron, excessive concentration of manganese is undesirable in most of the industrial water supplies. Particularly industrial processes such as paper manufacturing, brewing, dyeing and textile processing need water free from manganese. In cooling systems, manganese causes corrosion reactions in copper alloy tubing. Manganese in boiler water may also accelerate corrosion.

Among the methods available for the determination of manganese, the persulfate method is the best suited and is applicable to all types of waters.

PERSULFATE METHOD

This method is applicable to raw waters, potable waters, treated waters and waste waters.

Principle

The soluble manganese compounds are oxidized to permanganate by persulfate in acid solution. The resulting color of the permanganate solution is measured either visually or photometrically.

$$2MnSO_4 + 5(NH_4)_2S_2O_8 + 8H_2O + 10HNO_3 \rightarrow \underset{\text{Permanganic acid}}{HMnO_4} + 12H_2SO_4 + 10NH_4NO_3$$

Interferences

Chloride and ferric ions interfere. Chloride as NaCl up to 100 mg is eliminated by the addition of mercuric sulfate. Ferric iron is eliminated by the addition of phosphoric acid which form soluble complexes.

Reagents

1. Manganese stock solution: Dissolve 1.000 g high purity (99.9%) manganese metal in 100 ml 7% sulfuric acid. Cool and make up to 1000 ml in a volumetric flask. Pipet 50ml of solution into a 1000ml volumetric flask and make up to the mark with distilled water.

1.0 ml = 0.05 mg Mn

Alternatively, dissolve 144mg potassium permanganate, $KMnO_4$ in 100 ml distilled water. Add 25 ml dil.sulfuric acid (20%), reduce the permanganic acid by dropwise addition of hydrogen peroxide (H_2O_2 – 20 volume) and transfer to a 1000 ml volumetric flask and make up to the mark with distilled water.

2. Manganese working standard solution: Pipet 100 ml of the stock solution (1ml = 0.05 mg Mn) and make up to 500 ml with distilled water.

1.0 ml = 0.01 mg Mn

3. Special reagent: Add 400 ml conc. HNO_3 to 200 ml distilled water in a 2000 ml beaker. Dissolve 75g. mercuric sulfate, $HgSO_4$ in it. Add 200 ml 85% phosphoric acid, H_3PO_4 and 35 mg silver nitrate, $AgNO_3$ and dilute the cooled solution to 1000 ml.

4. Ammonium persulfate, $(NH_4)_2S_2O_8$ crystals.

Procedure

Clear samples

1. Place appropriate volumes of manganese standard solution in 250 ml beakers. Dilute each to 50 ml with distilled water. Include a beaker containing 50 ml distilled of water as the blank.
2. Place 100 ml or suitable aliquot of the sample in a 250 ml beaker.
3. To the blank, standards and sample add 5 ml special reagent.
4. Heat the beakers on a hot plate and when the contents are reduced to 40 ml, add 1g. ammonium persulfate while boiling. Continue the boiling for one minute*. Then cool the beakers under tap.
5. Transfer to 50 ml Nessler tubes and compare visually. Photometric measurements may also be made by using either spectrophotometer at 545 nm or filter photometer with a suitable green filter. Prepare a calibration curve and find out the mg manganese equivalent to the observed optical density.

* Too long a boiling and slow cooling results in the loss of permanganate color. Hence should be cooled immediately.

Express the result as mg manganese (as Mn) per litre of the sample.

Samples containing suspended matter, turbidity and organic matter: To a suitable aliquot of the sample in an evaporating dish add 5 ml conc. HCl and evaporate to about 25 ml. Digest at the boiling point to bring dissolution of the particulate materials and evaporate to dryness. Cool and add 5 ml conc. H_2SO_4 and 10 ml conc. HNO_3. Mix thoroughly and evaporate until white dense fumes of sulfur trioxide appears. Add a further 10 ml of conc. HNO_3 and evaporate to white dense fumes. Cool and add 40 ml distilled water. Warm and filter through a sintered crucible and make up the filtrate to 100 ml with distilled water. From this made up solution, pipet suitable aliquots in a 250 ml beaker and dilute to about 50 ml. Add 5 ml special reagent. Proceed from Step No.4 of the procedure for clear samples.

CHAPTER - 39

MERCURY

The occurrence of elemental mercury in water is rare. Mercury compounds are widely employed in the manufacture of disinfectants, detonators, pigments and medical products. Discharge of these wastes may contribute mercury to the receiving waters. Mercury can also enter into a watercourse through agricultural pesticides, herbicides, and fungicides. Among the compounds of mercury found in water, methyl mercury and phenyl mercuric acetate are the most common.

Mercury can be determined by conventional dithizone method and cold vapor atomic absorption method. Dithizone colorimetric method is described below.

DITHIZONE COLORIMETRIC METHOD

This method is applicable to potable waters and mercury concentration down to 1.0 μg can be measured. However if waste waters are to be tested, wet digestion with special type of apparatus is necessary because of the volatility of mercury and its compounds.

Principle

Mercury ions are allowed to react with dithizone (in chloroform) at a pH of about 1 to form an orange red colored complex and the color is measured using a spectrophotometer.

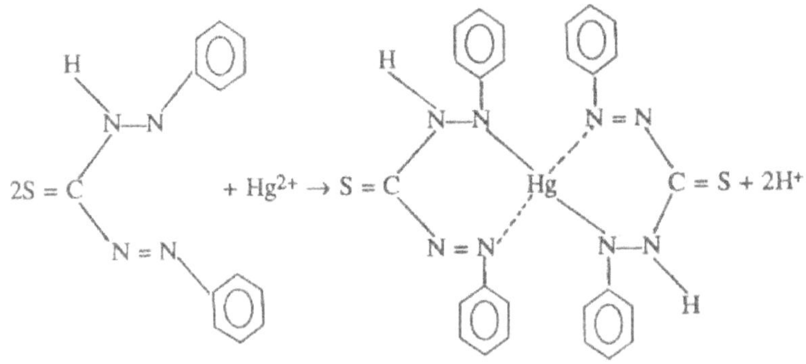

Dithizone Mercury dithizonate

Interferences

Copper is the major interference and is eliminated during the procedure by keeping the copper dithizonate in organic phase while mercury dithizonate in aqueous phase. The other interfering elements, silver, gold, palladium, platinum(divalent) are usually absent.

Reagents

Note: Use mercury free water (redistilled or deionized distilled water) for the preparation of all solutions and dilutions.

1. Mercury stock solution: Dissolve 135.4mg. mercuric chloride $HgCl_2$ in about 750 ml water. Add 1.5ml conc. HNO_3 and make up to the mark with distilled water.

$$1 \text{ ml} = 100\ \mu g \text{ Hg}$$

2. Mercury working solution: Pipet 10.0 ml mercury stock solution into a 1000 ml volumetric flask and make up to the mark with distilled water.

$$1 \text{ ml} = 1\ \mu g \text{ Hg}$$

Prepare only just before use.

3. Potassium permanganate solution 5%: Dissolve 5g. potassium permanganate $KMnO_4$ in 100 ml water

4. Sulfuric acid. conc.

5. Potassium persulfate solution 5%: Dissolve 5g. potassium persulfate, $K_2S_2O_8$ in 100 ml water.

6. Hydroxylamine hydrochloride solution 50%: Dissolve 50g. hydroxylamine hydrochloride $NH_2OH.HCl$ in 100 ml water.

7. Dithizone solution: Dissolve 6mg. dithizone in 1000 ml chloroform.

8. Sulfuric acid 0.25 N: Dilute 6.9 ml conc. H_2SO_4 to 1000 ml with water.

9. Potassium bromide solution, 40%: Dissolve 40g. potassium bromide, KBr in 100 ml distilled water.

10. Phosphate-carbonate buffer: Dissolve 150g. disodium hydrogen phosphate $Na_2HPO_4.12H_2O$ or 73.9g. $Na_2HPO_4.2H_2O$ and 38g. potassium carbonate (anhydrous) in 1000 ml water. Purify the buffer as follows:

Place the buffer in a 1500 ml separating funnel and extract with 10 ml portions of dithizone until the last extract remains green (bluish green). Remove the excess dithizone from the buffer by extraction with chloroform.

11. Sodium sulfate, Na_2SO_4 anhydrous.

Procedure

Clean all the glassware with nitric acid and then with distilled water.

1. Place 0.0, 1.0, 2.0, 4.0,..... 10.0 ml of mercury working solution into a series of 1000 ml beakers. Dilute each to 500 ml (or volume similar to sample).
2. Place 500 ml (or suitable aliquot of the) sample in a 1000 ml beaker.
3. To the blank, standards and sample add 1 ml potassium permanganate solution and 10 ml conc. H_2SO_4.
4. Stir well and boil (if the permanganate color decolorizes add more $KMnO_4$).
5. When boiling ceases, add carefully 5 ml of potassium persulfate solution and allow to cool for 30 minutes.
6. Discharge the pink color (present if any) by the addition of one or more drops of hydroxylamine-hydrochloride solution.
7. Transfer the contents of each beaker to individual 1000 ml separating funnels.
8. Extract the contents of each separating funnel, with 25 ml portions of dithizone until the last extract remains green (bluish green) as that of the original dithizone solution.
9. Collect the chloroform extracts in individual 250 ml separating funnels.
10. Wash the combined chloroform extracts by treatment with 50 ml 0.25N.H_2SO_4.
11. Now transfer the chloroform layer into another series of 250 ml separating funnel.
12. Add 50 ml of 0.25N.H_2SO_4 and 10 ml potassium bromide solution to each funnel and shake well. Now the mercury dithizonate will be in aqueous phase and other dithizonates (particularly of copper) will be in organic phase.
13. Discard the lower organic layer.
14. Wash the aqueous phase by shaking with a small volume of chloroform and discard the lower chloroform layer.
15. To each separating funnel add 12 ml phosphate – carbonate buffer and 10 ml dithizone solution. Shake well and allow the phases to separate.
16. Pass the organic layer containing mercury dithizonate through a layer of sodium sulfate (in a funnel) into a cuvette and measure the absorbance at 490 nm.
17. Prepare a calibration curve on a linear graph paper plotting absorbance (optical density) versus microgram of mercury.
18. From the calibration graph, find out the number of micrograms of mercury equivalent to the optical density of the sample.

Express the result as mg mercury (as Hg) per litre of the sample.

CHAPTER - 40
NICKEL

Nickel does not occur in natural water. Metal plating waste waters when discharged to water courses, contribute nickel. Increased use of stainless steel and other nickel containing alloys in water treatment plants, boilers etc., also contribute nickel. The determination of nickel is of much importance in water chemistry for high pressure boilers because of the following reasons:

(i) The presence of nickel indicates corrosion of nickel alloys.

(ii) It behaves similarly as copper and may be precipitated either as the oxide or as the metallic nickel.

The following dimethyl glyoxime method is one among the best methods for the determination of nickel and is more sensitive, being able to detect 0.001 mg/l.

DIMETHYL GLYOXIME METHOD

This method is applicable to all types of water and waste water.

Principle

Nickel reacts with dimethyl glyoxime in the presence of an alkaline oxidizing agent to form a characteristic red color complex which is measured visually or photometrically.

$$Ni_2+ + 2H_2DMG \rightarrow Ni(HDMG)_2 + 2H^+$$

Interferences

Iron, manganese and copper interfere. If they are present in concentration thrice that of nickel their interference is suppressed by the addition of citrate. If they are present in excess concentrations, nickel dimethylglyoxime complex alone is separated by extraction with chloroform. The chloroform layer is then acidified with dil.HCl to bring the nickel to the aqueous phase.

Reagents

1. Nickel stock solution: Dissolve 447.9 mg nickel sulfate, $NiSO_4.6H_2O$ in distilled water and make up to 1000 ml in a volumetric flask.

1.00 ml = 100 μg Ni

2. Nickel working solution: Pipet 10.0 ml nickel stock solution in to a 1000 ml volumetric flask and make up to the mark with distilled water.

1.00 ml = 1 μg Ni

3. Hydrochloric acid, approximately 0.05 N: Dilute 50 ml conc. HCl to 1000 ml with distilled water.

4. Sodium citrate solution: Dissolve 125g. sodium citrate, $Na_3C_6H_5O_7.2H_2O$ in 500 ml distilled water.

5. Iodide solution, approximately 0.05 N: Dissolve 20g. potassium iodide, KI in 5 ml distilled water. Dissolve in this solution 6.4g. iodine. Dilute the solution to 1000 ml.

6. Dimethyl glyoxime solution: Dissolve 1g. dimethyl glyoxime in 100 ml conc.ammonia solution. Add 100 ml distilled water. Filter if necessary.

7. Additional reagents for the removal of interferences

- **(i)** **Dilute ammonia solution:** Dilute 10 ml conc. ammonia solution (Sp. gr. 90) to 500 ml with distilled water.
- **(ii)** **chloroform**

Procedure

1. Place appropriate volumes of nickel working solution covering the range up to 20 μg in to a series of 50 ml volumetric flasks (or 50 ml Nessler tubes). Include a 50 ml volumetric flask (or 50 ml Nessler tube) as the blank.
2. Place a suitable aliquot of the sample containing not more than 20 μg in 50 ml volumetric flask (or 50 ml Nessler tube). If the sample contains organic matter it must first be destroyed by acid digestion (See Chapter – 9, "PRELIMINARY ACID DIGESTION OF SAMPLE"). Then an aliquot of the digested sample has to be neutralized and taken in the volumetric flask. (If the sample contains small amounts of iron, manganese and copper, nickel may be determined as per the procedure given below. If it contains excessive amounts, a separate procedure has to be adopted).
3. To the blank, standards and sample add 20 ml 0.5N.HCl.
4. Then add the following reagents in order with mixing after each addition :
   (i) 10 ml sodium citrate solution,
   (ii) 2 ml iodine solution, and
   (iii) 4 ml dimethyl glyoxime solution.
5. Make up to 50 ml with distilled water and allow to stand for 20 minutes.

6. Compare the color visually or photometrically using a spectro -photometer at 470 nm or filter photometer with a suitable blue filter taking water as the reference. Prepare a calibration curve and find out the mg nickel equivalent to the observed optical density.

Express the result as mg nickel (as Ni) per litre of the sample.

CHAPTER - 41

NITRATE

Nitrate generally occurs in trace quantities in surface waters but may attain high levels in some ground waters. Application of fertilizers to lands also contribute nitrate to ground waters. Waste waters from chemical–fertilizer manufacturing plants contribute nitrate to receiving waters.

Nitrate is an important consideration with regard to potable waters. It has been reported that nitrate nitrogen concentration above 20 mg/l may cause 'methemoglobinemia' in infants, a disease characterized by blood changes.

Regarding industrial water supplies nitrates are reported to be injurious to dyeing of wool and silk fabrics. They are also harmful to fermentation processes and cause disagreeable tastes to beer.

Nitrate is best determined by the following phenol disulfonic acid method.

PHENOL DISULFONIC ACID METHOD

Principle

Nitrate reacts with phenoldisulfonic acid and on subsequent reaction with alkali yields an yellow colored compound, the intensity of which is measured and the nitrate concentration is determined.

Interference

Chloride interferes seriously. It may be eliminated by the addition of silver sulfate solution.

Reagents

1. Phenol disulfonic acid reagent: Dissolve 25 g pure white phenol in 150 ml conc.H_2SO_4. Add 75 ml fuming sulfuric acid (15% free SO_3). Stir it well and keep it on a boiling water bath for 2 hrs. After cooling, transfer to an amber bottle.

2. Potassium hydroxide solution, 12N: Either potassium hydroxide or sodium hydroxide may be used. Dissolve 673 g KOH in distilled water and dilute to 1000 ml.

(or)

Sodium hydroxide solution, 12N: Dissolve 480g. NaOH in distilled water and dilute to 1000 ml.

Note: Alkali solutions should not be stoppered with glass stoppers. Use only rubber stoppers.

3. Nitrate stock solution: Dissolve 722 mg potassium nitrate, KNO_3 (previously dried at 105°C for 24hr) in distilled water and make up to 1000 ml in a volumetric flask.

1 ml = 0.1 mg N = 0.443 mg NO_3

4. Nitrate working standard solution: Place 50 ml of stock nitrate solution in a porcelain dish and evaporate to dryness by keeping it on a water bath. Add 2 ml phenol disulfonic acid reagent and swirl the sides of the dish with the reagent. Transfer quantitatively to a 500 ml volumetric flask. Rinse the dish several times with distilled water and make up to the mark with distilled water.

1 ml = 0.01 mg N = 0.0443 mg NO_3

5. EDTA solution: Dissolve 50g. EDTA in 20 ml distilled water to form a paste. Add 6 ml conc. ammonia solution and mix well to dissolve the paste.

Procedure

(i) Pipet a suitable aliquot of the sample (20 ml, 10 ml, 5 ml, or 2 ml) in a porcelain evaporating dish.

(ii) Evaporate to dryness on a water bath.

(iii) Add 1 ml phenol disulfonic acid to the dish. Swirl the sides of the dish with the reagent. Add a little distilled water.

(iv) Transfer quantitatively to a 50 ml Nessler tube. Rinse the dish thrice with distilled water and add the rinsings to the tube.

(v) Add dropwise 12 N potassium hydroxide or 12 N sodium hydroxide with vigorous mixing until maximum yellow color is developed (About 5 to 6 ml is required). If any turbidity is formed during the addition of alkali, add EDTA solution dropwise with stirring until the turbidity is dissolved.

Preparation of standards

(i) Pipet out 0.0, 1.0, 2.0, 3.0,.... 12.0 ml of working standard nitrate solution in to 50 ml Nessler tubes.

(ii) To each of the tube add 2 ml of 12N potassium hydroxide or 12 N sodium-hydroxide solution and mix thoroughly.

(iii) Dilute to the 50 ml mark with distilled water.

Compare the color of the samples with that of the standards.

Calculation

$$\text{Nitrate (as N), mg/l} = \frac{\text{matching std} \times 0.01 \times 1000}{\text{ml. sample taken for determination}}$$

$$\text{Nitrate (as } NO_3\text{), mg/l} = \text{Nitrate (as N), mg/l} \times 4.43$$

ELECTRODE METHOD

Nitrate can also be determined by ion selective electrode method.

Follow the manufacturers' instructions and prepare a calibration curve using known standards. From this graph, find out the nitrate concentration of the sample.

CHAPTER - 42
NITRITE

Nitrites are generally formed in water due to bacterial action on ammonia and organic nitrogen. Since nitrites are readily oxidized to nitrates, they are seldom present in significant concentrations in surface or other natural waters. Nitrite may also enter into a water supply through discharge of nitrite treated cooling waters.

The presence of nitrite in water indicates organic pollution (i.e) the nitrogenous organic matter is undergoing oxidation or nitrification and that the process is not complete. In some cases, nitrates are also reduced to nitrites.

Water containing nitrite is undesirable for industrial purposes, particularly for brewing and dyeing of wool and silk fabrics.

Nitrite is best determined by the following diazotization method.

DIAZOTIZATION METHOD

Principle

Sulfanilic acid is diazotized by nitrite and the diazocompound is coupled with α-naphthylamine hydrochloride to form a reddish purple azo dye at a pH 2.0 to 2.5. See the reaction given below.

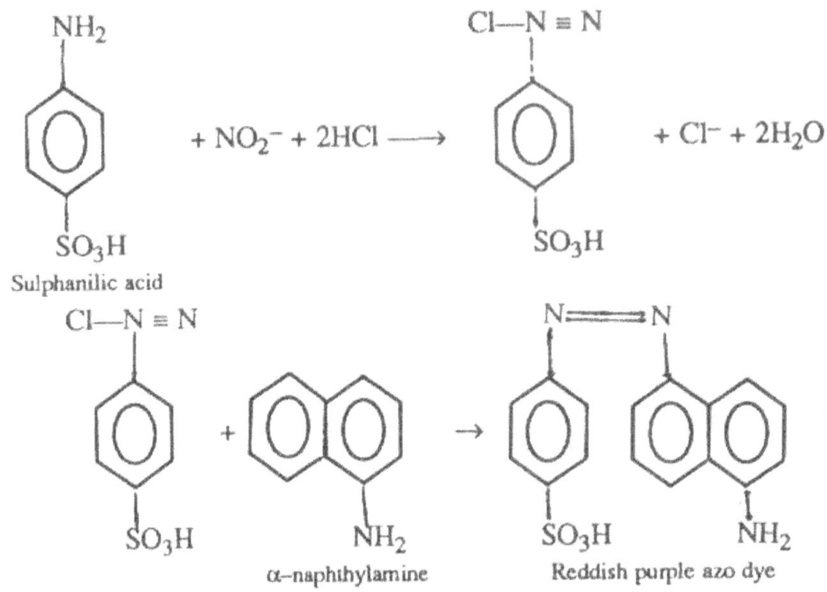

Interferences

Strong reducing agents, strong oxidizing agents and amines interfere. Color and turbidity interfere, but the interference may be eliminated or suitable corrections may be applied.

Reagents

Use nitrite-free distilled water for the preparation of reagents.

1. Sulfanilic acid solution: Dissolve completely 600mg. sulfanilic acid in 70 ml hot distilled water. Cool and add 20 ml conc. HCl. Dilute to 100 ml with distilled water.

2. α-naphthylamine hydrochloride solution: To 50 ml distilled water in a beaker add 1 ml conc. HCl. Dissolve in it 600 mg α-naphthylamine- hydrochloride. Dilute to 100 ml with distilled water. Store in a refrigerator.

3. Sodium acetate buffer solution, 2M: Dissolve 16.4g. sodium acetate $NaC_2H_3O_2$ or 27.2g. sodium acetate trihydrate $NaC_2H_3O_2.3H_2O$ in distilled water and dilute to 100 ml.

4. Nitrite stock solution: Dissolve exactly 493mg. sodium nitrite, $NaNO_2$ in distilled water and make up to 1000 ml in a volumetric flask. Prepare freshly.

1.00 ml = 100 μg Nitrite as N

5. Nitrite intermediate solution: Pipet 10.0 ml stock nitrite solution and make up to 100 ml.

1.00 ml = 10 μg Nitrite as N

Prepare daily.

6. Nitrite working solution: Pipete 5.0 ml nitrite intermediate solution into a 100 ml volumetric flask and make up to the mark.

1.0 ml = 0.5 μg Nitrite as N

Prepare this solution freshly.

7. Aluminium hydroxide suspension: Dissolve 125g. aluminium potassium sulfate, $AlK(SO_4)_2.12H_2O$ or aluminium ammonium sulfate $AlNH_4(SO_4)_2.12H_2O$ in 1000 ml distilled water in a 2-litre beaker. Warm the beaker to 60°C and add slowly with constant stirring, 55 ml conc. ammonia solution. Allow it to stand for about 1 hour. Transfer it to a large bottle and wash the precipitate by mixing and decantations with successive additions of distilled water until free from ammonia, chloride, nitrite and nitrate.

In the final stage decant the clear supernatant as far as possible.

Procedure

1. Pipet 0.1, 0.2, 0.4, 0.6, 0.8, 1.0, 1.5, 2.0, 2.5, 3.0, 4.0, 5.0 ml nitrite working solution in to 50 ml Nessler tube and make up to the mark with distilled water. Include a Nessler tube as blank. Place 50 ml distilled water in it.
2. Transfer 50 ml of the clear sample (neutralized to pH 7) or an aliquot diluted to 50 ml in a Nessler tube.
3. Add 1.0 ml sulfanilic acid to blank, standards and sample and mix thoroughly. After 10 minutes add 1.0 ml α-naphthylamine-hydrochloride solution and 1.0 ml sodium acetate buffer solution and mix thoroughly.
4. After 10 minutes compare the color visually or measure the absorbance using a spectrophotometer or a filter photometer at 520 nm.

Prepare a calibration curve and find out μg nitrite nitrogen, equivalent to the optical density.

Express the result as mg nitrite nitrogen (as N) per litre of sample. If it has to be expressed in terms of nitrite, then

$$mg/l\ NO_2 = mg/l\ nitrite\ nitrogen \times 3.29.$$

Colored and turbid samples: If the sample is colored or contains suspended solids, place 100 ml sample in a 250 ml beaker. Add 2 ml aluminium hydroxide suspension and stir well. Cover it with a watch glass and allow it to stand for a few minutes. Decant the supernatant and filter through a filter paper, in to a 50 ml Nessler tube and proceed as described above starting from Step no.3.

Chlorinated samples: Chlorinated samples may contain nitrogen trichloride which may impart a false red color. This effect can be minimized by adding α-naphthylamine hydrochloride solution first and then sulfanilic acid solution.

CHAPTER - 43

NITROGEN (AMMONIA)

Ammonia generally arises from the aerobic or anaerobic decomposition of nitrogenous organic matter. The presence of ammonia is often regarded as indicative of fresh pollution by nitrogenous wastes.

To understand the significance of free ammonia present in natural waters, it is necessary to consider the source from which the water is derived. Urine of man and animals yields large quantities of ammonium carbonate and hence sewage is rich in free ammonia. Consequently, water polluted by sewage contains more free ammonia than albuminoid ammonia. In other words, whenever the free ammonia content of a water is higher than the albuminoid ammonia, the water should be suspected of sewage pollution.

Deep well waters usually contain only traces of organic matter and ammonia. If an excess of ammonia and organic matter are present then it should be suspected of pollution through seepage.

The presence of more than traces of ammoniacal nitrogen in river waters (which are used as drinking water sources) is undesirable. Waters with high ammonia content have a high chlorine demand and require much longer contact periods for satisfactory sterilization.

As regards industrial waters, ammonia is often used in water treatment to control or adjust pH. For example, to counter the effect of carbonic acid that causes corrosion in condensate systems, ammonia is used. However if the condensate is used as boiler feed water, ammonia is objectionable.

DISTILLATION AND NESSLERIZATION METHOD

Principle

The sample after pH adjustment to 7.4 with phosphate buffer is distilled and the distillate containing ammonia is treated with Nessler reagent. The yellow brown color produced is compared with that of standards.

$$2(2KI.HgI_2) + 2NH_3 + 3KOH \rightarrow (NH_2)_2HgOH.HgI + 7KI + 2H_2O$$

Interferences

Sulfate interferes, but can be eliminated by adding lead carbonate to the sample before distillation. Volatile organic substances such as formaldehyde interfere and are eliminated by boiling at low pH, after which the normal distillation is followed. Aliphatic amines and aromatic-amines, organic chloramines, ketones, aldehydes and alcohols also interfere.

Reagents

1. **Ammonia-free distilled water:** Prepare ammonia-free distilled water by passing distilled water through a strong cation exchanger or add 0.1 ml. conc. H_2SO_4 per litre of distilled water and redistill.

2. **Phosphate buffer solution:** Dissolve 14.3g. potassium dihydrogen- phosphate, KH_2PO_4, and 68.8 g. dipotassium hydrogen phosphate, K_2HPO_4 in little ammonia-free distilled water and dilute to 1000 ml.

3. **Sodium sulfite solution:** Dissolve 0.2g. sodium sulfite, Na_2SO_3 in ammonia-free distilled water and dilute to 200 ml. Prepare freshly.

4. **Nessler reagent:** Dissolve 100g. mercuric iodide. HgI_2, and 70 g potassium iodide, KI in about 100 ml. ammonia-free distilled water. Add this solution slowly with constant stirring to a cool solution of 160g. sodium hydroxide NaOH in 500 ml. water. Dilute to 1 litre. Keep it in a rubber stoppered pyrex bottle in dark.

(or)

Dissolve 35g. potassium iodide, KI, and 12.5 g mercuric chloride, $HgCl_2$, in about 700 ml distilled water and add a saturated aqueous solution of mercuric chloride with stirring until a slight permanent red precipitate is formed. (About 40 to 50 ml would be required). Then mix with a solution (cooled to room temperature) of 120 g sodium hydroxide in 150 ml water. When the mixture has cooled down, transfer to a 1000 ml measuring cylinder, add a further 1 ml saturated mercuric chloride solution and mix well. Finally dilute to the 1 litre mark and shake again.

5. **Ammonium stock solution:** Dissolve 3.819 g. ammonium chloride, NH_4Cl (dried at 100°C) in ammonia-free distilled water and make up to 1 litre in a volumetric flask.

1.0 ml. = 1.0 mg N or 1.22 mg NH_3

6. **Ammonium intermediate solution:** Pipet 10.0 ml. stock ammonium solution in to a 1000 ml. volumetric flask and make up to the mark with ammonia-free distilled water.

1 ml. = 0.01mg N or 0.0122 mg NH_3

7. **Ammonium working solution:** Place 125.0 ml. intermediate ammonium solution and make up to 500 ml. in a volumetric flask.

1 ml. = 0.0025 mg N or 0.003 mg NH_3

Procedure

Preliminary distillation: Use for distillation a pyrex flask of 1000–1500 ml. capacity attached to a vertical condenser as shown in figure 43–1.

Place appropriate volume of the sample in a beaker. If the sample is unpolluted water, take 500ml or 250ml. If the sample is sewage or other type of waste water take 100 ml. or 50 ml. or 25 ml. according to the strength. In all cases make up the volume to 500 ml. with ammonia-free distilled water.

Dechlorinate the sample if it contains any chlorine with sodium sulfite solution. Use 1 ml. solution to remove 1 mg/l of residual chlorine in 500 ml. sample.

If necessary, neutralize the sample to approximately pH 7 with 1N sodium hydroxide solution or 1N sulfuric acid. Transfer the contents of the beaker to the distillation flask.

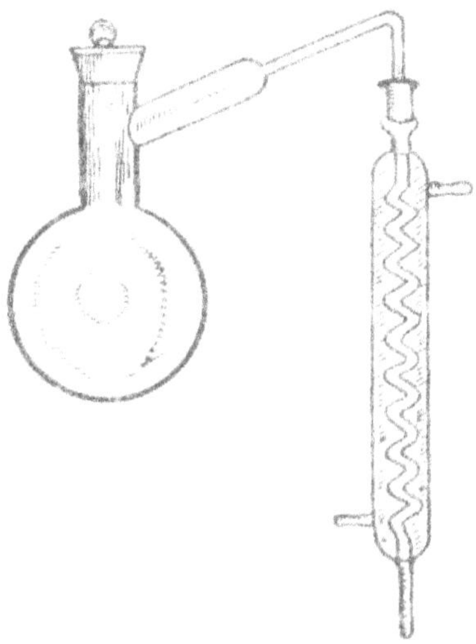

Fig. 43–1. Distillation apparatus for ammonia nitrogen, albuminoid nitrogen and phenol determination*

* Reproduced with the kind permission of American Public Health Association, from their publication "STANDARD METHODS FOR THE EXAMINATION OF WATER AND WASTE WATER", 20th edition, 1998.

Add 10 ml. phosphate buffer solution (If the sample contains calcium above 250 mg/l add an additional 10 ml. buffer solution for each 250 mg in the sample and adjust to pH 7.4 with acid or base). Immediately start distillation without delay. (See note). Distill at the rate of 6-10 ml/minute and collect the distillate in a volumetric flask until the last distillate shows no ammonia by testing with Nessler's reagent. Compensate the loss of water during distillation by adding ammonia-free distilled water to the distillation flask.

Note: The whole distillation assembly should be made free from ammonia prior to the sample distillation, using 500 ml. ammonia free distilled water and 10ml phosphate buffer solution with a few glass beads to prevent bumping.

Color comparison

Pipet in to a series of 100 ml. Nessler tubes, 2.0, 4.0, 6.0, 8.0, 10.0, 12.0 and 15.0 ml. standard ammonium solution. (1 ml. = 0.0025mg N). Include a Nessler tube as blank. Pipet in to a 100 ml. Nessler tube an aliquot of the distillate. (If the sample is unpolluted water, transfer completely 100 ml. of distillate). Make up the contents of all the Nessler tubes to 100 ml. with ammonia-free distilled water. Add 2.0 ml. Nessler reagent to each tube and mix thoroughly. Compare visually the color of the sample with that of the standards after 10 minutes. If photometric measurement is made, measure the absorbance or transmittance in a spectrophotometer at 400 to 500 nm using a light path of 1 cm or longer or filter photometer equipped with a violet filter.

Calculation

mg/l Ammonia nitrogen (as N) =

$$\frac{\dfrac{\text{matching. std in mg}}{\text{vol. of aliquot of the distillate in ml}} \times \text{Total vol. of the distillate in ml.}}{\text{vol. of sample taken for distillation}} \times 1000$$

CHAPTER - 44

NITROGEN (ALBUMINOID)*

Polluted waters and sewage contain nitrogen in the form of organic compounds and urea. Part of this organic nitrogen is readily converted into ammonia on oxidation with alkaline potassium permanganate solution. This quantity of ammonia in addition to the free ammonia represents the albuminoid nitrogen.

Albuminoid nitrogen gives a measure of the relatively easily decomposable organic matter. Organic nitrogen and albuminoid nitrogen are of particular value in determining the strength of nitrogenous organic wastes.

DISTILLATION AND NESSLERIZATION METHOD

Following the distillation of ammonia nitrogen, the addition of a strongly alkaline potassium permanganate solution to a water sample releases additional ammonia and is made to react with Nessler's reagent.

Reagents

In addition to all the reagents listed in the estimation of ammonia nitrogen, prepare the following reagent.

Alkaline potassium permanganate solution: Dissolve 8g. potassium permanganate, $KMnO_4$ in about 500 ml. ammonia free distilled water, in a 2 litre beaker. Add 144 g. sodium hydroxide, NaOH and dilute to about 1250 ml. and concentrate to 1000 ml. by keeping it on a hot plate.

Procedure

After distillation of free ammonia, continue the distillation adding ammonia-free distilled water and 50 ml. alkaline potassium permanganate solution per 500 ml. of the contents of the

* Usually albuminoid nitrogen is determined only on water samples and not on polluted water and waste water. Instead, organic nitrogen content of them are determined.

flask. Collect the distillate in a standard flask. Continue the distillation until the last distillate shows no ammonia by testing with Nessler's reagent.

Transfer appropriate aliquot of the distillate into a 100 ml. Nessler tube. Make up to the mark with ammonia-free distilled water. Simultaneously prepare standards as described in "Ammonia Nitrogen", compare the color of the sample with that of the standards either visually or photometrically.

Calculation

$$mg/l\ Albuninoid\ nitrogen\ (as\ N) = \frac{\frac{matching.\ std\ in\ mg}{vol.\ of\ aliquot\ of\ the\ distillate\ in\ ml} \times Total\ vol.\ of\ the\ distillate\ in\ ml.}{vol.\ of\ sample\ taken\ for\ distillation} \times 1000$$

When interpreting the results, the free ammonia and albuminoid ammonia should be considered together, since the relative proportion is more important than the actual quantities. Generally the free ammonia of crude sewage is greater than albuminoid ammonia. Hence if a natural or treated water contains more free ammonia than albuminoid ammonia, the water may be suspected of polluted with sewage. However it has to be remembered that decaying vegetable matter such as dead leaves etc. contribute more albuminoid ammonia.

CHAPTER - 45

NITROGEN (ORGANIC)

Organic nitrogen is the organically bound nitrogen in nitrogenous organic matter such as proteins, amino acids, organic bases and urea. Albuminoid nitrogen is a fraction of the organic nitrogen which is easily decomposed when treated with alkaline potassium permanganate solution. Organic nitrogen and albuminiod nitrogen are of particular value in determining the strength of nitrogenous organic wastes.

Organic nitrogen is the difference between total kjeldahl nitrogen and ammonia nitrogen. Hence deduction of ammonia nitrogen from total kjeldahl nitrogen gives the amount of organic nitrogen. It may also be determined by digestion of the sample after removal of free ammonia with subsequent distillation and titration using 0.02N. H_2SO_4 as described for total kjeldahl nitrogen.

Express the result as mg nitrogen (as N) per litre of the sample.

CHAPTER – 46

NITROGEN (TOTAL KJELDAHL)*

Total Kjeldahl Nitrogen is the sum of ammonia nitrogen and organic nitrogen. This does not include nitrite nitrogen and nitrate nitrogen. The classical Kjeldahl method is used to determine the total nitrogen content.

KJELDAHL METHOD

Principle

The nitrogen of the organic matter is converted into ammonium sulfate when treated with sulfuric acid in the presence of copper sulfate catalyst. An excess of alkali is then added (to liberate the ammonia from ammonium sulfate) and distilled and the distillate is either treated with Nessler reagent or titrated with standard sulfuric acid after absorbing in boric acid solution.

$$2NH_3 + 2H_3BO_3 \rightarrow 2NH_4^+ + 2H_2BO_3^-$$

$$H_2SO_4 \rightarrow 2H^+ + SO_4^{2-}$$

$$2NH_4^+ + 2H_2BO_3^- + 2H^+ + SO_4^{2-} \rightarrow (NH_4)_2SO_4 + 2H_3BO_3$$

Reagents

1. Sulfuric acid, conc.

2. Copper sulfate solution 10%: Dissolve 10 g. copper sulfate, $CuSO_4.5H_2O$ in 100 ml. distilled water.

3. Potassium sulfate crystals.

4. Phenolphthalein indicator solution: See Chapter – 11, "ALKALINITY".

5. Sodium hydroxide solution, 50%: Dissolve 100g. NaOH in 200 ml. distilled water. Place it in a rubber stoppered bottle.

6. Boric acid solution, 2%: Dissolve 10g. boric acid H_3BO_3 in ammonia-free distilled water and dilute to 500 ml.

---

* Usually the total kjeldahl nitrogen is not determined on nmpolluted water samples. In those cases determination of albuminoid nitrogen is sufficient.

7. Mixed Indicator solution: Dissolve 200 mg methyl red indicator in 100 ml 95% ethyl or isopropyl alcohol. Dissolve 100 mg methylene blue in 50 ml 95% ethyl or isopropyl alcohol. Combine the two solutions. Prepare monthly. End point with this indicator is the color change from pale green to lavender.

Alternatively, methyl red indicator can also be used. It can be prepared by dissolving 50 mg methyl red in 100 ml ethyl alcohol. End point with this indicator is the appearance of red color.

8. Standard sulfuric acid solution, 0.02 N: See Chapter – 11, "ALKALINITY".

Procedure

Digestion

Place 100 ml. or an appropriate volume of the sample in a Kjeldahl flask. Dilute to about 200 ml. with ammonia-free distilled water. Add 10 ml. conc. H_2SO_4 and 1 ml. copper sulfate solution. If the organic matter is hard to destroy, add 20 ml. conc. H_2SO_4 and 5 g. potassium sulfate. Add a few glass beads and boil under hood until the solution becomes clear. Then digest for an additional 30 minutes. Allow to cool.

Distillation

Transfer the contents of the flask carefully in to a distillation flask and dilute to about 300 ml. Make the solution in the flask alkaline with sodium hydroxide using phenolphthalein indicator. Start the distillation after immersing the tip of the condenser in 50 ml. boric acid solution in a conical flask. Collect 200 ml. distillate.

Titration

Add 0.5 ml. mixed indicator solution to the distillate. Titrate against 0.02N sulfuric- acid. The end point is the color change from pale green to lavender (If methyl red is used as indicator, the color change is from yellow to red). Conduct a blank also, starting from the digestion step to final titration.

Calculation

$$mg / l \text{ Total kjeldahl nitrogen (as N)} = \frac{ml.\ 0.02N.H_2SO_4 \text{ for sample} - ml.\ 0.02N.H_2SO_4 \text{ for blank} \times 0.28 \times 1000}{ml. \text{ sample taken for determination}}$$

CHAPTER - 47

ODOR

Odor determination is useful in many respects. For example, it may be used as a check on the quality of raw and treated waters, in determining the effectiveness of deodorising procedures, and as a means of tracing contamination or leaks in industrial processes.

No water can be quite satisfactory for domestic and industrial purposes if it possesses any odor. Disagreeable odors and tastes in water are due to the presence of a variety of objectionable substances particularly living microscopic organisms or decaying vegetation including algae, actinomycetes, bacteria, fungi and weeds. Sewage and industrial wastes also contribute tastes and odors to receiving waters. Where heavy pollution by sewage or industrial effluents occur, the odor may be more marked and offensive.

The commonest form of objectionable odor in streams and other water bodies is attributed to the presence of small amounts of sulfides. The extent of odor depends upon the pH of water. The lower the pH the higher will be the amount of hydrogen sulfide produced.

The presence of inorganic and organic compounds of nitrogen, sulfur and phosphorus, and the putrefaction of protein, and other organic materials present in sewage and in trade wastes are causes of odor in polluted rivers. Many of the industrial effluents containing organic substances (such as phenol, alcohols, esters, aldehydes, ketones and sulfur compounds) possess characteristic odors and are imparted to the receiving waters.

The common odors of surface waters in lakes, rivers and impounding reservoirs may be described as “earthy”, “weedy”, “musty”, “decaying vegetable”, etc. Large volumes of water in lakes and raw water storage reservoirs are affected by algae growth. Many algae have characteristic peculiar odors and the odors are transmitted to the water.

For example, the protozoa dinobryon imparts a fishy odor to water. The algae oscillatoria and rivularia produce a ‘mouldy’ odor and the algae anabaena a strong grassy odor. It is suggested that liberation of traces of essential oils from algae is the reason for odor production. Micro organisms of certain types are reported to be responsible for odors in water supplies. Growths of iron and sulfur bacteria may cause offensive odors. The presence of colloidal vegetable matter in surface water gives off a faint flavor which is generally described as "Peaty" or "Vegetable".

Mineral matters such as clay and sand when present in finely divided state give a faint earthy taste and odor. These odors are sometimes observed in the first filtrates through new sand filters or freshly washed sand filters. Water from newly constructed wells and reservoirs also contain this odor.

Apart from all, chlorination is one of the most important causes of odor and tastes. In addition to the chlorinous taste and odor produced by chlorination, 'Chlorphenol' and 'iodoform' taste and odors are also produced.

Qualitative determination of odor

Collect samples in glass bottles and determine the odor as soon as possible after sample collection. Do not use plastic containers for collection of samples for odor tests.

Place 250 ml of the sample in a 500 ml wide mouth earlenmeyer flask at room temperature and sniff the odor. This is the cold odor quality. Some odors may be missed at room temperature and it is better to determine the hot odor quality also. Place 250 ml sample in a 500 ml stoppered conical flask and heat to about 58 – 60°C. Remove the stopper and sniff the odor. Report the nature and type of odor present. Report the temperature also at which observations are made. A list showing qualitative descriptions of odor is as follows:

1.Aromatic	9.Septic
2.Chemical	10.Earthy
3.Chlorinous	11.Peaty
4.Hydrocarbon	12.Grassy
5.Medicinal	13.Musty
6.Sulphuretted	14.Moldy
7.Disagreeable	15.Vegetable
8.Fishy	

The intensity of odor is expressed in terms of Threshold Odor Number (TON). For details refer, "STANDARD METHODS FOR THE EXAMINATION OF WATER AND WASTE WATER", 20th edition, 1998, published by American Public Health Association.

CHAPTER - 48

OIL AND GREASE

The presence of even very low concentrations of oil and greasy matter in raw water sources is undesirable due to the complications produced by them in the treatment processes such as coagulation, and sedimentation. If any of the free or emulsified oil reaches the sand filters or ion exchange beds, it will coat the grains and decrease the effectiveness of filtration and also interfere with the back washing.

The presence of oil in boiler feed water may cause foaming, priming, overheating of tubes resulting in bulging or even bursting. It is to be mentioned that cooling waters should also be free from oil otherwise biological slimes may occur. Most of the industrial processes require only oil free water, because of the possibility of production of unwanted tastes and odors in the manufactured products due to the oil.

The determination of oil and grease includes all the substances that are extractable by the specified solvent. Generally, the substances extractable are, oils, fats and waxes. It is to be noted that the results obtained indicates only the non-volatile fraction of these materials. Determination of the volatile fraction is beyond the scope of this book.

SOLVENT EXTRACTION METHOD

Principle

Oils, grease, and other extractable matters are dissolved in suitable solvent and separated from the aqueous phase. The solvent layer is then evaporated and the residue is weighed as oil and grease.

Reagents

1. Sulfuric acid 1 + 1: Add carefully 250 ml. conc. H_2SO_4 to 250 ml. distilled water and cool.
2. Petroleum ether: Boiling point 35° to 60°C.

Procedure

1. Place 1000 ml. or appropriate volume of the sample in a suitable separating funnel.

2. Add 5 ml. sulfuric acid per litre of sample.
3. Rinse the sample bottle with 15 ml. petroleum ether and add the rinsings to the separating funnel. Add a further 25 ml. ether to the funnel and shake vigorously for 2 minutes.
4. Draw the aqueous phase in to a clear container and transfer the ether layer through a filter paper (Whatman No 40) containing enough sodium sulfate (anhydrous) wetted with petroleum ether, to a clean, tarred evaporating dish.
5. Continue the extraction twice and add the ether extracts to the evaporating dish after passing through a cone of anhydrous sodium sulfate Na_2SO_4 to remove traces of water (if any).
6. Evaporate the ether in the dish until about 10 ml. remains in the flask. Dry on a water bath.
7. Cool in a desiccator and weigh.

Calculation

$$mg / l \ Oil\ and\ grease = \frac{mg.\ residue\ in\ the\ distilling\ flask}{ml.\ sample\ taken\ for\ determination} \times 1000$$

While expressing the result, specify the solvent used.

Note: This value represents only the nonvolatile oil and greasy matter. For volatile oils and hydrocarbons use "Sherrat method" (Ref.: J.G.Sherrat, "Analyst", 1956,81,518 and J.G.Sherrat, "Analyst", 1962,87,595).

CHAPTER - 49
OXYGEN (DISSOLVED)

As dissolved oxygen induces corrosion reactions, its determination in industrial waters is essential. The test for dissolved oxygen is not only carried out on industrial waters but also on polluted waters and industrial effluents. The test constitutes a means of controlling pollution of water courses. This test is an indicator of the purity achieved during treatments such as activated sludge process and oxidation ponds etc. Also it is the basis of BOD test.

Sample collection and preservation

Samples have to be collected in BOD bottles (See figure 49 -1). If volumes higher than 300 ml. are required, then they have to be collected in separate 1 litre pyrex or corning bottles. Fill the bottles without entrapment of any air. Special sampling devices such as displacement sampler (or) Kemmerer type samplers may be employed in collecting samples from depth. (For details of these samplers refer "ANNUAL BOOK OF ASTM STANDARDS" – part – 23, 1972").

The temperature of the sample should be recorded as soon as the sample is collected. It is also useful to note barometric pressure, especially for calculating the percent saturation of oxygen.

The DO should be determined in the field itself. If it is not possible, the DO may be fixed by the addition of 2 ml. of manganous sulfate solution and 2 ml. of alkali- iodide-azide reagent to the sample in the BOD bottle. (See procedure). The fixed sample has to be transported to the laboratory by keeping it in ice. It should also be kept in dark. The final titration should be performed within 6 hours – the time reckoned from the addition of above reagents.

Among the methods available for the determination of dissolved oxygen, iodometric and electrode methods are having wide range of usefulness. Both the methods are equally good and the selection of the method largely depends on the interferences present and accuracy desired.

IODOMETRIC METHOD

The basic Winkler method is used to determine DO. It has been modified however to remove the most common interferences. Among the modifications, the Alsterberg's modification of adding sodium azide to remove nitrite and Rideal-Stewart's modification of adding permanganate to eliminate the interference of ferrous iron are important.

Principle

When manganous sulfate is added to the sample containing alkaline potassium - iodide, manganous hydroxide is formed, which is oxidized by the dissolved oxygen of the sample to basic manganic oxide. On addition of sulfuric acid, the basic manganic oxide liberates iodine equivalent to that of dissolved oxygen originally present in the sample. The liberated iodine is titrated with a standard solution of sodium thiosulfate using starch as the indicator.

$$MnSO_4 + 2KOH \rightarrow Mn(OH)_2 + K_2SO_4$$

$$2Mn(OH)_2 + \underset{\text{(Dissolved oxygen)}}{O_2} \rightarrow \underset{\substack{\text{(Basic Manganic Oxide)}\\ \text{(brown in color)}}}{2MnO(OH)_2}$$

$$MnO(OH)_2 + 2H_2SO_4 \rightarrow \underset{\text{(Manganic sulfate)}}{Mn(SO_4)_2} + 3H_2O$$

$$Mn(SO_4)_2 + 2KI \rightarrow MnSO_4 + K_2SO_4 + I_2$$

$$2Na_2S_2O_3 + I_2 \rightarrow Na_2S_4O_6 + 2NaI$$

Interferences & Elimination

Nitrite is the major interfering ion. Ferrous and ferric iron, residual chlorine, oxidizable sulphur compounds such as sulfide, sulfites, thiourea and its derivatives, organic matter and suspended matter also interfere.

Elimination of Nitrites (Alsterberg modification)

Nitrites interfere during acidification with sulfuric acid. They react with potassium iodide and liberate iodine.

$$2KI + H_2SO_4 \rightarrow 2HI + K_2SO_4$$

$$2HNO_2 + 2HI \rightarrow 2H_2O + N_2O_2 + I_2$$

If a longer duration is allowed, the DO of the sample reacts with N_2O_2 liberating more iodine and a cyclic reaction occurs.

$$2N_2O_2 + 2H_2O + O_2 \rightarrow 4HNO_2$$

However nitrites are eliminated by the addition of sodium azide (incorporated in the alkali-iodide-azide reagent) in the presence sulfuric acid as follows:

$$2NaN_3 + H_2SO_4 \rightarrow 2HN_3 + Na_2SO_4$$
$$HNO_2 + HN_3 \rightarrow N_2O + H_2O + N_2$$

Elimination of Ferrous salts and Nitrites (Rideal-Stewart modification)

Addition of potassium permanganate in presence of acid suppresses the interference due to ferrous salts, and nitrites. Excess of permanganate is removed by reaction with potassium oxalate.

Elimination of Ferric compounds

Interference by ferric compounds are suppressed by the addition of phosphoric acid, in place of sulfuric acid.

Elimination of Suspended Solids

Samples containing suspended solids absorb appreciable quantities of iodine in acid solution. Suspended matter can be flocculated by the use of alum solution and the supernatant liquid may be used for DO determination (See Procedure).

Elimination of Sulfites, Thiosulfates and Polythionates

When sulfites, thiosulfates are present, a preliminary oxidation with alkaline sodium hypochlorite is necessary. (See procedure).

Elimination of Residual Chlorine

If the sample contains any residual chlorine, it has to be removed by reaction with sodium sulfite / sodium thiosulfate solution. (See procedure).

Apparatus

Generally BOD bottles of 300 ml. capacity are used. This bottle is made of a hard resistance glass with low alkali content. It has a flared lip for making water seal and penny head glass stopper pointed to eliminate trapping of air bubbles. Each bottle has a frosted shield marking. Both bottle and stopper have the same number. (See Figure 49–1).

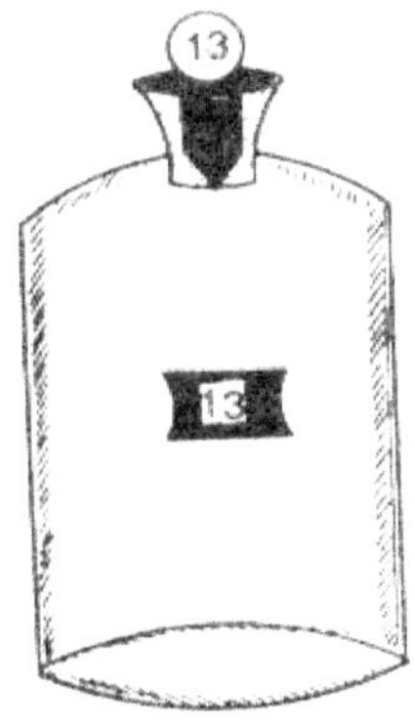

Fig 49–1. BOD bottle

Reagents

1. Manganous sulfate solution: Dissolve 91.0g. manganous sulfate mono hydrate $MnSO_4.H_2O$ in distilled water. Filter if necessary. Dilute to 250 ml. with distilled water. 1 ml. of this solution when treated with 50 ml. of acidified potassium iodide solution should not liberate iodine.

2. Alkali-iodide-azide reagent

(i) Dissolve 175g. potassium hydroxide (or 125g. sodium-hydroxide) and 37.5 g potassium iodide (or 33.7g. sodium-iodide) in distilled water and dilute to 250 ml.

Note: Sodium and potassium salts may be used interchangeably.

(ii) Dissolve 2.5 g sodium azide in 10 ml. distilled water.

Pour the azide solution to the alkali-iodide solution and mix well. This solution should not give color with starch solution when diluted and acidified.

3. Conc. sulfuric acid

4. Phosphoric acid: 85 to 90%

5. Sodium thiosulfate solution, 0.1N: Dissolve 24.82g. sodium thiosulfate, $Na_2S_2O_3.5H_2O$ in boiled and cooled distilled water and make up to 1000 ml. in a volumetric flask. Standardize it against standard potassium dichromate solution (0.1N).
(See Chapter–51, "PERMANGANATE VALUE").

6. Sodium thiosulfate solution, 0.025N: Dilute appropriate volume of 0.1 N sodium thiosulfate solution (about 250 ml. required) to 1000 ml. with distilled water in a volumetric flask.

1.0 ml. 0.025 N thiosulfate ≡ 0.2 mg D.O.

7. Starch solution : See Chapter – 51, "PERMANGANATE VALUE"

Procedure

The following procedure is applicable to domestic and industrial waters and also to sewage effluents.

1. Collect the sample in a 300 ml. B.O.D bottle with utmost care. (For the collection of samples for DO and BOD determinations special sampling devices are preferred. Refer "STANDARD METHODS FOR THE EXAMINATION OF WATER AND WASTE WATER" – 20th edition 1998).
2. Add 2 ml. manganous sulfate solution, followed by the addition of 2 ml. alkali-iodide-azide reagent. The tip of the pipet should be below the surface of the liquid.
3. Stopper the bottle without entrainment of air and mix by inverting the bottle atleast 10 times. (See the note below).

   Note: (Now a brown precipitate of basic manganic-oxide will be formed. Instead of brown precipitate if a white precipitate is formed, the sample is considered to be devoid of oxygen).
4. Allow the precipitate to settle completely leaving a clear supernatant liquid.(with sea water and other saline waters the precipitate settles only with difficulty).
5. Carefully remove the stopper and add 2 ml. conc.H_2SO_4* by the sides of the bottle.
6. Stopper the bottle and mix thoroughly until dissolution is complete.
7. Measure 203 ml. of the solution from the bottle (which corresponds to 200 ml. of the original sample) into a conical flask of 500 ml. capacity.
8. Titrate immediately with 0.025N. sodium thiosulfate solution using starch as the indicator.

Note: Before transporting the sample to the laboratory, the steps from 1to 3 have to be followed to fix the oxygen. Within a few hours it has to be acidified and titrated.

Calculation

If the thiosulfate solution is exactly 0.025 N, then,

$$mg/l \text{ Dissolved oxygen} = ml. \text{ thiosulfate} \times 1$$

$$ml/l \text{ of DO at } 0^{\circ}C \text{ and 760 mm pressure} = mg/l \text{ DO} \times 0.70$$

* If the sample contains ferric iron, use phosphoric acid instead of sulfuric acid. Phosphoric acid may also be used instead of sulfuric acid on all occasions.

Samples containing excess Nitrite (5 to 40 mg/l)

First add 0.2 ml. 3% aqueous solution of sodium azide to the sample and then proceed from Step No.2.

Samples containing Ferrous iron

Samples such as mine drainage waters and dairy waste waters usually contain significant amounts of ferrous iron. It can be eliminated as follows:

To the sample collected in a 300 ml. BOD bottle, add 0.7 ml. conc. H_2SO_4 keeping the tip of the pipet below the surface of the liquid. Add sufficient amount of potassium permanganate solution (630 mg $KMnO_4$/100 ml.) such that the liquid after mixing retains a red-violet tinge for about 20 minutes. Add carefully potassium oxalate solution (2g $K_2C_2O_4.H_2O$/100 ml.) using a graduated 1 ml. pipet. Take care to avoid large excess of potassium oxalate solution. Mix well and allow to stand in the dark. (The reaction occurs more readily in the dark). After the color of the permanganate has been discharged, add 2 ml. manganous sulfate solution. Add 3 ml. alkali–iodide–azide reagent instead of 2 ml. Proceed from Step No.3.

Since potassium permanganate and potassium oxalate solutions contribute to the volume, measure 205 ml. of the final solution instead of 203 ml. and titrate with 0.025N sodium thiosulfate.

Samples containing Suspended matter

Collect the sample in 500–1000 ml. bottle with usual precautions as for regular DO samples. Add 10 ml. alum solution [10g. $AlK(SO_4)_2.12H_2O$/100 ml.] add 2 ml. conc. ammonia solution. Stopper and mix well by inverting. Allow to settle for about 10 minutes. Siphon the clear supernatant into a 300 ml. BOD bottle until it overflows. Stopper it carefully and proceed from Step No.2.

Samples containing Sulfite, Thiosulfate and Polythionates

Industrial waste waters such as paper mill wastes usually contain sulfite, thiosulfate and polythionates. Siphon two portions of 300 ml. sample into two BOD bottles. To the sample in one bottle add 2N. sodium hypochlorite (NaOCl) solution (0.2 to 0.4 ml. is enough) to oxidize the sulfur compounds to sulfates. Mix by inversion for 20 to 30 seconds. Add 1 ml. potassium iodide solution (1 N) and then acidify with 1 ml. or more of H_2SO_4 (1 + 9). Mix by inversion. Neutralize the liberated iodine with 0.1N. sodium sulfite solution using 0.2 ml. starch indicator. Calculate the quantity of sodium hypochlorite required to neutralize the oxidisable substances and add it to the sample in another BOD bottle. Allow four minutes for completion of reaction and then proceed from Step No.2 of the procedure.

Samples containing Chlorine and Hypochlorites

Proceed as described for samples containing sulfites, but start from addition of potassium iodide solution. Add the amount of sodium thiosulfate (or sulfite) required to neutralize chlorine or hypochlorites to the sample in another BOD bottle. Proceed from Step No.2 of the procedure.

Samples containing sulfides and excess organic matter

Electrometric method is preferable.

ELECTRODE METHOD

An alternate method for determining dissolved oxygen in water/waste water is the membrane electrode method. The main advantage of electrode method is that it suffers little / no interference. It does not require any reagents.

Dissolved oxygen membrane electrode is manufactured by many firms. Instructions offered by the manufacturers' have to be followed scrupulously. The sensitivity of the electrode has to be checked up frequently using water of known oxygen concentration.

CHAPTER - 50

PARTICLE COUNTING AND SIZE DISTRIBUTION

Particle counting is an instrumental method introduced in recent times. The instruments, named particle counters provide information about the size and number of particles in a specified volume of water. Particle counting and size distribution analysis provides a basic idea of the makeup of natural water, treated waters and process waters. Hence variations in treatment processes could be identified immediately. It is particularly useful in evaluating the filtration efficiency. Particle counter sensors are installed at the inlet and outlet of filters such that the number and size of particles are identified which provides a means of detecting filter problems very early before they become serious.

Measurement of particle size is carried out by electronic measurement devices. Very sensitive instruments capable of producing both size and number concentration information on particulate dispersions are readily available. In most of the instruments particles pass through a sensing zone where they are measured individually. Instruments produce an electronic pulse (voltage, current or resistance), proportional to the characteristic size of the particle. The instruments responses (pulse height, width or area) are classified by magnitude and counted in each class to yield the particle size distribution. Three different types of instruments are used. They are, electrical sensing zone instruments, light-blockage instruments and light scattering instruments. Among these, some of them can be set up for either continuous-flow or batch sampling. Others can be used only for batch analysis. The manufacturers' instructions have to be followed scrupulously to obtain better results. For more details, refer 'STANDARD METHODS FOR THE EXAMINATION OF WATER AND WASTE WATER' – 20th edition, 1998, published by American Public Health Association.

CHAPTER - 51

PERMANGANATE VALUE (OR) OXYGEN ABSORBED (OR) TIDY'S TEST

This test determines the amount of oxygen absorbed by the sample from potassium - permanganate, in acid solution. The test gives some indication of the nature of oxidisable substances present in the sample. Though the test is an empirical one, it is useful as a quick means to find out the pollution potential of waste waters and is useful in laboratories with limited facilities. When it is not feasible to conduct biochemical oxygen demand test due to interference by toxic substances, this test is of much use. It is also used to determine the Mc Gowan strength of polluted waters, sewage and industrial effluents.

Principle

Waters and waste waters when acidified with sulfuric acid and mixed with potassium permanganate, absorb oxygen. The amount of oxygen absorbed can be calculated from the quantities of potassium permanganate added and the amount left after the reaction.

$$4KMnO_4 + 6H_2SO_4 \rightarrow 2K_2SO_4 + 4MnSO_4 + 6H_2O + 5O_2 \uparrow$$

$$O_2 + 4HI \rightarrow 2H_2O + 2I_2$$

$$I_2 + 2Na_2S_2O_3 \rightarrow Na_2S_4O_6 + 2NaI$$

Reagents

1. Potassium permanganate stock solution, 0.1250N: Dissolve 4 to 4.2g. of potassium - permanganate, $KMnO_4$ in about 900 ml. distilled water in a 2 litre beaker. Cover it with a watch glass and heat the solution to boiling; boil gently over a small flame for 15 - 30 minutes and allow the solution to cool to room temperature. Keep the solution overnight and filter through a funnel containing a plug of purified glass wool. Collect the filtrate in a 1000 ml. volumetric flask cleaned with chromic acid mixture. Make up to the mark with distilled water and standardize it against 0.1250 N sodium oxalate solution as follows. Place the permanganate solution in an amber bottle and keep it in dark.

Standardization of potassium permanganate solution

(i) Dissolve 2.0938g. sodium oxalate $Na_2C_2O_4$ in distilled water and make up to 250 ml. in a volumetric flask. This is exactly 0.1250 N.

(ii) Place 25 ml. of this solution in a 500 ml. conical flask and add 150 ml. 2N sulfuric acid (prepared by mixing 50 ml. conc. H_2SO_4 and 950 ml. distilled water).

(iii) Titrate with the potassium permanganate solution rapidly at room temperature to the persistence of a faint pink color. Allow to stand until the solution becomes colorless. Then heat to 55-60°C and complete the titration until a faint pink color persists for 30 seconds.

Calculate the normality of potassium permanganate solution.

1 ml. of exactly 0.1250N.$KMnO_4 \equiv$ 1.0 mg oxygen

2. Potassium permanganate working solution 0.0125N: Dilute appropriate volume of 0.1250 N potassium permanganate solution ten times such that the resulting solution is exactly 0.0125 N.

1 ml. of exactly 0.0125N.$KMnO_4 \equiv$ 0.1 mg oxygen

3. Stock sodium thiosulfate solution 0.1250N: Dissolve 31.03g. sodium-thiosulfate, $Na_2S_2O_3.5H_2O$ in distilled water (boiled and cooled) and make up to 1000 ml. in a volumetric flask with recently boiled and cooled distilled water. Standardize it against standard 0.1250 N potassium dichromate solution as follows.

Standardization of sodium thiosulfate using standard potassium dichromate

(i) Dissolve 6.13g. potassium dichromate, $K_2Cr_2O_7$ previously dried at 105°C for 2 hrs in distilled water and make up to 1000 ml. in a volumetric flask. This is exactly 0.1250N solution.

(ii) Place 100 ml. boiled and cooled distilled water in a 500 ml. conical flask.

(iii) Add 3g. potassium iodide, 2.0g. sodium bicarbonate and shake until the salts dissolve.

(iv) Add 6 ml. conc. HCl and pipet 25.0 ml. 0.1250 N potassium dichromate solution in to the flask.

(v) Cover the flask with a watch glass and keep it in dark for 5 minutes.

(vi) Rinse the watch glass into the flask and dilute the solution in the flask to 250 ml. with boiled and cooled distilled water, titrate the liberated iodine with the sodium thiosulfate solution in the burette. When the solution acquires yellowish green color, add 1ml starch indicator solution. Now the color changes to greenish blue. Rinse the sides of the flask and continue the titration until one drop changes the color from greenish blue to light green.

Calculate the normality of sodium thiosulfate solution.

1.0 ml. of exactly 0.1250 N sodium thiosulfate ≡ 1.0 mg oxygen.

4. Sodium thiosulfate working solution 0.0125 N: Dilute appropriate volume of 0.1250N solution ten times with boiled and cooled distilled water such that the resulting solution is exactly 0.01250 N.

1.0 ml. of exactly 0.01250 N sodium thiosulfate ≡ 0.1 mg oxygen.

5. Sulfuric acid 1 + 3: Add carefully 250 ml. conc. H_2SO_4 into 750 ml. distilled eater with constant stirring and cool.

6. Potassium iodide solution, 10%: Dissolve 10g. potassium iodide, KI in 100 ml. distilled water. Keep it in an amber bottle. (when it becomes colored, reject it).

7. Starch indicator solution: Dissolve 1g. starch in a little water. Stir it with a glass rod to make it as a thin paste. Pour this paste in about 100 ml. boiling distilled water and boil for 2 minutes and cool.

Procedure

Water

Rinse two 400 ml. stoppered glass bottles first with sulfuric acid and then with distilled water. Place 250 ml. of the well mixed sample in one bottle. Add 10 ml. sulfuric acid (1 + 3) followed by 10 ml. of 0.01250 N potassium permanganate solution. (The volume of potassium permanganate added may be such that half of it remains after 4 hrs. Accordingly 20 ml. or 30 ml. may be added depending upon the nature of the sample). Prepare a blank also in the other bottle using 10 ml. sulfuric acid and same volume of potassium permanganate solution used for sample. Keep the sample and blank at 27°C in dark for 4 hrs. At the end of 4 hrs. add 1 ml. potassium iodide solution and titrate with 0.01250N. sodium thiosulfate solution using starch as indicator.

$$\underset{\text{(Permanganate Value)}}{mg/l\ oxygen\ absorbed} = \frac{(ml\ 0.01250N.thio\ for\ blank - ml\ 0.01250N.thio\ for\ sample) \times 0.1 \times 1000}{ml.\ sample\ taken\ for\ determination}$$

Note 1: The amount of residual chlorine normally present in treated waters does not interfere with this test. But however excessive amounts of residual chlorine interfere and hence the sample should be suitably dechlorinated with sodium sulfite before the determination is made.

Note 2: The Tidy's figure obtained above indicates an empirical sum of both the organic and inorganic matter. the amount of inorganic matter alone can be obtained as described below.

Oxygen absorbed in 3 minutes: Place 250 ml. of the well mixed sample or a suitable portion of the well mixed sample in a 400 ml. glass stoppered bottle. Add 10 ml. sulfuric acid solution. Add 10 ml. 0.01250N. potassium permanganate solution and rotate the bottle for effective mixing. Exactly after 3 minutes (after the addition of potassium permanganate) add 1.0 ml.

potassium iodide solution and titrate with 0.01250N. sodium thiosulfate solution. Calculate the mg/l oxygen absorbed as described above.

Inorganic substances such as nitrite, sulfide and ferrous iron react with potassium permanganate and oxidized within 3 minutes. Thus the 3 minutes Tidy's figure is approximately the amount of inorganic substances only. In examining potable waters, this determination is rarely required.

Sewage and other waste waters: Adopt the same procedure as for water, but use suitable smaller volumes of the sample and higher volumes 0.1250 N potassium permanganate solution. Titrate with 0.1250N. sodium thiosulfate solution.

Calculation

If sodium thiosulfate solution is exactly 0.1250N, then mg/l oxygen absorbed.

$$= \frac{(\text{ml } 0.1250\text{N.thiosulfate for blank - ml } 0.1250\text{N.thiosulfate for sample}) \times 1 \times 1000}{\text{ml. sample taken for determination}}$$

Note: (If N/80 permanganate is used instead of N/8 solution, usually a lower oxygen absorbed figure is obtained. According to the "Ministry of Health report on Methods of Sewage Analysis", if N/80 permanganate is used, the value obtained should have to be multiplied with 1.6 to obtain an approximately correct result).

Interpretation of the result: (See BOD also). A water sample can be regarded as of satisfactory quality when the tidy's figure is in the range of 0.5 – 1.0 ppm. If it exceeds 2 p.p.m. investigation of the source is necessary. For peaty waters, the tolerance limit is 4 p.p.m.

The Tidy's value is useful in assessing the strength of sewage and septic tank liquors. According to McGowan, domestic sewages and septic tank liquors can be classified as follows:

Oxygen absorbed from 0.1250N potassium permanganate (in p.p.m)	Classification
Less than 80	Weak
100 to 120	Average
170 to 250	Strong
250 or more	Very strong

CHAPTER - 52

PHENOLS

Phenols and phenolic compounds are not present in natural waters and are contributed by the effluents from oil refineries, gas works, coke plants and chemical plants. Phenols impart a characteristic objectionable odor to receiving waters. Even very small concentration (0.001 mg/l) produce disagreeable tastes and odors in water particularly during chlorination.

The following 4-amino antipyrine colorimetric method is best suited for the determination of phenols.

DISTILLATION & COLORIMETRIC DETERMINATION

The presence of varied type of inorganic substances present in polluted waters and effluents interfere with the colorimetric procedure, and samples should be subjected to preliminary treatment, distillation and extraction. The phenol concentration is then determined by 4–amino antipyrine colorimetric method.

Collection & Preservation of samples: Collect 1000 ml. of sample in a glass bottle. Acidify to pH below 4 using H_3PO_4 (1 + 3) (to eliminate interferences due to alkaline conditions) and add 1 g. copper sulfate (per litre) to precipitate sulfides as cupric sulfide and to prevent biological growths etc. Keep the sample at 5° to 10^oC. Start the analysis within 24 hrs.

Removal of oxidizing agents: If oxidizing agents such as chlorine are present (detected by acidification and addition of KI and starch) remove them by adding sodium arsenite.

Removal of sulfur compounds: Bring the pH of the sample below 5 with H_3PO_4 (1 + 9) and aerate briefly by stirring prior to the addition of copper sulfate.

DISTILLATION

The phenols are separated from the other non-volatile impurities by distillation. As the rate of voltalization of phenols is gradual, the volume of the distillate must be equal to that of sample being distilled.

Procedure

Use an all glass distillation apparatus (See Chapter - 43 "NITROGEN (AMMONIA)").

1. Place 500 ml. sample in the distillation flask (if the sample is known to contain enough phenols, smaller aliquots may be taken and volume of distillate may be reduced accordingly).
2. Add 5 ml. copper sulfate solution (10%) and bring the pH below 4 using H_3PO_4 (1 + 9) (omit these additions if the sample is already preserved).
3. Start distillation and collect 450 ml. distillate. Cool the flask and add 50 ml. distilled water and continue the distillation until the total volume of the distillate is 500 ml.

[If distillate is turbid, acidify with 1 + 9 H_3PO_4 to pH below 4 and add 5 ml. $CuSO_4$ solution (10%) and distill as described above. If the second distillate is also turbid, acidify 500 ml. of original sample using 1N.H_2SO_4 and methyl orange indicator and add 150 g. NaCl. Extract with 40 ml. chloroform. Extract further with 25 ml. chloroform for four times combine all the chloroform extracts in a second separating funnel and extract with 2.5N sodium hydroxide solution (100 g. NaOH/litre) first with 4.0 ml. and then with 3.0ml. volumes two times. Combine the alkaline extracts, heat on a water bath until chloroform is removed. Cool and dilute to 500 ml. and proceed distillation as described above starting from Step No.2].

Note–1: Appropriate smaller volumes of regents may be used in cases where smaller volumes of samples are taken for distillation.

Note–2: Diethyl ether is recommended in place of chloroform as it eliminates emulsion problem and the use of NaCl. Chloroform is preferred because of the hazards in handling ether.

4-AMINO ANTIPYRINE COLORIMETRIC METHOD

This method is applicable to most of the phenolic compounds excepting para cresol and similar para substituted phenols.

Principle

Phenols react with 4-amino antipyrine at a pH of 10.0±0.2 in the presence of potassium ferricyanide to form a yellow red antipyrine dye.

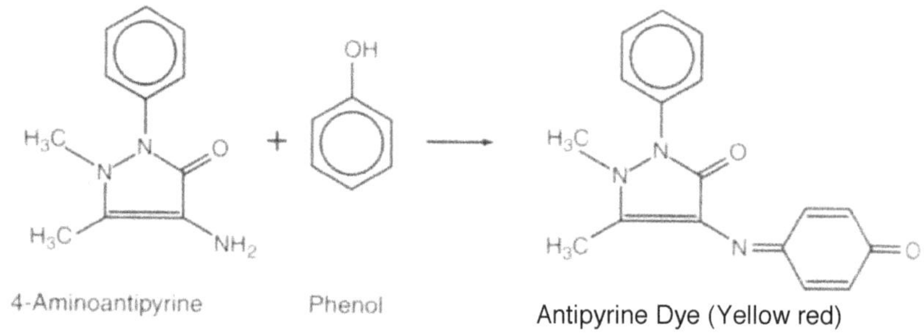

Interferences

All the interferences are eliminated or reduced to a minimum level by preliminary treatment and distillation.

Reagents

1. Phenol stock solution: Dissolve 1.0 g. phenol in distilled water and make up to 1000 ml. in a volumetric flask. Standardize* this solution every time as described below.

Pipet 25.0 ml phenol stock solution into an iodine flask and dilute to 100 ml. Pipet 20.0 ml. bromate - bromide solution (0.5N). (prepared by dissolving 6.96 g. potassium bromate and 37.5g. potassium bromide in 500 ml) into the flask. Add 10 ml 1 + 3 sulfuric acid solution and stopper the flask immediately. Seal the neck of the flask with 10 ml. 10% potassium iodide solution. Keep the flask in the dark for one hour. Wash the sealing solution into the flask and add a further 10 ml of 10% potassium iodide solution. Titrate the liberated iodine with 0.25N. sodium thiosulfate using starch indicator. Conduct a blank simultaneously using 100 ml. distilled water and 20 ml. bromate-bromide solution.

Calculate the amount of phenol present in one litre as follows:

$$mg / l\ phenol = \frac{ml.\ thio\ used\ for\ blank - ml.\ thio\ used\ for\ back\ titration \times 3.921 \times 1000}{25}$$

2. Phenol intermediate solution: Dilute appropriate volume of phenol stock solution to 1000 ml. such that 1.0 ml. of the diluted solution contains 10 μg phenol.

3. Phenol working solution: Pipet 50.0 ml of phenol intermediate solution to 500 ml. with distilled water.

1.00 ml. = 1 μg phenol

* Standardization itself constitutes one method for the determination of phenols if present in higher concentrations.

4. Ammonia-ammonium chloride buffer: Dissolve 16.9 g. ammonium chloride, NH_4Cl in 143 ml conc. ammonia solution and dilute to 250 ml. with distilled water. Place it in a tightly stoppered bottle.

5. 4–Amino antipyrine solution: Dissolve 2g. 4–amino- antipyrine in distilled water and dilute to 100 ml. with distilled water. Prepare daily.

6. Potassium ferricyanide solution: Dissolve 8g. potassium ferricyanide $K_3Fe(CN)_6$ in distilled water and dilute to 100 ml. This solution may be kept for one week.

7. Chloroform

8. Sodium Sulfate, anhydrous, granular

Procedure

1. In to a series of 1000 ml separating funnels place 0.0, 2.0, 5.0, 10.0, 20.0 .... 50.0 ml of phenol working solution and dilute each to 500 ml with distilled water.
2. Place 500 ml. distillate in 1 litre separating funnel (lower aliquots may also be used if phenol concentration is more, but dilute to 500 ml. with distilled water).
3. To the blank, standards and sample distillate, add 10 ml ammonia-ammonium chloride buffer solution (Addition of 1 ml. of this buffer to 50 ml. volume brings the pH to 10±0.2).
4. Add 3.0 ml of 4–amino antipyrine solution. Mix immediately and add 3.0 ml. potassium ferricyanide and again mix immediately.
5. Allow to stand for 3 mts. and extract with 25 ml chloroform if 5 cm. cell is used or with 50 ml. if 10 cm cell is used.
6. Pass through fritted-glass funnel containing 5g. layer of sodium sulfate and collect the dried extract in 50 ml Nessler tubes.
7. Measure the optical density of sample and standards at 460nm setting blank at 100% transmittance and prepare a calibration curve. Find out the µg. equivalent of phenol from the curve. Visual comparison can also be made.

Express the result as mg. phenolic substances [as phenol (C_6H_5OH)] per litre of sample.

CHAPTER - 53

PHOSPHATE

Phosphate may occur in surface or ground waters as a result of leaching from minerals or ores, from agricultural run off, as a constituent of boiler waters and cooling waters, as a result of discharge of industrial wastes, and as a major element of municipal sewage due to the utilization of synthetic detergents.

Phosphates are the widely employed chemicals to prevent scale formation and to inhibit corrosion, particularly in boiler feed waters and cooling waters. It is reported that polyphosphates are detrimental in that they interfere with coagulation, flocculation, and limesoda treatment of water.

Phosphorus occurs in natural waters and waste waters in the form of various phosphates. They are commonly classified as orthophosphates condensed phosphates (meta-, pyro-, and polyphosphates), and organically bound phosphate. These various forms of phosphates may occur in soluble form or in particulate form.

According to "Standard Methods", the phosphates which respond to the colorimetric tests without preliminary hydrolysis or oxidative digestion of the sample are considered as "Orthophosphates".

The phosphate fraction which is obtained along with orthophosphates on mild acid hydrolysis is known as condensed phosphate. This fraction includes meta, pyro and polyphosphates. The sum of the ortho and condensed phosphates is the total inorganic phosphates.

The phosphate fraction obtained on acid digestion along with ortho and acid hydrolysable phosphates is known as the 'organic' or 'organically bound' phosphates.

Phosphate determination consists of two general procedural steps:

1. Conversion of various forms of phosphates to orthophosphate and
2. Determination of ortho phosphate.

Among the methods available for the determination of orthophosphate, stannous - chloride method is recommended here.

ORTHOPHOSPHATES
(Stannous Chloride Method)

Principle

Ammonium molybdate reacts with phosphate to form molybdophosphoric acid which is reduced to a blue colored complex 'molybdenum blue' by the addition of stannous chloride.

$$PO_4^{3-} + 12(NH_4)2MoO_4 + 24H^+ \rightarrow (NH_4)_3PO_4.12MoO_3 + 21NH_4^+ + 12H_2O$$

$$(NH_4)_3PO_4.12MoO_3 + Sn^{2+} \rightarrow (Molybdenum\ blue) + Sn^{4+}$$

Interferences

Arsenate, fluoride, sulfide, thiosulfate, thiocyanate and excess molybdate cause negative interferences. Sulfides may be removed by oxidation with bromine water.

Reagents

1. **Phenolphthalein indicator solution:** See Chapter – 11, "ALKALINITY".
2. **Sulfuric acid – nitric acid solution:** Carefully add 75 ml conc. H_2SO_4 to about 150 ml distilled water and cool. Add 1 ml conc. HNO_3 and dilute to 250 ml with distilled water.
3. **Ammonium molybdate solution:**
   (a) Dissolve 25g. ammonium molybdate, $(NH_4)_6Mo_7O_{24}.4H_2O$ in about 200 ml distilled water.
   (b) Add carefully 280 ml conc. H_2SO_4 to 400 ml distilled water and cool.
   Add the molybdate solution to the diluted acid and dilute to 1000 ml.
4. **Stannous chloride solution:** Dissolve 2.5g. fresh stannous chloride, $SnCl_2.2H_2O$ in 100 ml glycerol and heat in a water bath. Mix by stirring with a glass rod. This reagent is stable and requires no special storage.
5. **Phosphate stock solution:** Dissolve 439mg anhydrous potassium dihydrogen phosphate, KH_2PO_4 in distilled water and make up to 1000 ml in a volumetric flask. Add two drops of toluene as a preservative.

   $$1.0\ ml = 100\ \mu g\ P = 306\ \mu g\ PO_4$$

6. **Phosphate standard solution:** Pipet 10.0 ml phosphate stock solution into 1000 ml volumetric flask and make up to the mark with distilled water. Prepare freshly.

$$1.0 \text{ ml} = 1 \ \mu g \text{ P} = 3.06 \ \mu g \text{ PO}_4$$

Procedure

Important Note: Do not use synthetic detergents containing phosphates for cleaning glassware. Use only acid solutions.

1. Place 100 ml or a suitable aliquot of the sample containing not more than 20 μg P in a 100 ml Nessler tube. (The sample should be free from color and turbidity). Add 1 drop of phenolphthalein indicator. If any pink color appears, destroy it by adding one or 2 drops of sulfuric-nitric acid solution. If more than 5 drops are required, take a smaller aliquot, add phenolphthalein, discharge the pink color and dilute to 100 ml.
2. Into a series of 100 ml Nessler tubes, pipet appropriate volumes of phosphate working solution covering the range up to 20 μg P. Dilute to 100 ml. Include a Nessler tube containing 100 ml distilled water as the blank.
3. To the blank, standards and sample add 4.0 ml ammonium molybdate solution and 0.5 ml stannous chloride solution, mixing after each addition.
4. After 10 minutes but before 12 minutes, measure the color using a spectrophotometer at 690 nm.
5. Prepare a calibration curve and find out the number of micrograms of P equivalent to the observed optical density of the sample.

Express the result as mg phosphate as P per litre of sample.
If it has to be expressed in terms of PO_4 multiply by 3.06

INORGANIC PHOSPHATES
(Orthophosphates + Condensed phosphates)

To 100 ml sample or an aliquot diluted to 100 ml, add 1 drop of phenolphthalein indicator. If any pink color appears discharge by the addition of sulfuric-nitric acid solution. Then add 1 ml excess of sulfuric acid – nitric acid solution and boil* gently for about 90 minutes. Distilled water may be added during boiling to keep the volume between 25 and 50 ml. Cool and neutralize with 6N sodium hydroxide solution (240g NaOH/1000ml) using phenolphthalein indicator. Make up the volume to 100 ml. Determine the phosphate as described in "Stannous chloride method".

* Instead of boiling, the sample containing sulfuric acid-nitric acid solution may be auto claved at 15-20 psi for 30 minutes. (Ref : James W. McCoy M, "CHEMICAL ANALYSIS OF INDUSTRIAL WATER").

The result obtained is the sum of total inorganic phosphates (i.e.) the sum of orthophosphates and condensed (meta-, pyro-, and poly) phosphates. If the concentration of condensed phosphates only is needed subtract the orthophosphate value from the total inorganic phosphate value.

TOTAL PHOSPHATES

(Orthophosphates+Condensed phosphates+Organic phosphates)

Organic matter must first be destroyed by wet oxidation. Perchloric acid digestion, sulfuric acid-nitric acid digestion, and persulfate digestion are the three recommended procedures. The choice of the digestive procedure largely depends on the nature of the sample.

Perchloric acid digestion: Acidify a suitable aliquot of the well mixed sample placed in an evaporating dish with conc. HNO_3 using methyl orange indicator (until red color appears) and add 5 ml excess of conc. HNO_3. Evaporate on a water bath or hot plate to 15 – 20 ml. Transfer to a 125 ml conical flask using 5 ml conc. HNO_3. Add 5 ml conc. HNO_3 add 10 ml per chloric acid, $HClO_4$ (70-72%). Add a few boiling chips and evaporate gently on a hot plate until white dense fumes of $HClO_4$ just appear. If the solution is not clear and 10 ml more conc. HNO_3 and continue the evaporation. Then cool the digested sample and neutralize with 6N-NaOH solution using phenolphthalein indicator. Transfer to a 100 ml volumetric flask (filter if necessary) and make up to the mark with distilled water. Determine the phosphate content of the solution as described in "Stannous chloride method".

Sulfuric acid-nitric acid digestion: Place a suitable aliquot of the well mixed sample in a micro-kjeldahl flask and add 1 ml conc. H_2SO_4 and 5 ml conc. HNO_3. Digest the sample to a volume of 1 ml and continue the digestion until the solution becomes colorless. Cool and add about 20 ml distilled water and neutralize with 1N sodium hydroxide solution(40g NaOH/1000 ml) using phenolphthalein indicator. Transfer to a 100 ml volumetric flask (filtering if necessary) and make up to the mark with distilled water. Determine the phosphate content of this made up solution as described in "Stannous chloride method".

Persulfate digestion: To 100 ml or a suitable aliquot of the sample diluted to 100 ml, add 1 drop of phenolphalein indicator. If pink color appears discharge it by the addition of 3 + 7 sulfuric acid solution and add 1 ml in excess followed by 15 ml of 5% solution of potassium persulfate (prepared freshly). Boil gently for about 90 minutes keeping the volume to be about 50 ml with the addition of distilled water, or auto clave at 15 – 20 psi for 30 minutes. Cool and neutralize with 1N sodium hydroxide solution using phenolphthalein indicator. Transfer to a 100 ml volumetric flask and make up to the mark with distilled water. Determine the phosphate concentration of this made up solution using the procedure described in "Stannous chloride method".

The result obtained indicates the concentration of total phosphates. If the concentration of organic phosphates is required, subtract the total inorganic phosphate from the total phosphate.

CHAPTER - 54

pH VALUE

pH is a term used universally to express the intensity of the acid or alkaline condition of a solution. It is a measure of hydrogen ion concentration, or more precisely, the hydrogen ion activity.

pH is defined as the "logarithm" (base 10) of the reciprocal of the hydrogen-ion concentration". Thus if $[H^+] = 10^{-6}$ moles per litre, then pH = 6. pH is an important factor in water chemistry, since it enters into the calculation of acidity and alkalinity and processes such as coagulation, disinfection, softening and corrosion control.

The pH of a raw water used for domestic purposes is important in that it affects taste, efficiency of chlorination and treatment processes such as coagulation. The killing power of chlorine diminishes with increasing pH values and consequently it is advantageous to apply chlorine to a water with a pH value of 7 or less. Knowledge of pH is essential in the selection of coagulants for water purification. For example, aluminium sulfate is more effective at pH 6.7, whereas ferrous sulfate coagulates well at a higher pH; ferric sulfate is effective over a wide pH range. Measurement of pH and its control is not only essential for water purification processes but is also important for many industrial processes too. Besides, maintaining correct pH range is extremely important in waste water treatment.

pH in conjunction with total salinity, temperature, calcium content and total alkalinity is used to determine whether a water is corrosive in nature or having scale forming tendencies. Langelier Saturation Index and the Ryznar Stability Index based on the actual pH values and theoretical pH values are the widely employed parameters to measure the particular water to be corrosive or to form scale. For further details, see Chapter – 35.

pH can be measured electrometrically or colorimetrically. Electrometric methods are by far the most accurate and suffers with little or no interference. Though the colorimetric methods require less expensive apparatus, it suffers from serious interferences such as color, turbidity, oxidants, reductants, and colloidal matter. Hence colorimetric methods are suitable only for rough estimation.

ELECTROMETRIC METHOD

pH meters are the widely employed instruments for the electrometric measurement of pH. A pH meter is a high-impedance electrometer calibrated in terms of pH. In principle, electrometric determination of pH involves the measurement of the electromotive-force (EMF) of a cell comprising an indicator electrode responsive to hydrogen ions and a reference electrode both immersed in the test-solution. The indicator electrode commonly used is glass electrode and the reference electrode is calomel electrode. At present, 'combined electrode' comprising both reference and glass electrode as a single electrode unit is available. Measurement of pH values above 10 and at high temperature is best made with special glass electrodes designed for such purposes.

Various types of pH meters are available commercially. They range from battery-operated portable units to highly precise instruments. pH meters with an accuracy of 0.005 pH unit are available. But for most of the purposes a pH meter with an accuracy of 0.05 unit is sufficient.

pH meters are usually standardized with standard buffer solutions, the preparation of some of them are given below.

The points to be observed during the preparation of standard buffer solutions are:

1. Use distilled water having a specific conductance less than 2 micromhos at 25°C and a pH 5.6 to 6.0.
2. The distilled water used for the preparation of borax and phosphate solutions should be boiled and cooled to expel the carbon dioxide and to produce a pH of 6.7 to 7.8.
3. Care should be taken to make complete dissolution of salts, otherwise the pH may be incorrect.
4. Store the buffer solutions in polythene containers.

pH 4* Buffer solution: Dissolve 1.012g. anhydrous potassium hydrogen phthalate $KHC_8H_4O_4$, in distilled water and make up to 1000 ml. in a volumetric flask.

* at 25°C

pH 7* Buffer solution: Dissolve 1.361g. anhydrous potassium dihydrogen phosphate, KH_2PO_4 and 1.420g. anhydrous disodium hydrogen phosphate Na_2HPO_4 (both of which have been dried at 110°C to 130°C for 2 hrs) in distilled water and makeup to 1000ml in a volumetric flask.

pH 9* Buffer solution: Dissolve 3.81g. sodium borate decahydrate (borax) $Na_2B_4O_7.10H_2O$ in distilled water and make up to 1000 ml.

Commercial buffers are also available in tablet, powder and concentrated solution forms. These must be dissolved and made up to the stated volume with distilled water.

Measurement of pH

Follow the manufacturers' instructions to operate the pH meter and determine the pH of the sample of water or waste water. Before measurement of pH of the sample, standardize the pH meter, using a standard buffer solution of pH near that of the sample to be tested. Check the electrode response occasionally by measuring the pH of another standard buffer solution with a different pH.

The following points may be found useful:

(i) Wash the electrodes thoroughly with distilled water and then with the sample before the measurement of pH.

(ii) Allow the system to stabilize before making the final reading.

(iii) When samples containing oil or grease are to be determined remove them by filtration and then measure the pH. If any only film sticks to the electrode, it can be removed with a soft tissue soaked in a suitable solvent or detergent followed by thorough washing with distilled water.

(iv) If the pH exceeds 10, sodium ions also respond. In that case use a special glass electrode. If no such electrode is available make suitable corrections in the measurement made with the ordinary glass electrodes. Usually manufacturers' of pH meter supply such correction charts, along with the instrument.

(v) If glass electrodes are dried out, soak them in N/10 HCl for about 10 to12 hrs or according to the manufacturers' instructions.

(vi) Care should be taken while filling the solution in the reference electrode such that no air bubbles are entrapped in the solution or in the liquid junction.

---

* at 25°C

The pH of most natural waters falls within the range of 4 to 9. Most of the waters are slightly alkaline due to the presence of carbonates and bicarbonates. The desirable pH range for drinking water is 7.0 to 8.5.

Based on the pH value, natural waters can be divided into three distinct classes:

(a) Those which contain carbonates with or without bicarbonates. They do not have free carbonic acid. The pH of these waters is 8.3 or above 8.3

(b) Those which contain no carbonates but contain bicarbonates and carbonic acid. The pH value of these waters ranges from 4.5 to 8.3 Most of the natural waters fall under this category.

(c) Those which contain free acid in addition to carbonic acid. They do not contain carbonates or bicarbonates. The pH value of these waters is 4.5 or below 4.5.

CHAPTER - 55

POTASSIUM

Though potassium ranks seventh among the elements in the order of abundance, its concentration in most waters is trivial. Potassium is an essential nutritional element, but in excessive amounts it acts as a cathartic. It is reported that foaming may be caused in boilers by more than 50 mg/l of potassium and sodium in the water. Potassium is best determined by flame photometric method.

FLAME PHOTOMETRIC METHOD

Follow the manufacturers' instructions scrupulously Calibrate the instrument with potassium standard solutions which are prepared as follows.

Potassium stock solution: Dissolve 1.907 g potassium chloride, KCl dried at 110°C in little de-ionised distilled water and make up to 1000 ml in a volumetric flask.

1.0 ml = 1.0 mg K.

Make dilutions as required for use with deionised distilled water. Determine potassium content of the sample using flame photometer with usual precautions. (refer Chapter–63, "SODIUM").

Express the result as mg potassium (as K) per litre of the sample.

ION-SELECTIVE ELECTRODE METHOD

Ion selective electrode for potassium determination is available now. The potassium electrode membrane has good selectivity for potassium ions. Other ions (sodium, ammonium, hydrogen, rubidium and cesium) only slightly interfere.

Follow the manufacturers' instructions to find out the potassium content of a water sample using this ion selective electrode.

CHAPTER - 56

RESIDUAL CHELANT (EDTA)

Residual Chelant is the uncombined chelant remains in boiler water after reaction with hardness. Determination of residual chelant is useful in controlling the dosage of chelant. It is also used as an indirect method to determine the average total hardness of feed water. Needless to say that the accurate determination of hardness values less than 1 ppm (as $CaCO_3$) is always difficult. Besides, hardness varies in feed water because of leakage from softeners, contamination of condensate by raw water and other sources of hardness. For this reason, it is convenient to control chelant dosage by measuring the residual chelant in the boiler water.

In this chapter, separate procedures for the determination of residual chelant (EDTA) and combined chelant (EDTA) are given. In addition, calculating the average total hardness of feed water from the values of concentration of chelants is also described.

RESIDUAL CHELANT (EDTA) DETERMINATION BY BACK TITRATION METHOD

Principle: The test is carried out by adding a known excess of standard magnesium chloride solution to the boiler blowdown water and titrating the excess magnesium chloride with standard EDTA solution from which the value for residual chelant is derived.

Reagents

1. **Ammonia – ammonium chloride buffer:** See Chapter – 31, "HARDNESS (TOTAL)".
2. **Eriochrome Black-T indicator:** See Chapter – 31, "HARDNESS (TOTAL)"
3. **Standard Magnesium Solution:**

Dissolve 64.46 mg magnesium metal turnings in 50ml distilled water containing 1 ml HCl (conc). Evaporate to dryness, then dissolve in little water, quantitatively transfer to a 1000 ml volumetric flask and make upto the mark with distilled water.

1.0 ml ≡ 1.0 mg EDTA (Disodium salt)

4. **Standard EDTA solution:**

Dissolve 1.000 g ethylene diaminetetra acetate dehydrate disodium salt, $Na_2H_2C_{10}$ - $H_{12}O_8N_2.2H_2O$ in distilled water and make upto 1000 ml in a volumetric flask.

1.0 ml = 1.0 mg EDTA (Disodium salt)

Procedure

1. Place 100 ml or suitable aliquot of boiler blowdown water in a conical flask.
2. Add 10.0 ml standard magnesium solution and mix well.
3. Add 2 ml ammonia–ammonium chloride buffer solution and mix again.
4. Add 1 drop of Eriochrome Black-T indicator and mix well.
   (If the solution is not wine red at this point it indicates that more residual chelant is present. Therefore an additional 5.0 ml or 10.0 ml standard magnesium solution has to be added.)
5. Titrate with standard EDTA titrant until the color of the solution changes from wine red to blue. Record the titre value and calculate the quantity of residual chelant as shown below.

Calculation

mg/l Residual Chelant (EDTA Disodium salt)

$$= \frac{\text{(ml. standard magnesium solution - ml. standard EDTA solution)}}{\text{ml. boiler blow down water}} \times 1000$$

TOTAL CHELANT (EDTA) DETERMINATION BY COLORIMETRIC METHOD

Determination of total chelant is useful in determining the average total hardness of feed water. The following colorimetric method using xylenol orange and ziroconyl chloride is more precise.

Principle

Ziroconium reacts with the chelant EDTA that has combined with the calcium and magnesium and also with the residual chelant. A known volume of the zirconyl oxychloride solution is added to boiler water which after reaction with the chelant (both combined and residual) is allowed to react with xylenol orange solution and the color of the resulting solution is measured.

Reagents

1. Standard EDTA solution [1.0 ml = 1.0 mg EDTA (Disodium salt)]

See the section, "Residual Chelant Determination"

2. EDTA Working standard solution

Pipet 10.0 ml of standard EDTA solution into a 100 ml volumetric flask and make upto the mark with distilled water.

1.0 ml = 0.1 mg EDTA (Disodium salt)

3. Xylenol orange solution

(i) Dissolve 200 mg xylenol orange in 85 ml conc.Hcl.

(ii) Dissolve 25 g hydroxylamine hydrochloride, NH_2OH. HCl in 100 ml distilled water.

Mix these two solutions in a 250 ml volumetric flask and dilute to the mark with distilled water. Allow to stand overnight. Filter if necessary.

4. Zirconyl chloride stock solution: Dissolve 1.800 g. zirconyl oxychloride, $ZrOCl_2.8H_2O$ in 500 ml water containing 65 ml HCl (conc). Dilute to 1000 ml with distilled water in a volumetric flask.

0.1 ml = 0.50 mg Zr

5. Zirconyl chloride working solution : Place 10.0 ml of the stock solution in a 500 ml volumetric flask. Add 10 ml conc. HCl and mix well. Dilute to the mark with distilled water.

1.0 ml = 0.01 mg Zr

Prepare freshly

Procedure

(A) Preparation of calibration graph

1. Place 1.0 ml, 2.0 ml, 3.0 ml, 4.0 ml and 5.0 ml of EDTA working standard solution into a series of 50 ml volumetric flasks.
2. Bring the total volume of the solution in each flask to 35 ml with distilled water.
3. To each flask add 5.0 ml xylenol orange reagent and mix well.
4. Add 5.0 ml Zirconyl chloride working solution to each flask and mix thoroughly.
5. Dilute to the mark with distilled water and again mix well.
6. Prepare a blank simultaneously as per the procedure given below.
7. Place 35 ml distilled water in a 50 ml volumetric flask, add 5.0 ml xylenol orange, mix well and dilute to the mark with distilled water and mix thoroughly.
8. Allow the blank and standard solutions to stand for 30 minutes.
9. Set the transmittance of the blank at 100 percent at a wavelength of 535 nm. Measure the transmittance of the standards and plot a calibration curve on a semilogarithmic graph paper relating transmittance of standard EDTA solutions to their concentration.

(B) Sample

Place 25 ml or a suitable lower aliquot of the filtered blowdown water in a 50 ml volumetric flask and dilute to 35 ml with distilled water. Add 5.0 ml xylenol orange solution and mix well. Add 5.0 ml zirconyl chloride working solution and mix thoroughly. Makeup to the mark with distilled water, mix well and allow to stand for 30 minutes. Then measure the transmittance and read the corresponding EDTA concentration from the calibration curve.

$$mg/l\ EDTA\ Disodium\ salt = \frac{mg\ EDTA\ disodium\ salt \times 1000}{ml.\ sample\ taken\ for\ determination}$$

CALCULATION OF TOTAL HARDNESS FROM THE VALUES OF TOTAL EDTA AND RESIDUAL EDTA

By measuring the total chelant (EDTA) and residual chelant (EDTA) concentrations, it is possible to calculate the average total hardness of the feed water. Calculation of total hardness is illustrated below by a worked out problem.

Problem 56 – 1

On analysis, a boiler water shows a total EDTA of 10 ppm and a residual EDTA of 2 ppm. Calculate the average total hardness of the feed water, if the boiler is operated at 5% blowdown.

Total EDTA (Disodium salt) = 10 ppm

Residual EDTA (Disodium salt) = 2 ppm

∴ Combined EDTA = (10 – 2) = 8 ppm

1 ppm EDTA disodium salt chelates = 0.2686 ppm hardness (as $CaCO_3$)

∴ 8 ppm EDTA disodium salt chelates = 0.2686 x 8

= 2.15 ppm hardness (as $CaCO_3$)

Boiler is operated at 5% blow down.

∴ The average total hardness of the feed water

$$= 2.15 \times \frac{5}{100}$$

= **0.1075 ppm (as $CaCO_3$)**

Rounded to = **0.11 ppm (as $CaCO_3$)**

CHAPTER - 57

RESIDUAL CHLORINE

Since chlorination is the widely employed method of disinfection, the presence of chlorine is common with potable and industrial waters. Many industrial waters are chlorinated in order to control bacteria and slime forming organisms. It has to be remembered that the residual chlorine in the water used for food canning, may cause tastes in the canned foods or frozen products or it may corrode cans. Chlorine produces undesirable tastes and odors, especially in the presence of even traces of organic substances such as phenol. It is also reported that chlorine present in water used for metal plating may affect the smoothness and brightness of the deposits.

In cooling towers, continuous contact with free chlorine may result in the deterioration of lumber. Further, large concentrations of chlorine can accelerate corrosion particularly when oxygen is present. Hence determination of chlorine in industrial waters is highly essential.

Chlorine in water may be present as free available chlorine (in the form of hypochlorous acid (HOCl) or hypochlorite ion (OCl^-) or both), and/or as combined available chlorine (chloramines such as mono chloramines NH_2Cl, dichloramine $NHCl_2$ and nitrogen trichloride (NCl_3) and other chloro derivatives). Total chlorine is the sum of free and combined available chlorine residuals.

Two methods for the determination of residual chlorine are given here. The iodometric method is considered to be the standard one. It is however best suited to samples containing chlorine more than 1 mg/l only. The ortho tolidine method can be applied with precision to samples containing less than 1 mg/l chlorine. This colorimetric procedure is widely used as a field method also. Both the methods are suitable for potable water, swimming pool water, cooling water, moderately polluted water and for water treatment plant effluents.

Note: The residual chlorine content of water and particularly industrial effluent samples decreases after collection. This effect is more with rising temperature and in the presence of organic matter. Hence, the residual chlorine test should be conducted without delay, and preferably in the absence of sunlight.

IODOMETRIC METHOD

Principle

Both free available chlorine and combined available chlorine liberates free iodine on reaction with potassium iodide and the liberated iodine is titrated with standard sodium thiosulfate solution using starch as indicator. The reaction is preferably carried out at a pH of about 3 to 4.

$$2KI + Cl_2 \quad - \; 2KCl + I_2$$

$$I_2 + 2Na_2S_2O_3 \rightarrow Na_2S_4O_6 + 2NaI$$

Interferences

Nitrite ions interfere seriously. Ferric iron in excess of 2 mg/l, interferes. Manganic ions are also reported to interfere. These interferences however are tolerable if the pH of the sample is adjusted between 4.5 and 8.0 by the addition of acetic acid. (See modified procedure).

Reagents

1. Acetic acid, conc. (glacial)

2. Potassium iodide, crystals

3. Starch indicator solution: See Chapter - 52 "PERMANGANATE VALUE".

4. Sodium thiosulfate solution, 0.1N: Dissolve 24.82g. sodium thiosulfate $Na_2S_2O_3.5H_2O$ in boiled and cooled distilled water (oxygen free) and make up to 1000 ml. in a volumetric flask. Standardize it against standard potassium dichromate solution (0.1N) as per the procedure given below. Preserve it by adding chloroform 3 drops/litre.

Standardization of Sodium Thiosulfate

(i) Dissolve 1.226 g potassium dichromate in little distilled water and make up to 250 ml. in a volumetric flask.

(ii) Place 100 ml. boiled and cooled distilled water (i.e. oxygen free) in a 500 ml. conical flask. Add 3 g potassium iodide and 2 g sodium bicarbonate. Shake until the salts dissolve.

(iii) Add 6 ml. conc. HCl followed by 25.0 ml. standard potassium dichromate solution (0.1 N) and cover the flask with a glass. Allow to stand in dark for 5 minutes.

(iv) Rinse the watch glass and dilute with 300 ml. distilled water. Titrate the liberated iodine with the sodium thiosulfate solution. When the solution becomes yellowish green, add 2 ml. starch solution. Rinse the sides of the flask and continue the titration until one drop changes the color from greenish blue to light green.

(v) Calculate the normality of sodium thiosulfate solution.

$$\text{Normality of sodium thiosulfate} = \frac{25 \times 0.1}{\text{ml. sodium thiosulfate}}$$

5. Sodium thiosulfate solution, 0.0025N: Dilute appropriate volume of 0.1N sodium thiosulfate to give 0.0025N solution.

1.0 ml. 0.0025N sodium thiosulfate ≡ 0.08865 mg.Cl_2. Prepare daily.

Procedure

Place 500 ml. or suitable lower aliquots of the sample in a conical flask. Add 0.5g to 1.0 g potassium iodide crystals and 5 ml. acetic acid. Mix and immediately titrate with 0.0025N sodium thiosulfate until the color of the liberated iodine is nearly discharged. Add 1 ml. starch solution and titrate until the blue color is discharged.

Calculation

$$\text{mg/l chlorine} = \frac{\text{ml. 0.0025N sodium thiosulfate} \times 0.0886 \times 1000}{\text{ml. sample taken for estimation}}$$

Modified procedure when nitrite, manganese and ferric iron (more than 2 mg/l) are present.

Adjust the pH of the sample to lie between 4.5 and 8.0 by the addition of acetic acid or sodium acetate as necessary(Excess acidity must be avoided). Add 0.5 to 1.0 g potassium iodide crystals. Titrate immediately with 0.0025N thiosulfate solution as described above.

ORTHOTOLIDINE METHOD

Principle

Orthotolidine reacts in acid medium with chlorine to form a yellow color product and the yellow color is compared with standards. (Reaction - see next page). Residual chlorine color dises are available commercially. They are available in many ranges. The ranges 0.1 to 1.0 mg/l chlorine and 0.2 to 2.0 mg/l chlorine are preferable.

Interferences

Iron above 0.3 mg/l, (manganic) manganese above 0.01 mg/l and nitrite* above 0.1 mg/l interfere by producing additional yellow color. However, use of arsenite reagent enables the determination of the color produced by chlorine and its derivatives only (see procedure).

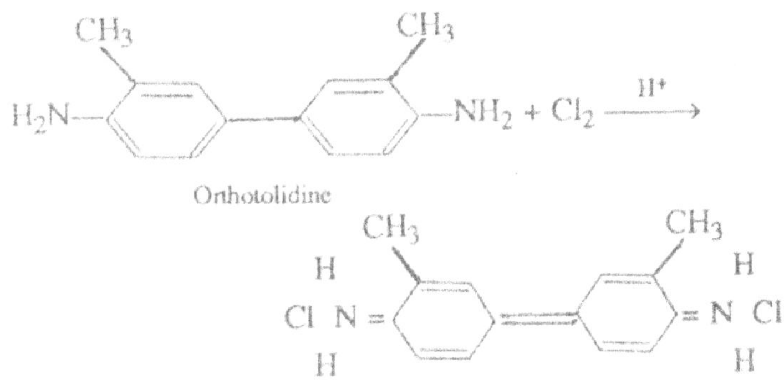

Reagents

1. Orthotolidine solution: Dissolve 1.35 g orthotolidine hydrochloride in 500 ml. distilled water. (do not use orthotolidine base). Add this solution with constant stirring to a solution containing 350 ml. distilled water and 150 ml. conc. HCl.

(**Caution:** Toxic-take care to avoid ingestion).

2. Sodium arsenite solution: Dissolve 5.0g. sodium arsenite $NaAsO_2$ in distilled water and dilute to 1000 ml.

(**Caution:** Toxic-take care to avoid ingestion).

Procedure

Free chlorine reacts instantaneously and the maximum color is developed at once, while the color development by combined chlorine takes a little time, and is temperature dependent. Usually a period of 3 to 5 minutes is allowed and a temperature of 20°C is preferred. The color development is preferably done in dark.

---

* Usually nitrite will not co-exist with free chlorine. However it may exist in the presence of chloramines.

Volume of sample and reagents: If 10 ml. sample is used, add 0.5 ml. OT solution, if 15 ml. is used, add 0.75 ml. and if the sample volume is 50 ml. use 2.5ml. OT solution. Use the same volume of arsenite solution as orthotolidine. Either comparator tubes or nessler tubes may be employed. Use standard color glass disc for comparison.

Estimation of Total available chlorine and interferences: To a comparator tube or nessler tube containing orthotolidine reagent add suitable volume of the sample and mix well immediately. Compare the color with the standards in exactly 5 minutes. The color produced is due to the total available chlorine and total amount of interfering ions. Record the result as T. (when interferences are known to be absent this reading represents the amount of total available chlorine).

Estimation of Free available chlorine and interferences: To a comparator tube or nessler tube containing orthotolidline solution, add appropriate amount of sample, mix quickly and add immediately sodium arsenite solution and once again mix well immediately. Compare the color as rapidly as possible. The color produced is due to the free available chlorine and interfering ions. Record the result as F.

Estimation of interferences: To a comparator tube or nessler tube containing sodium arsenite solution, add appropriate volume of sample. Mix well quickly and add orthotolidine solution immediately. Mix well once again and compare the color immediately. Record the results as I_1. compare the color once again with the standards in exactly 5 minutes and record the result as I_2. The colors produced immediately and in 5 minutes are due to the interfering substances only.

Calculation

Total available residual chlorine (as Cl_2) $= T - I_2$

Free available residual chlorine (as Cl_2) $= F - I_1$

Combined available residual chlorine (as Cl_2) $= (T - I_2) - (F - I_1)$

CHAPTER - 58

RESIDUAL SODIUM CARBONATE (RSC)

Residual Sodium Carbonate (RSC) is a term used when a water is tested for its suitability for irrigation purpose. It is the excess amount of carbonate and bicarbonate above the sum of calcium and magnesium content of water [all the values being expressed in milli equivalent (meq) per litre].

$$RSC = (CO_3^{2-} + HCO_3^{-}) - (Ca^{2+} + Mg^{2+})$$

If the total hardness (calcium + magnesium) exceeds the value of alkalinity (both of them are expressed as $CaCO_3$), there is no residual sodium carbonate. Only when the alkalinity exceeds total hardness, there will be residual sodium carbonate.

Waters containing less than 1.25 meq/l residual sodium carbonate are probably safer, those containing 1.25 - 2.5 meq/l are marginal; and those with more than 2.5 meq/l are not suitable for irrigation.

Problem 58–1

The results of analysis of a water sample are given below. Calculate the RSC value of that water.

Phenolphtalein alkalinity (as $CaCO_3$), mg/l	*= 0*
Methyl Orange alkalinity (as $CaCO_3$), mg/l	*= 148*
Total hardness (as $CaCO_3$)	*= 136*
Calcium (as $CaCO_3$), mg/l	*= 42*
Magnesium (as $CaCO_3$),mg/l.................	*= 94*

Since there is no phenolphthalein alkalinity, there is no carbonate. All the alkalinity is present as bicarbonate only.

Bicarbonate alkalinity is 148 mg/l. It is expressed in terms of $CaCO_3$. Therefore the first step is the conversion of calcium carbonate to bicarbonate. For converting calcium carbonate, multiply the alkalinity value by a factor of 1.22

$$148 \times 1.22 = 180.56 \text{ mg/l}$$

Now the mg/l unit has to be converted to meq/l, for which it has to be multiplied by 0.01639.

$$\text{Bicarbonate, meq/l} = 180.56 \times 0.01639 = 2.96$$

Calcium is expressed in terms of $CaCO_3$. Hence it has to be converted in terms of Calcium.

$$\therefore \text{mg/l Calcium (as Ca)} = \text{mg/l Calcium (as } CaCO_3) \times 0.4$$
$$= 42 \times 0.4 = 16.8$$

Magnesium is expressed in terms of $CaCO_3$. Hence it has to be converted in terms of magnesium.

$$\therefore \text{mg/l Magnesium (as Mg)} = \text{mg/l Magnesium (as } CaCO_3) \times 0.24$$
$$= 94 \times 0.24 = 22.56$$

In the third stage the mg/l unit of Calcium and Magnesium have to be converted to meq/l.

$$\text{Calcium, meq/l} = \text{mg/l, Calcium} \times 0.0499$$
$$= 16.8 \times 0.0499$$
$$= 0.84$$

$$\text{Magnesium, meq/l} = \text{mg/l Magnesium} \times 0.08226$$
$$= 22.56 \times 0.08226 = 1.86$$

$$\therefore \text{Residual Sodium Carbonate} = (0 + 2.96) - (0.84 + 1.86)$$
$$= (2.96 - 2.70) = \mathbf{0.26 \text{ meq/l}}$$

RSC Value of the water is 0.26 meq/l. It is below 1.25 meq/l and the water can be used for irrigation.

Problem 58–2

Following is the results of analysis of a sample of water. Find out the RSC value of the water.

Phenolphthalein alkalinity, (as $CaCO_3$),mg/l	*= 0*
Methyl orange alkalinity (as $CaCO_3$), mg/l	*= 364*
Calcium (as $CaCO_3$), mg/l	*= 68*
Magnesium(as $CaCO_3$),mg/l	*= 88*

Since there is no phenolphthalein alkalinity, there is no carbonate. All the alkalinity is present as bicarbonate only.

Bicarbonate alkalinity is 364 mg/l. It is expressed in terms of $CaCO_3$. Therefore, the first step is the conversion of calcium carbonate to bicarbonate. For converting calcium carbonate, multiply the alkalinity value by a factor of 1.22

$$364 \times 1.22 = 444 \text{ mg/l}$$

Now, the mg/l unit has to be converted to meq/l. for which it has to be multiplied by 0.01639

$$\text{Bicarbonate, meq/l} = 444 \times 0.01639 = 7.27$$

In the second stage, Calcium hardness (as $CaCO_3$) and Magnesium hardness (as $CaCO_3$)have to be converted to Calcium (as Ca) and Magnesium (as Mg) respectively.

$$\text{mg/l, Calcium (as Ca)} = 68 \times 0.4 = 27.2$$
$$\text{mg/l, Magnesium (as Mg)} = 88 \times 0.24 = 21.12$$

In the third stage, the mg/l unit of calcium and magnesium have to be converted to meq/l

$$\text{Calcium, meq/l} = \text{mg/l, calcium} \times 0.0499$$
$$= 27.2 \times 0.0499 = 1.36$$

Magnesium, meq/l $= \text{mg/l magnesium} \times 0.08226$
$= 21.12 \times 0.08226 = 1.74$

$\therefore$ Residual Sodium Carbonate $= (0 + 7.27) - (1.36 + 1.74)$
$= 7.27 - 3.10 =$ **4.17 meq/l**

Since the RSC value of this water exceeds 2.5 meq/l, it is unsuitable for irrigation.

CHAPTER - 59

SELENIUM

Selenium has a toxic effect on man and animals comparable with that of arsenic, giving rise to similar symptoms. Selenium also has been suspected of causing dental carries in man and of being carcinogenic agent. The selenium concentration of most drinking waters is less than 10μg/l. Water contaminated with wastes may contain higher concentrations of selenium.

DIAMINO BENZIDINE METHOD

Principle

This method of selenium determination is based on its reaction with diaminobenzidine to produce an yellow color. A stable yellow dipehnylpiazselenol is formed at pH 2 to 3 and is quantitatively extracted by toluene at pH 5 and higher.

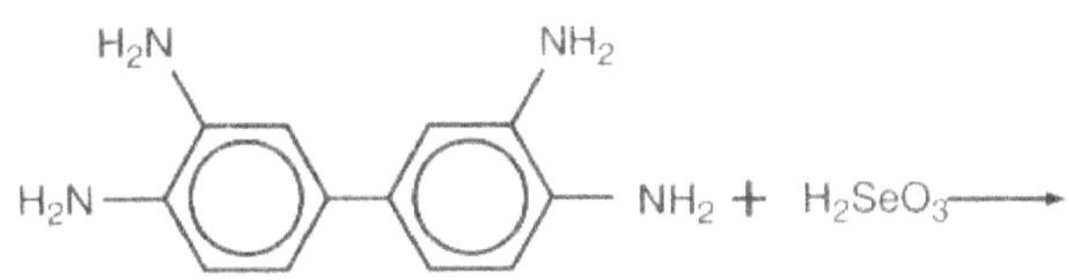

Diaminobenzidine

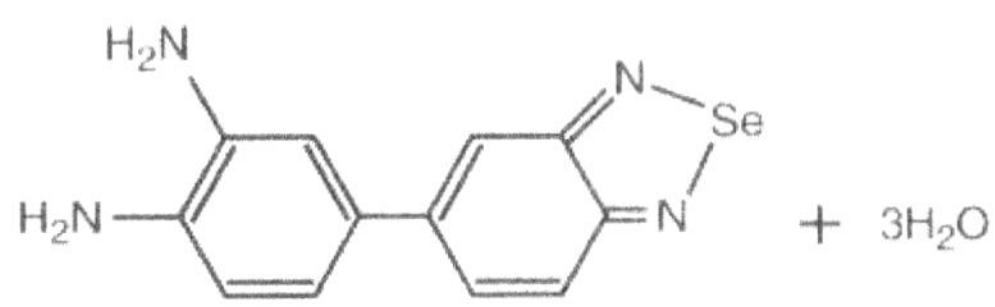

Diphenyl piazselenol

Interference

Oxidizing agents must be absent. Ferric ion also interfere but its interference is eliminated by the addition of oxalic acid.

Reagents

1. Stock Selenium solution: Dissolve 2.190g. sodium selenite, Na_2SeO_3 (or 1.9114 of Sodium hydrogen selenite) in water containing 10ml conc. HCl and dilute to 1000ml with water.

1ml = 1.00mg Se (IV)

2. Intermediate Selenium solution: Dilute 10.0ml stock selenium solution to 1000ml with distilled water.

1ml = 0.01mg Se

3. Working Standard Selenium solution: Dilute 10.0 ml intermediate selenium solution to 100ml with distilled water.

1ml = 0.001mg Se

4. Formic acid (H.COOH) 11.5%: Add 11.5ml formic acid to 88.5ml distilled water.

5. Sodium hydroxide 4% : Dissolve 4g. NaOH in 100ml distilled water.

6. Ethylene diamine tetra acetic acid, disodium salt, 8%: Dissolve 8g. Ethylene diamine tetra acetic acid disodium salt in 100ml. distilled water.

7. 3-3'–Diamino benzidine 0.5%: Dissolve 0.5g. 3-3'–Diamino benzidine tetrahydrochloride ($C_{12}H_{18}Cl_4N_4.2H_2O$) in 100ml. distilled water. If any insoluble matter remains in the solution filter through whatman No. 4 filter paper. Prepare the solution freshly.

Caution: Handle this reagent with extreme care

8. Ammonium Hydroxide solution 1 + 1: Mix 50ml. conc. Ammonia solution with 50ml. distilled water. Keep it in a tightly stoppered bottle.

9. Toluene, AR Grade

Procedure

As the selenium concentrations in water are generally lower, samples have to be concentrated prior to determination. Place a suitable volume of the sample (not less than 1000 ml.) in a silica dish and acidify with 1ml. conc. HCl and concentrate on a water bath. Makeup the volume to 50ml with distilled water. Use this concentrated sample for determination.

1. Place 0.0, 1.0, 2.0, 4.0, 8.0 ml of selenium working solution into a series of 100ml. Nessler tubes. Dilute each to 50ml. with distilled water.
2. Place 50ml. of the prepared sample in a 100ml. Nessler tube.
3. To the blank, standards and sample add 0.2ml. EDTA solution. Mix well. Allow 5 minutes for the reaction.
4. Add 2ml. formic acid and mix well.

5. Adjust the pH to 2 to 3 by the addition of sodium hydroxide solution. (for blank and standards about 1ml. 1N. NaOH is required)
6. Add 2 ml. of 3 – 3' Diaminobenzidine solution and allow it to stand for 50 minutes for color development.
7. Raise the pH to 6 to 7 by the addition 1 + 1 ammonium hydroxide (About 0.3 to 0.5 ml is required)
8. Transfer the contents of each Nessler tube to 250ml separating funnels.
9. Extract the color with 10 ml. toluene and read the absorbance at 420mμ against a reagent blank.

Find out the microgram equivalent of selenium from the absorbance value.

Express the result as mg Selenium (as Se) per litre of the sample.

CHAPTER - 60

SILICA

Silica is a common constituent of natural waters and treated waters. Silicates are normally added to water in a number of uses. They are primarily used as water conditioners and corrosion inhibitors. Silica is of special importance in steam generation as it forms troublesome scales of silica and complex silicates in the boiler, under certain conditions. Moreover some of it is carried over with the steam and deposits are formed on the blades of steam-turbines resulting in pluggage of small passages and imbalance of the rotor. Hence silica levels must be kept below 0.005 mg/l for very high pressure boilers. Not only in boilers but the characteristic glassy scales of silica lower the efficiency of heat transfer in heat exchangers also.

Silica can also form deposits in cooling water systems. However it is significant only when it is present in excess amounts or if the pH value and magnesium content are sufficiently high.

Measurement of silica is extremely useful for feed waters of high pressure boilers and steam condensates. Besides, determination of silica provides an excellent method of checking the efficiency of demineralizers and to detect the organic fouling of anion exchange resins. Silica is one of the impurities that comes out first from the exchanger unit when the exchange capacity of a demineralizer is exhausted. Thus the presence of silica in ion exchanger effuluent clearly indicates the exhaustion of resins and need for regeneration. Organic fouling of anion exchange resin is also detected by the slow increase of residual silica content of demineralized water.

Two methods* for the determination of silica are given below. The molybdosilicate method is rapid and used as a routine method for higher concentrations of silica. To determine very low concentration of silica as in high pressure boiler feed waters, molybdenum blue method is employed.

---

* Both the methods described here are applicable to raw, feed and boiler waters. However it should be indicated that colloidal silica is not detected by both the methods. Therefore blowdown ratio calculated from silica values may be erroneously high, as part of the colloidal silica is unreactive in the feed water, but transformed to reactive form under boiler conditions. That is why the concentration of silica in feed water is less while in boiler water it is more – which is supposed to be the correct one.

MOLYBDOSILICATE METHOD

This method is applicable for the determination of non colloidal silica in natural water, industrial waters such as boiler water and cooling water and industrial waste waters that are not too colored or turbid.

Principle

Ammonium molybdate reacts with silica at a pH 1.2 to form a yellow colored complex. The intensity of the yellow color is proportional to the concentration of silica in the sample.

$$\underset{}{SiO_4^{4-}} + \underset{\text{Ammonium molybdate}}{12(NH_4)_2MoO_4} + 24HCl \rightarrow \underset{\beta\text{-Silico-12-Molybdate Anion (Yellow)}}{SiMo_{12}O_{40}^{4-}} + 24NH_4Cl + 12H_2O$$

Interferences

In addition to the yellow color produced by silica, phosphate also reacts with ammonium molybdate and gives yellow color. However this yellow color can be eliminated by adding oxalic acid. The addition of oxalic acid simultaneously reduces the interference from tannin also. Large amounts of iron, color and turbidity also interfere, but may be compensated if photometric measurements are made. (See procedure for colored and turbid samples).

Glass containers and reagents may also contribute to silica. Hence, keep all the reagent solutions in polyethylene bottles and use only reagents with low silica content.

Reagents

Store all the reagents in polythene containers.

1. **Hydrochloric acid, 1 + 1:** To 100 ml distilled water in a beaker add 100ml conc.HCl and mix well. Place it in a polythene bottle.
2. **Ammonium molybdate solution, 10%:** Dissolve 10g. ammonium molybdate, $(NH_4)_6Mo_7O_{24}.4H_2O$ in about 50 ml distilled water by gentle warming and stirring. Dilute to 100 ml and filter if necessary. Adjust the pH to 7-8 with (silica free) ammonium or sodium hydroxide.

3. **Oxalic acid solution, 10%:** Dissolve 10 g oxalic acid, $H_2C_2O_4$ in distilled water and dilute to 100 ml.
4. **Artificial standard solutions**
   (a) **Potassium chromate solution:** Dissolve 0.630g. potassium chromate, K_2CrO_4 in distilled water and make up to 1 litre in a volumetric flask.
   (b) **Borax solution, 1%:** Dissolve 10g. sodium borate decahydrate, $Na_2B_4O_7.10H_2O$ in 1000 ml.distilled water.

Procedure

Clear Samples

Place 50.0 ml of the sample or suitable aliquots in 50 ml Nessler tube. Add 1.0 ml 1 + 1 HCl and 2.0 ml ammonium molybdate solution. Mix well and allow to stand for 5 to 10 minutes. Add 1.5 ml oxalic acid solution and mix thoroughly.

Compare the color with that of the standards prepared simultaneously after 2 minutes but before 15 minutes, measuring time from the addition of oxalic acid.

Preparation of Standards

Prepare the standards using artificial standard solutions according to the table given below.

Value in silica mg	Volume of Potassium chromate solution in ml	Volume of Borax solution in ml	Volume of water in ml
0.00	0.0	25	30
0.10	1.0	25	29
0.20	2.0	25	28
0.40	4.0	25	26
0.60	6.0	25	24
0.80	8.0	25	22
1.00	10.0	25	20

Colored and Turbid samples

Choose a suitable volume of the sample and place the same volume of sample in two 50 ml Nessler tubes and make up to the mark with distilled water. To one tube add 1 + 1 HCl, 2.0 ml ammonium molybdate solution and after 5 to 10 minutes 1.5 ml oxalic acid solution as described above. To the other tube, add 1.0 ml 1 + 1 HCl and after 5 to 10 minutes 1.5 ml oxalic acid solution. Determine the transmittance of these two solutions and deduct the transmittance of the sample to which no ammonium molybdate has been added. The value represents the color due to the silica only.

Note: For photometric measurements do not use the artifical color standards. Use only the standard solution prepared from pure silica. For preparation, see Molybdenum Blue Method. Observe the transmittance of standards and samples at 410 nm using a light path of 1 cm or longer.

'MOLYBDENUM BLUE' or 'HETEROPOLY BLUE' METHOD

This method is applicable to steam condensate, feed waters for high pressure boilers and nuclear reactors which contain only very low concentrations of silica.

Principle

Molybdosilicic acid is formed by reaction between silica and ammonium molybdate at pH about 1.2, which on reduction with a solution containing sodium sulfite, sodium bisulfate, and 1-amino-2 naphthol-4 sulfonic acid yields a blue complex "molybdenum blue" and is measured colorimetrically.

$$\underset{\beta\text{-Silico-12-Molybdate Anion (Yellow)}}{SiMo_{12}O_{40}^{4-}} + \underset{\text{from reducing agent}}{4e^-} \rightarrow \underset{\text{Silico-12-Molybdate Anion or Molybdenum blue}}{SiMo_{12}O_{40}^{8-}}$$

Reagents

Keep all the reagents in polythene containers.

1. **Hydrochloric acid, 1 + 1.**

2. **Ammonium molybdate solution:** See "Molybosilicate method".

3. **Oxalic acid solution 10%:** See "Molybdosilicate method"

4. **Reducing solution:** Dissolve 1.0g. sodium sulfite, Na_2SO_3 and 500 mg 1-amino-2-naphthol-4-sulfonic acid* in 50 ml water. Warm if necessary. Add this solution to a solution of 30g. sodium bisulfite $NaHSO_3$ in 150 ml distilled water. Filter into a polyethylene container. Discard the solution when it becomes dark. Keep the solution in a refrigerator.

5. **Stock silica solution:** Weigh 1.000g. pure silica in a platinum crucible and cover the silica with 5g. anhydrous sodium carbonate. Cover the crucible with the platinum lid and heat to red

* If the amino naphthol sulfonic acid is incompletely soluble, that material is not suitable for silica determination.

hot until the sodium carbonate begins to fuse. When the melt has subsided, increase the heating until the crucible is bright red and continue heating for further ten minutes. After the melt becomes clear, cool the crucible, add water to dissolve the melt and make up to 1000 ml in a volumetric flask.

$$1.0 \text{ ml} = 1.0 \text{ mg } SiO_2$$

6. Standard silica solution: Pipet 10.0 ml of the above stock solution in to a 1 litre volumetric flask and make up to the mark with distilled water.

$$1.0 \text{ ml} = 10\ \mu g\ SiO_2$$

Procedure

Pipet a suitable volume of the sample containing not more than 80 μg SiO_2 in to a 50 ml Nessler tube. Pipet appropriate volumes of standard silica solution covering the range from 10 to 80 μg SiO_2 in to 50 ml Nessler tubes. Include a Nessler tube as the blank. Make up the volume of the sample, standards and blank to 50 ml with distilled water.

1. To each tube add 1 ml 1 + 1 HCl, 2.0 ml ammonium molybdate solution and mix well. After 10 minutes add 1.5 ml oxalic acid solution and mix thoroughly. After five minutes from the addition of oxalic acid add 2.0 ml reducing solution and mix well. Allow to stand for five minutes.
2. After 5 minutes compare the color of the sample with that of the standards either visually or photometrically at 820 nm. (Prepare the calibration curve using a semilogarithmic graph paper).

Express the result as mg silica (as SiO_2) per litre of the sample.

CHAPTER - 61

SILT DENSITY INDEX (SDI) (OR) FOULING INDEX

Silt Density Index is a test used to measure the level of suspended solids in feed water for a reverse osmosis system. This test is also called "microporous membrane pluggage test" and "fouling index". SDI is used as a means of correlating the maximum level of suspended solids allowable in the feed water to fouling in reverse osmosis systems in order to maintain membrane performance. SDI values less than 4 is recommended for feeding RO systems without treatment (especially for spiral wound elements.) The method is based on determining the rate of pluggage when water is passed through a 0.45 micron membrane filter at a constant applied pressure (i.e.) 30 lb/in^2. Although SDI is not absolute, it is by far the best method by which a good correlation can be made between the SDI and the rate of membrane fouling.

Equipment Required

The test equipment consists of a ball valve, a pressure regulator and a pressure gauge followed by a membrane filter assembly with holder to hold a 47 mm diameter filter disc.

Procedure

1. Connect the water supply to be tested to the filter assembly and adjust the pressure regulator to provide a constant pressure of 30 lb/in^2 (2.1 bar).
2. Place a 0.45 micron filter disc on the back-up plate of filter holder. Seal with 'O' ring and replace the top half of the filter holder and bolt loosely.
3. Bleedout the trapped air by cracking the ball valve. Close the valve and tighten the bolts.
4. Open the valve and record the time taken for the collection of 500 ml water (t_i).
5. Continue the flow through the filter for an elapsed time of upto 15 minutes (T).
6. At the end of this period, record the time required to collect a second 500 ml sample (t_f).
7. Disconnect the apparatus, take the filter and place in a plastic bag to save it for future reference or for analysis of the deposit to determine the type of material that has caused fouling.

Calculation

$$Silt\ Density\ Index = \frac{(1-\frac{t_i}{t_f})}{T} \times 100$$

Important Note

For accurate measurements, t_f should be no more than four times the value of t_i. If higher than this, the elapsed time (T) between readings should be reduced. Changes in sample amounts and times may be considered for very high SDI values. If the time required to obtain a second 100 ml sample is greater than 60 seconds, pluggage will be about 90% and it is not necessary to continue the test.

Limitations

1) Pore size of the membrane used in this test is 0.45 micron while the pore size of a reverse osmosis membrane is less than 0.002 microns.
2) In this apparatus, all flow goes through the filter leaving suspended solids as a deposit over the filter. In RO, there is a parallel flow across the membrane which carry away suspended solids.
3) Fouling in a RO system is also caused by the concentration and precipitation of salts rather than by colloidal and suspended matter present in the feed water. This test does not take into account such effects of concentration.

CHAPTER - 62
SILVER

Silver is uncommon to natural waters. However catadyn method of disinfecting water contributes silver in minor quantities to water. Silver in drinking water is highly undesirable as it can cause argyria , a permanent blue grey discoloration of the skin and eyes that imparts a ghostly appearance.

The following colorimetric method using cupric diethyl dithio carbamate is more accurate and easy to carry out.

COLORIMETRIC METHOD

Principle

The method is based upon the decolorization of cupric diethyl carbamate (in carbon tetra chloride) by silver ion. Silver displaces copper from cupric diethyl dithiocarbamate to form a colorless silver compound.

Interferences

Heavy metal ions usually interfere. However their action is suppressed by adjusting the pH to between 4 and 5 with sodium citrate before extraction with carbon tetrachloride.

Reagents

1. Silver stock solution: Dissolve 0.1575g. silver nitrate, $AgNO_3$ in distilled water and make up to 1000 ml in a volumetric flask.

1.0 ml = 0.1 mg Ag.

Store it in an amber bottle.

2. Silver working solution: Pipet 10.0 ml of silver stock solution in to a 1000 ml volumetric flask and make up to the mark with distilled water.

1.0 ml = 0.001 mg Ag.

Prepare freshly.

3. Cupric diethyl dithiocarbamate reagent

(i) **Copper standard solution:** Dissolve 0.393g. copper sulfate, $CuSO_4.5H_2O$ in distilled water and make up to 1000 ml with distilled water.

1.0 ml = 0.1 mg Cu

(ii) **Sodium diethyl dithiocarbamate solution:** Dissolve 0.2g sodium diethyl - dithiocarbamate in 100 ml water and filter.

(iii) **Cupric diethyl dithiocarbamate reagent:** Into a 100 ml separating funnel, pipet 5.0 ml copper standard solution. Add 25 ml water and 1.0 ml sodium diethyl dithiocarbamate, solution and mix well. Shake with 10 ml carbon tetra chloride for about one minute. Allow the layers to separate and run the organic layer through a filter paper in to a dry 500 ml measuring cylinder. Rinse the funnel with two 5 ml portions of carbon tetra chloride and add the rinsings to the extract in the measuring cylinder. Dilute to 300 ml with carbon tetra chloride .Keep it in dark in an amber bottle. This solution will keep for atleast two weeks.

Procedure

1. Place 1.0 ml 2.0 ml,... 6.0 ml of silver standard working solution in to a series of 100 ml separating funnels. Dilute to 25 ml with distilled water.
2. Include a separating funnel containing 25 ml distilled water as the blank.
3. Place 250 ml of the sample in a silica dish and evaporate to 25 ml. Quantitatively transfer to a 100 ml separating funnel.
4. To the blank, standards and sample, add exactly 10 ml cupric diethyl- dithiocarbamate reagent from a burette*. Add 5 ml sodium citrate solution and shake vigorously for two minutes.
5. Allow the layers to separate and run the lower layer in to the cuvette of a spectrophotometer or filter photometer.
6. Measure the optical densities of blank, standards and sample in a spectrometer at 435 nm or in a filter photometer with a suitable violet filter, using carbon - tetrachloride in the reference cell.
7. Plot a calibration curve and find out the number of micrograms of silver equivalent to the optical density.

Express the result as mg Silver (as Ag) per litre of the sample.

* Do not keep the cupric diethyl dithiocarbamate reagent for more than 10 minutes as the color will fade due to exposure to light.

CHAPTER - 63
SODIUM

Sodium is present in most natural waters from negligible to appreciable concentrations. Sodium is an important consideration with regard to irrigation waters only. Regarding drinking and industrial purposes it is of minor importance only. However, it has to be mentioned that excessive amounts of sodium in drinking water is harmful to persons suffering from cardiac, renal, and circulatory diseases. Sodium when present as sodium chloride in concentrations more than 500 mg/l makes the water unpalatable and causes appetite disturbances. According to Larson and King[*], sodium chloride and other sodium salts in water are strong contributors to the rate of corrosion of iron and steel. It is also reported that, if the concentration of sodium plus potassium in boiler waters exceeds 50 mg/l, foaming troubles may occur.

Determination of sodium is sometimes used to indicate the purity of water; for example, in steam condensates, the concentration of sodium indicates whether there is any carry over of water from boiler system in to the steam. Likewise, the completeness of cation exchange can be checked by determining sodium concentration in demineralizer effluents.

Sodium can be determined using either a flame photometer or ion selective electrode. The flame photometer method is presented here.

FLAME PHOTOMETRIC METHOD

The flame photometric method is by far the simplest and most rapid method for the determination of sodium. Even very small amounts of sodium in large proportion of water can be determined by this method.

Reagents

Store all reagent solutions in polythene containers only.

[*] Larson, T.E and King R.M, "CORROSION BY WATER AT LOW FLOW VELOCITY", JAWWA, 46 (1954).

1. Deionized distilled water: Prepare by passing distilled water through a mixed bed of ion-exchange resins. Use only deionized distilled water for the preparation of solutions and dilutions.

2. Sodium stock solution: Dissolve 2.5422g. sodium chloride, NaCl (Previously dried at 140°C for 1 hour) in little deionized distilled water and make up to 1000 ml in a volumetric flask.

1.0 ml = 1.0 mg Na

Make dilutions as required for use.

Procedure

Follow the manufacturers' instructions for operating the instrument. The following points will however be useful:

1. The reproducibility and accuracy of the instrument should be checked constantly. For this, a standard may be included for test every time the instrument is operated.
2. Sample should be free from suspended impurities. If the sample contains any suspended matter, filter it.

Express the result as mg sodium (as Na) per litre of sample.

ION-SELECTIVE ELECTRODE METHOD

Ion selective electrodes capable of measuring sodium ion concentration in water are available. Suppliers of the electrodes provide necessary instructions for the determination. They have to be followed scrupulously.

PERCENT SODIUM

In agricultural practice, sodium concentration is expressed in terms of 'Percent Sodium'. It can be defined as the percentage of the sodium concentration of a water to the total cationic concentration. It is calculated as follows:

$$\text{Percent sodium (as Na)} = \frac{100\,A}{(A + B + C + D)}$$

where A = Sodium content (as Na) in meq/l*,
B = Calcium content (as Ca) in meq/l,
C = Magnesium content (as Mg) in meq/l,
D = Potassium content (as K) in meq/1.

Conversion of mg/l to meq/l :

Calcium 1 mg/l	= 0.04990 meq/l
Magnesium 1 mg/l	= 0.08226 meq/l
Sodium 1 mg/l	= 0.04349 meq/l
Potassium 1 mg/l	= 0.02557 meq/l

CALCULATION OF PERCENT SODIUM-EXAMPLES

Problem 63-1

On analysis, a water is found to contain sodium (as Na) 69 mg/l, calcium (as Ca) 64 mg/l, magnesium (as Mg) 41.3 mg/l and potassium (as K) 4 mg/l. Find out the percent sodium content of the water.

$$\text{Percent sodium (as Na)} = \frac{100\,A}{(A+B+C+D)}$$

The values are given in mg/l. Convert them to meq/l.

$$\begin{aligned} A &= 69 \times 0.04349 &&= 2.9573 \\ B &= 64 \times 0.04990 &&= 3.1936 \\ C &= 41.3 \times 0.08226 &&= 3.3973 \\ D &= 4 \times 0.02557 &&= \underline{0.1023} \\ & && \underline{9.6505} \end{aligned}$$

$$\therefore \text{Percent sodium (as Na)} = \frac{2.9573}{9.6505} \times 100 = \mathbf{30.64}$$

Rounded to **30.6%**

* $\text{milliequivalent / litre} = \dfrac{\text{mg/l}}{\text{Equivalent wt.}}$

Problem 63–2

The calcium hardness and magnesium hardness of a water sample are 108 mg/l (as CaCO3) and 124 mg/l (as CaCO3) respectively. If the sodium and potassium contents are 122 mg/l (as Na) and 2 mg/l (as K) respectively, find out the percent sodium (as Na)

Since the value of calcium hardness is expressed in terms of $CaCO_3$ convert it as Ca by multiplying with 0.40 and likewise the value of magnesium hardness from $CaCO_3$ to Mg by multiplying with 0.24.

Then convert all the mg/l value to meq/l value.

$$A = 122 \times 0.04349 = 5.3057$$

$$B = 108 \times 0.4 \times 0.04990 = 2.1557$$

$$C = 124 \times 0.24 \times 0.08226 = 2.4480$$

$$D = 2 \times 0.02557 = \underline{0.0511}$$

$$\underline{9.9605}$$

$$\textit{Percent sodium (as Na)} = \frac{100\,A}{(A + B + C + D)}$$

$$\therefore \textit{Percent sodium (as Na)} = \frac{5.3057}{9.9605} \times 100 = 53.26$$

Rounded to **53.3%**

Calcium and magnesium in the proper proportions maintain soil in good condition of tilth and structure, while the opposite is true when sodium predominates. In most normal soils (in arid and semi arid regions), calcium and magnesium are the principal cations held by the soil in replaceable or exchangeable forms. When water containing high amounts of sodium is used for irrigation it replaces calcium and magnesium from the soil, resulting in the break down of granular structure of soil. This apart, various changes takes place resulting in the scaling of soil pores and a decrease in soil permeability. With further increase in the sodium percentage, the soil continues to deteriorate and its pH increases to the level of alkali soils.

It has been widely recommended that the percent sodium in irrigation water should not exceed 50 - 60, in order to avoid the deleterious effects on soil as mentioned above.

SODIUM ADSORPTION RATIO

It is a parameter calculated from the sodium, calcium and magnesium ion concentrations of the water intended for irrigation to assess the sodium hazard.

$$Sodium\ Adsorption\ Ratio\ (SAR) = \frac{Na}{\sqrt{\frac{1}{2}\ (Ca+Mg)}}$$

where the ionic concentrations are expressed in meq/l.

Problem 63–3

On analysis, a water is found to contain calcium (as Ca) 123mg/l, magnesium (as Mg) 65mg/l and sodium (as Na) 81mg/l. Calculate the SAR of the water.

All the values are expressed in mg/l.
First they have to be converted to corresponding meq/l by dividing them by their equivalent weights.

$$meq/l\ Ca = \frac{mg/l}{Eq.\ wt.\ of\ Ca} = \frac{123}{20} = 6.15$$

$$meq/l\ Mg = \frac{mg/l}{Eq.\ wt.\ of\ Mg} = \frac{65}{12} = 5.42$$

$$meq/l\ Na = \frac{mg/l}{Eq.\ wt.\ of\ Na} = \frac{81}{23} = 3.52$$

$$SAR = \frac{Na}{\sqrt{\frac{1}{2}\ (Ca+Mg)}}$$

$$= \frac{3.52}{\sqrt{\frac{1}{2}\ (6.15+5.42)}} = \frac{3.52}{\sqrt{\frac{1}{2}\ (11.57)}} = \frac{3.52}{\sqrt{5.785}}$$

$$= \frac{3.52}{2.405}$$

= **1.46**

Problem 63-4

The analysis of a water shows Calcium (as $CaCO_3$) 68 mg/l, magnesium as ($CaCO_3$) 88 mg/l and sodium (as Na) 198 mg/l. Find out the SAR of the water.

Calcium is expressed as $CaCO_3$. Hence it has to be converted to calcium.

$$Calcium\ (as\ CaCO_3) \times 0.4 = Ca$$

$$68 \times 0.4 = 27.2$$

Similarly magnesium is also expressed as $CaCO_3$. Hence it has to be converted to magnesium.

$$Magnesium\ (as\ CaCO_3) \times 0.24 = Mg$$

$$88 \times 0.24 = 21.12$$

All the ionic concentrations are expressed in terms of mg/l. For finding out SAR they have to be converted to corresponding meq/l by dividing them by their equivalent weights.

$$meq/l\ Ca = \frac{mg/l}{Eq.\ wt.\ of\ Ca} = \frac{27.2}{20} = 1.36$$

$$meq/l\ Mg = \frac{mg/l}{Eq.\ wt.\ of\ Mg} = \frac{21.12}{12} = 1.76$$

$$meq/l\ Na = \frac{mg/l}{Eq.\ wt.\ of\ Na} = \frac{198}{23} = 8.60$$

$$SAR = \frac{Na}{\sqrt{\frac{1}{2}(Ca+Mg)}}$$

$$= \frac{8.6}{\sqrt{\frac{1}{2}(1.36+1.76)}} = \frac{8.6}{\sqrt{\frac{1}{2}(3.12)}} = \frac{8.6}{\sqrt{1.56}}$$

$$= \frac{8.6}{1.249}$$

= **6.88**

A high SAR value affects the plants. Besides, it affects clay minerals with ion exchange capacities (causing the trace metals essential for crops to be regenerated off the clay particle). The permeability of clay soils is also reported to be reduced. Another report indicates that waters with a SAR value greater than 10 will present an appreciable sodium hazard in fine-textured soil having high cation-exchange capacity, especially as the salinity increases. However, there is no definite limit for SAR. It depends on the crops cultivated. For example, the SAR limits for certain fruits may be four whereas grain and alfalfa may tolerate an SAR as high as 18.

CHAPTER - 64

SODIUM SLIP DETERMINATION

Sodium slip determination is useful in checking the functioning of the strong acid cation exchanger unit. It is highly useful to detect the exhaustion of bed as well as improper bed and/or operating conditions.

A water when passed through a strong acid cation exchanger unit will exchange all its cations to hydrogen. Hydrogen combines with the anions of the water resulting in the formation of following acids: carbonic acid, sulfuric acid, hydrochloric acid and nitric acid. Carbonic acid decomposes immediately to carbondioxide and water. The remaining acidity is termed as 'Free Mineral Acidity (FMA)'. Normally this should be equal to (or just less than) the theoretical amount (i.e.) 'Equivalent Mineral Acidity' (EMA) of the water. If the difference is significant, it may be surmised that cation leakage (or slip) is occurring. The cation normally leaks out is sodium. The sodium leakage normally permitted is 1 – 2 ppm; any increase over this amount will tend to affect the quality of demineralized water. Sodlium slip can be easily calculated as shown below.

1) Calculate the Equivalent Mineral Acidity (EMA) of feed water as per the procedure given in Chapter – 29.

2) Determine the Free Mineral Acidity (FMA) of the decationized water as per the procedure given in Chapter – 10, under the section, "Free Mineral Acidity".

3) Sodium slip can then be calculated by using the following formula.

Sodium Slip = EMA – FMA

(All values are expressed in terms of calcium carbonate)

Problem 64–1

The analysis of a feed water shows the following results.

Sulfate (as $CaCO_3$) mg/l	=	*124*
Chloride (as $CaCO_3$) mg/l	=	*288*

Nitrate (as $CaCO_3$) mg/l = 28

The effluent from Strong Acid Cation unit shows a Free Mineral Acidity of 420 mg/l (as $CaCO_3$) when titrated with a standard alkali. Find out whether there is any sodium leak.

Equivalent Mineral Acidity(EMA)	=	*(Sum of the anions that form strong acids with hydrogen ions expressed in terms of $CaCO_3$)*
	=	*(124 + 288 + 28)*
	=	*440 mg/l (as $CaCO_3$)*
Free Mineral Acidity (FMA)	=	*420 mg/l (as $CaCO_3$)*
Sodium Leak	=	*EMA – FMA*
	=	*440 – 420*
	=	***20 mg/l (as $CaCO_3$)***

Since the value of FMA is significantly lower than EMA, there is certainly a sodium leak, with a quantity equivalent to 20 mg/l as $CaCO_3$.

CHAPTER - 65

SOLIDS

'Solids' is a term applied to all matter except the water contained in liquid material and thus the definition of solids refers to the matter that remains as residue upon evaporation and drying at a definite temperature.

'Total solids' is the residue that includes both dissolved solids and suspended solids.

The amount and nature of dissolved and undissolved matter occurring in liquid material vary considerably. Potable waters contain mostly inorganic mineral matter in dissolved condition and small amounts or none of organic matter, whereas waste waters such as sewage and industrial effluents contain considerable amount of undissolved matter. Determination of dissolved and undissolved matter is made with filtered and unfiltered portions of samples. The undissolved matter is usually referred to as "Suspended solids".

The common practice of determination of solids is to evaporate a suitable volume of the sample in a tared dish and drying at 105°C followed by weighing. The increase in weight represents the amount of solids.

For evaporation, dishes of 50 to 250 ml capacity, are employed. Dishes made up of various materials are available. The Table 65-1 will be useful in selecting a dish for particular use.

Drying Temperature

Regarding potable waters, dissolved solids only is of important concern. Hence it is enough to dry the residue at 103°–105°C. However water of crystallization of inorganic salts and mechanically occluded water are not completely removed at this temperature. The advantage of drying at this temperature is that loss of organic matter by volatilization is much less. The recommended temperature for potable waters is 179°–181°C at which almost all the mechanically occluded water and most of the water of crystallization are removed. However, organic matter is reduced by volatilization though not completely destroyed. Bicarbonates are converted to carbonates which in turn are also decomposed partially to oxides or basic salts.

Some chloride and nitrate salts also may be lost. Inspite of all these limitations, drying water samples at 179°–181°C yield values for dissolved solids which conform more closely to that obtained by summation of individually determined mineral salts. It is left to the analyst to choose the drying temperature best suited to the type of water under examination. However it should be noted that if the sample of water contains considerable amount of organic matter or its pH is above 9 the drying temperature should be 179-181°C. When reporting the results indicate the drying temperature also.

TABLE 65–1

VARIOUS TYPES OF DISHES USED FOR SOLIDS DETERMINATION

S. No.	Sample	Dish
1.	Water	Platinum (In the absence of Platinum, Stainless steel may be used).
2.	Sewage	Porcelain, Silica
3.	Industrial Effluents	Porcelain

Regarding sewage and other waste waters, generally the temperature 103°-105°C is preferred. Volatile solids are determined separately using a temperature of 550°C.

TOTAL SOLIDS

Total solids

For water samples, use platinum dish or stainless steel dish. For sewage and industrial effluent samples, use porcelain dish.

Clean the evaporating dish and subject to a preliminary drying in an oven at the same temperature intended for the residue. Cool in a desiccator and weigh. Place it on a water bath and measure into it 100 ml or suitable aliquots of the well mixed sample. Evaporate to dryness and further dry at 103°-105°C or 179°-181°C for one to two hours in an oven. Then cool it in a desiccator and weigh.

Calculation

$$mg / l \text{ Total solids} = \frac{mg. \text{ Total solids} \times 1000}{ml. \text{ sample for determination}}$$

Express the result as "Total Solids on drying at - °C" in terms of mg/l and to the nearest whole number. For results exceeding 1000 mg/l, report only three significant figures.

Total fixed solids

(Usually this determination is not made on water samples).

Keep the dish containing the residue (obtained from total solids test) in a muffle furnace at 550°C for 1 hr. Any organic matter and volatile substances present will be volatilized at this temperature and the remaining is the total fixed solids.

Express the result as mg total fixed solids per litre of sample.

Total volatile solids

$$mg/l \text{ Total volatile solids} = mg/l \text{ Total solids} - mg/l \text{ Total fixed solids.}$$

DISSOLVED SOLIDS

The determination of dissolved solids is useful in checking the correctness of analysis by comparing the result obtained with the sum of individual constituents.

In natural waters, the dissolved solids consist mainly of bicarbonates, carbonates, sulfates, chlorides, nitrates and possibly phosphates of calcium, magnesium, sodium and potassium with traces of iron, manganese and other substances. The amount of dissolved solids present in water is a consideration for its suitability for domestic use. In general, waters with a total solids content less than 500 mg/l are the most desirable.

Dissolved solids in industrial waters are undesirable for many reasons. They form scales, cause foaming in boilers, accelerate corrosion and interfere with the color and taste of the many finished products.

Total dissolved solids

Centrifuge or filter a suitable volume of sample through a Whatman No.42 or equivalent filter paper or through a glass fibre filter paper. Evaporate the filtered sample in a tared dish. Use platinum dish or stainless steel dish for water samples and porcelain dish for sewage and industrial effluents. Dry the residue at 103 – 105°C or at 179 – 181°C. The increase in weight of the dish equals the Total Dissolved Solids. It may also be obtained by the difference between the total solids and total suspended solids.

Express the result as "Total dissolved solids on drying at – °C" in terms of mg/l.

Fixed dissolved solids

Ignite the residue obtained from "Total Dissolved Solids" by keeping it in a muffle furnace at 550°C for 1 hr. Cool and weigh.

Express the result as mg "Fixed Dissolved Solids" per litre of sample.

Volatile dissolved solids

$$\text{mg/l Volatile dissolved solids} = \text{mg/l Total Dissolved Solids} - \text{mg/l Fixed Dissolved Solids}$$

SUSPENDED SOLIDS

The undissolved matter present in a water/waste water is usually referred to as suspended solids. No specific limit has been described for suspended solids in waters used for domestic and industrial purposes, but it is indirectly controlled by the limits on turbidity (5 units). The suspended solids determination is particularly useful in the analysis of sewage and other waste waters and is as significant as BOD determination. It is used to evaluate the strength of domestic waste waters and to determine the efficiency of treatment units.

The suspended matter is determined by filtering or centrifuging the sample, drying the residue and determining its weight by difference.

Total suspended solids

(i) Filtration method: Use a Gooch crucible containing asbestos mat for filtration. Prepare asbestos mat in the Gooch crucible as follows:

To 10g. of good quality asbestos in a 250 ml beaker add 100 ml distilled water and stir well. Decant the supernatant water containing the impurities of asbestos. Repeat this process atleast 10 times and then mix the asbestos with 1000 ml distilled water. Pour 25 ml of the well mixed asbestos solution into a Gooch crucible of 30 ml capacity with small perforations. Apply suction and dry the crucible. Place the crucible in an oven at 105°C. If 'Fixed Suspended Solids' and 'Volatile suspended solids are to be determined, place the crucible in a muffle furnace at 550°C. Cool and weigh.

Note: The thickness of the asbestos mat should be about 2-3 mm and should not exceed 3mm.

Filter a suitable volume of the sample through the crucible by applying suction. The filtration should be carried out at a lower vacuum than that used for be preparation of the mat. Wash the filter with distilled water to remove the soluble salts. Care should be exercised to prevent loss of fine asbestos particles during filtration and washing. Dry the crucible in an oven at 103-105°C. the increase in weight equals the total suspended solids.

Express the result as mg "Total Suspended Solids" per litre of sample.

Note: Instead of using Gooch crucible, glass micro fibre filter papers can be used, but they are not suitable for determining volatile suspended solids. Alternatively, the suspended solids may be calculated as the difference between total solids and dissolved solids.

(ii) Centrifugation method: A known volume of the well shaken sample (usually about 50 – 200 ml) is centrifuged using special conical-end centrifuge tubes. After centrifugation, the separated liquid is poured off and the residue in the tubes is washed with distilled water, re-centrifuged, the separated water is again removed and the suspended matter in the centrifuge tube is transferred by means of a water jet to a tared platinum or silica or porcelain dish and dried at 105°C. It should be remembered that centrifugation method gives slightly lower results than that obtained by filtration method.

Note: The liquid obtained after separation from the suspended matter by filtration or centrifugation may be used for the determination of dissolved solids.

Fixed suspended solids

Keep the dish containing the total suspended solids in a muffle furnace at 550°C for 1 hr. Cool and weigh.

Express the result as mg Fixed Suspended solids per litre of sample.

Volatile suspended solids

mg/l Volatile Suspended Solids = mg/l Ttoal Suspended Solids – mg/l Fixed Suspended solids

SETTLEABLE SOLIDS

The term settleable solids is applied to solid in suspension that will settle under quiescent conditions, due to the influence of gravity. The coarser suspended solids with a specific gravity greater than that of water, only will settle.

Determination of settleable solids is usually carried out on waste waters only and is useful in determining the design of primary settling tanks in plants employing biological treatment process. It is also used to determine the efficiency of sedimentation tanks.

For the determination of settleable solids an Imhoff cone is necessary. It consists of an inverted glass cone of 1 litre capacity having the apex part graduated in millitres (see Figure 65.1).

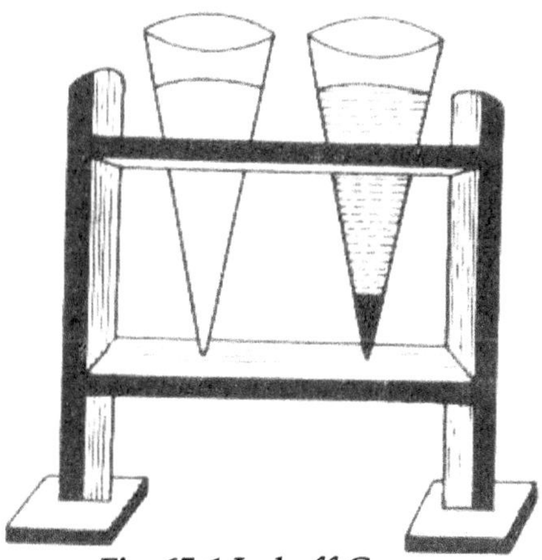

Fig. 65-1 Imhoff Cone

Fill in an Imhoff cone to the 1-litre mark with the thoroughly mixed sample and allow to settle. After 45 minutes, gently dislodge the material adhering to the sides of the cone using a glass rod or by spinning the cone. Record the volume of the sludge that has settled to the apex after one hour or other specified time related to the works condition.

Express the result as ml settleable solids per litre of sample per unit time.

CHAPTER - 66
SULFATE

Sulfate occurs naturally in water as a result of leachings from gypsum and other common minerals. Sulfate ions are relatively abundant in natural waters ranging from a few to several thousand mg/l. The sulfate content of municipal and industrial water supplies is usually increased during clarification, as alum (aluminium sulfate) is added as a coagulant.

High amounts of sulfate imparts a bitter taste to the water. Sulfate as magnesium sulfate causes laxative effects to children particularly in hot weather or climates. Sulfate in industrial waters causes potential scaling problems. The State Water Resources Control Board of California has set the following threshold limit for sulfate in industrial water supplies.

Industrial Process	SO_4 in mg/l	$CaSO_4$ in mg/l	$MgSO_4$ in mg/l	Na_2SO_4 in mg/l
Brewing	......	100-500	100-200	100
Carbonated beverages	250	......	......	......
Ice making	......	300	130-300	300
Daily	60	......	......	......
Sugar making	20	......	......	......
Textiles	100	......	......	......

The determination of sulfates is useful in the estimation of dissolved solids. It is useful for regulation of allowable sulfate concentrations in boiler water. The determination of sulfate is especially important with regard to cooling waters to control or to increase the addition of sulfuric acid which is added to reduce alkalinity and control pH value. Similarly it is important in oil field applications where two or more waters from different sources are mixed.

Two methods for the determination of sulfate are given here. The gravimetric method is applicable to waters containing normal to high concentration of sulfates. Turbimetric method is applicable to low concentration of sulfates.

GRAVIMETRIC METHOD

This method is precise one and applicable to all types of water and waste water.

Principle

Sulfate is precipitated as barium sulfate in the presence of hydrochloric acid. The precipitated barium sulfate is filtered, dried, ignited and weighed as $BaSO_4$.

$$BaCl_2 + SO_4^{2-} \longrightarrow BaSO_4\downarrow + 2Cl^-$$

Interferences

The gravimetric estimation of sulfate suffers from many interferences. Suspended matter, silica, nitrate, sulfite and barium chloride precipitant, lead to high results. Alkali metal sulfates yield low results. Chromium and iron also cause low results. Strict adherence to the procedure minimizes the interferences.

Reagents

1. Methyl red indicator: Dissolve 100mg. methyl red sodium salt in 100 ml distilled water.

2. Hydrochloric acid, 1+ 1.

3. Barium chloride, 5%: Dissolve 25g. barium chloride dihydrate $BaCl_2.2H_2O$ in 500 ml distilled water. (If any turbidity appears on standing, reject the solution).

4. Silver nitrate-nitric acid reagent (to test chloride): Dissolve 8.5g. silver nitrate and 0.5 ml conc. HNO_3 in 500 ml distilled water.

Procedure

1. Measure in to a 400 ml beaker appropriate volume of clear sample* containing 8 to 60 mg sulfate (usually 200 ml). Dilute the sample to 200 ml with distilled water. (If the sulfate content of the sample is very low, take higher volumes of the sample and evaporate on a water bath to 200 ml. Turbidimetric method may be employed for these samples).
2. Adjust the pH of the sample to 4.5–5.0 with HCl using methyl red indicator (to orange color). Then add additional 2 ml HCl. The samples are acidified to eliminate the precipitation of $BaCO_3$ which might occur in highly alkaline waters maintained near the boiling temperature. (If the sample is already digested with conc. HCl to remove suspended impurities, omit this step).
3. Boil the solution for about one minute and add 10 ml hot barium-chloride solution slowly from a pipet with constant stirring.

* If the sample is turbid or containing suspended impurities, digest with 2ml conc. HCl and filter through Whatman No.42 filter paper, or alternatively the residue obtained on evaporation for total solids determination may be leached with 1+1 HCl and made up to 200ml after filtration and then precipitated.

4. Keep the beaker on a water bath to digest the precipitate at 80 – 90°C for atleast 2 hrs, preferably over night.
5. Filter the contents of the beaker through an ashless filter paper (Whatman No.42 or equivalent). Wash the precipitate with hot water until the washings give no opalescence with silver nitrate-nitric acid reagent.
6. Place the filter paper with the precipitate in a previously weighed platinum or silica crucible (see note) and heat in a low flame with its lid loosely covered. Gradually increase the heat until the paper chars. Do not allow the filter paper to flame as it may cause losses. After the charring is complete ignite at 800°C for 1 hr (with free access of air). Cool in a desiccator and weigh.

Note: If silica crucible is used pre-ignite it at 800°C for 1 hour. Cool in desiccator and weigh.

Calculation

$$mg/l\ Sulfate\ (as\ SO_4^{2-}) = \frac{mg\ BaSO_4}{ml.\ sample\ taken\ for\ determination} \times 411.5$$

Samples having high silica content: If the silica concentration is above 25 mg/l, evaporate the aliquot of the sample nearly to dryness in a platinum dish on a water bath. Add 1 ml 1 + 1 HCl and moisten the residue completely. Evaporate the dish once again to dryness. Complete the drying by keeping the dish in an oven at 180°C. (If any organic matter is present, char over the flame). Add 2 ml distilled water and 1 ml 1 + 1 HCl and evaporate to dryness on a water bath. Add 2 ml 1 + 1 HCl and dissolve the residue using hot distilled water. Filter and wash with hot distilled water and proceed from step No.3 of the above procedure.

TURBIDIMETRIC METHOD

Though the gravimetric determination of sulfate is more precise, it is a time consuming one. The turbidimetric method though less accurate is more rapid and is especially applicable to waters with sulfate concentrations up to 60 mg/l.

Principle

Sulfate ions are precipitated as barium sulfate crystals of uniform size in acid medium. Light absorbed by the precipitate is measured using a spectrophotometer or an absorptiometer.

Interferences

Color and suspended matter interfere. If both are small in comparison with sulfate concentration, correction can be applied. With waters containing large quantities of organic matter, satisfactory precipitation of barium sulfate may not be possible.

Reagents

1. Conditioning reagent: Dissolve 75g. sodium chloride in 300 ml distilled water and add 30 ml conc. HCl and 100 ml 95% ethyl alcohol (or isopropyl alcohol). Add 50 ml glycerol and mix well.

2. Barium chloride: Crystals, 20 to 30 mesh.

3. Standard sulfate solution: Dissolve 147.9mg. anhydrous sodium sulfate in distilled water and make up to 1000 ml.

$$1.0 \text{ ml} = 100\ \mu g\ SO_4^{2-}$$

Procedure

Standards

1. Measure in to a series of 250 ml conical flasks 5.0, 10.0, 15.0, 20.0, 25.0, 30.0, 35.0, 40.0 ml of standard sulfate solution and dilute to 100 ml.
2. Add 5.0 ml conditioning reagent and mix well using magnetic stirrer. The speed of stirring should be the same for both standards and samples.
3. While stirring, add about 0.5g. barium chloride crystals and continue to stir exactly for one minute.
4. Immediately after one minute measure the optical density using a spectrophotometer at a wavelength of 420 nm or an absorptiometer with a violet filter having a maximum transmittance near 420 nm using 40 to 50 mm cells with water in the reference cell. Take readings at intervals of 30 seconds for a period of 4 minutes and take the maximum reading for calculating sulfate concentration. (The maximum reading will be normally obtained after a period of 2 minutes from the completion of a stirring).
5. Carry out a blank determination on the reagents used.
6. Prepare a calibration graph relating net optical density to μg of sulfate (as SO_4^{2-}) using a semilogarithmic graph paper.

Samples

1. Measure in to a 250 ml conical flask a suitable quantity of the sample containing not more than 4 mg sulfate and dilute to 100 ml and proceed from step No. to 4 as for standards.

2. From the calibration graph, read the mg of sulfate equivalent to the optical density.

Note: If the sample is colored or turbid, conduct a sample blank using the same amount of sample and conditioning reagent, without the addition of barium chloride. Measure the optical density and deduct the value from optical density of the barium chloride added sample and find out the mg equivalent sulfate for the deducted value from the calibration graph. If the absorption due to color or colloidal matter of the sample is large in comparison with that due to the precipitated barium sulfate, the method is unsuitable.

Calculation

$$mg/l\ Sulfate\ (as\ SO_4^{2-}) = \frac{mg.\ SO_4 \times 1000}{ml.\ sample\ taken\ for\ determination}$$

CHAPTER - 67

SULFIDE

The presence of sulfide in natural water is uncommon. Contamination by industrial wastes and reduction of sulfates (by anaerobic organisms) may introduce sulfides to water. Decomposition of organic matter produces sulfides in sewage and it is a common constituent of many industrial wastes such as those from tanneries, paper mills, chemical plants, edible oil refineries and gas works.

Presence of sulfide in industrial waters is highly objectionable because of its offensive odor and corrosive nature. At many instances corrosion of boilers and cooling systems is attributed to sulfide. When present in high concentrations it may even attack the resistant alloys such as brass and stainless steel and cause deterioration.

Sulfide is best determined by the following colorimetric methylene blue method.

COLORIMETRIC METHYLENE BLUE METHOD

Principle

This colorimetric method determines total sulfide which includes dissolved H_2S, HS^-, S^{2-} ions and acid soluble metallic sulfides. Acid insoluble sulfides are not determined by this method. This method is based on the reaction between p-amino dimethyl aniline, ferric chloride and sulfide ion, resulting in the formation of methylene blue.

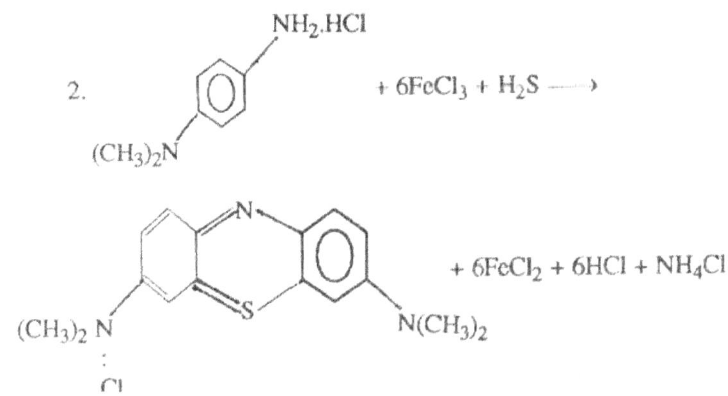

Methylene blue

Interferences

Strong reducing agents interfere but they occur only in insignificant concentrations.

Reagents

1. Sulfuric acid 1+1: Add carefully 500 ml. conc.H_2SO_4 to 500 ml. distilled water with constant stirring and cool.

2. Amine sulfuric acid stock solution: Add carefully 50ml. of conc.H_2SO_4 to 20 ml. distilled water and cool. Dissolve in this solution 27g. p-amino dimethy1 aniline oxalate (also called N, N dimthy1 p-phenylene diamine) and dilute to 100 ml. with distilled water. Store in an amber bottle.

3. Amine-sulfuric acid working solution: Dilute 12.5 ml. of amine–sulfuric acid stock solution to 500 ml with 1+1 sulfuric acid. Store in an amber bottle.

4. Ferric chloride solution: Dissolve 100g. ferric chloride, $FeCl_3.6H_2O$ in 40 ml. distilled water.

5. Diammonium hydrogen phosphate solution: Dissolve 100g. diammonium hydrogen - phosphate, $(NH_4)_2HPO_4$ in 200 ml. distilled water.

6. Methylene blue solution I: Dissolve 1.0g methylene blue (USP grade) in distilled water and make up to 1000 ml. in a volumetric flask. Standardise this solution with standard sulfide solution as follows:

Weigh 12 to 15g. pure sodium sulfide (white) and wash several times with distilled water to remove black sulfide adhering and dissolve the remaining in distilled water and make upto 1000ml in a volumetric flask.

To 100 ml distilled water in an iodine flask add a suitable aliquot (5 ml or 10 ml) of sulfide solution below the surface of water. Add a known excess of 0.025 N iodine solution (1 ml = 0.4 mg sulfide). Add 5 ml. conc. HCl, Stopper and shake well and back titrate with 0.025N sodium thiosulfate using starch indicator.

$$mg.\ sulfide\ present\ in\ the\ aliquot = \left(\underset{(0.025N)}{ml.\ iodine} - \underset{(0.025N)}{ml.\ thio} \right) \times 0.400$$

From the above sulfide solution, prepare a standard sulfide solution containing 1 mg. sulfide per 1 ml. To about 980 ml. distilled water in a 1000 ml. volumetric flask add 2 ml. of this standard sulfide solution such that the sulfide concentration of the resulting solution is 2 mg/l. Use this solution for standardization of methylene blue solution by the colorimetric procedure described below. Use 2 drops methylene blue and compare the color with that of sulfide solution. If the color of methylene blue is more than that of sulfide solution, methylene solution has to be diluted. If the intensity is less, small increment of dye (MB) has to be added such that the color of both the sulfide solution and MB are the same. Now the methylene blue solution I is standardized.

0.05 ml. (1 drop) of MB solution I ≡ 1.0 mg/l sulfide.

7. Methylene blue solution II: Pipet 10.0 ml. methylene blue solution into a 100 ml. volumetric flask and make up to the mark with distilled water.

0.05 ml (1drop) of methylene blue solution II ≡ 0.1mg/l sulfide.

Procedure

1. Place 7.5 ml. of the sample into two 10ml. comparator tubes. (125mm long and 15 mm outside diameter). Mark the tubes as A and B.
2. To tube A add 0.5 ml. amine-sulphuric acid working solution and 0.15 ml. (3 drops) ferric chloride solution. Mix immediately by inverting the tube only once. If any sulfide is present a blue color will appear. Allow three minutes for the complete color development.
3. To tube B, add 0.5 ml. 1 + 1 sulfuric acid and 0.15 ml. ferric chloride solution and mix.
4. After three minutes, add 1.6 ml. diammonium hydrogen phosphate solution to each tube to remove the color of the ferric chloride. Allow to stand for about 10 minutes.
5. If the sample seems to contain more than 5 mg/l sulfide, add methylene blue solution I, using a 1 ml pipet drop by drop to tube B until the color matches with that developed in tube A. If the sulfide concentration of the sample is less than 5 mg/l, add methylene blue solution II dropwise to tube B.

Note: If the sulfide concentration of the sample exceeds 20 mg/l, dilute the sample ten times and use the diluted sample for determination.

Calculation

(i) If the sample volume is 7.5 ml. and methylene blue solution I is used, then

$$mg/l\ sulfide = \frac{ml.\ Methylene\ blue\ I \times 1}{0.05}$$

(ii) If the sample volume is 7.5 ml. and methylene blue solution II is used, then,

$$mg/l\ sulfide = \frac{ml.\ Methylene\ blue\ II \times 0.1}{0.05}$$

CHAPTER - 68

SULFITE

The occurrence of sulfite in natural waters is rare. It is commonly found in boiler waters and cooling waters, as sodium sulfite is added to remove oxygen to prevent corrosion. Waste waters from paper and pulp industries also contain sulfites and contribute to receiving waters.

IODOMETRIC METHOD

Principle

Sulfite is oxidized with potassium iodate-iodide solution, under acidic conditions. From the amount of iodate-iodide solution required for oxidation, quantity of sulfite present is calculated.

$$KIO_3 + 5KI + 3H_2SO_4 \rightarrow 3K_2SO_4 + 3I_2 + 3H_2O$$

$$SO_3^{2-} + I_2 + H_2O \rightarrow SO_4^{2-} + 2HI$$

Interferences

Sulfide, ferrous iron and organic matter interfere seriously. Nitrate, if present will oxidize sulfite during acidification.

Reagents

1. **Sulfuric acid, 1+1:** Dilute 100 ml. conc. H_2SO_4 to 200 ml. with distilled water.
2. **Potassium iodate-iodide solution, 0.0125N:** Dissolve 445.8mg. potassium iodate, KIO_3 (previously dried at 120°C for several hours) in 200 ml. distilled water. Add 4.35g. KI and 310 mg $NaHCO_3$ and make up to 1000ml. in a volumetric flask.

   1.00 ml. 0.0125N potassium iodate-iodide solution ≡ 500μg SO_3.

3. **Starch indicator:** See Chapter - 51,"PERMANGANATE VALUE".

Procedure

Precaution: Conduct the determination as soon as possible after the sample has been collected.

1. Collect a fresh water sample, allowing only a little contact with air.

2. It is advisable that the temperature of the sample should not exceed 20° C.
3. If hot samples are collected, cool it in cooling apparatus as described in "Standard Methods".

Titration: Place 1 ml. 1+1 H_2SO_4 in a 250 ml. conical flask. Measure 50 ml. of water sample using a measuring cylinder and transfer to the flask. Add 1 ml. starch indicator and titrate with potassium iodate-iodide solution until a faint permanent blue color develops.

Calculation

$$mg/l\ sulfite\ (as\ SO_3^{2-}) = \frac{ml.\ potassium\ iodate\text{-}iodide \times 0.5 \times 1000}{ml.\ sample\ taken\ for\ titration}$$

If sulfite has to be expressed in terms of sodium sulfite, then,

$$mg/l\ Na_2SO_3 = mg/l\ SO_3 \times 1.57$$

CHAPTER - 69
SURFACTANTS (ANIONIC)

The increased use of synthetic detergents for general cleaning in houses and industries makes the possible presence of surface active agents in many water courses. Majority of the syndets contain anionic surfactants. Because of their unwanted effects, they are undesirable in water and a quantitative estimation of them in water is necessary.

LONGWELL AND MANIECE METHYLENE BLUE METHOD

This method is applicable to industrial waters, drinking waters, river waters, sewage effluents and other waste waters.

Principle

Methylene blue reacts with anionic surfactants to form a blue colored salt. The intensity of the blue color depends on the concentration of the anionic surfactants.

Interference

Sulfide interferes by reducing the methylene blue. However it can be oxidized by hydrogen peroxide.

Reagents

1. Alkaline phosphate solution 1%: Dissolve 5g. anhydrous sodium dihydrogen phosphate, NaH_2PO_4 in water. Adjust to pH 10 using sodium hydroxide and make upto 500 ml. in a volumetric flask.

2. Neutral methylene blue solution: Dissolve 350mg. methylene blue in distilled water and make up to 1000ml. in a volumetric flask .

3. Chloroform

4. Acid methylene blue solution: Dissolve 35mg. methylene blue in 500 ml. distilled water in a 1000 ml. volumetric flask and add 6.5 ml. conc. H_2SO_4 and make upto the mark with distilled water.

5. LAS (Linear alkyl sulfonate) stock solution: Dissolve 0.100g. manoxol OT (sodium dioctyl sulfo succinate) or any other reference LAS material in water and make up to 1 litre in a volumetric flask.

1.0 ml = 100μg Manoxol OT (or reference LAS material)

6. LAS standard solution: Pipet 10.0 ml. stock solution in to a 100 ml. volumetric flask and make up to the mark with distilled water.

1.0 ml = 10μg Manoxol OT (or reference LAS material)

7. Hydrogen peroxide: '20-volume'.

Procedure

1. Place a suitable volume of the sample containing 20-200μg surfactant in a separating funnel, (if the volume is below 100 ml, make up to 100 ml. with distilled water.) Add 10 ml. alkaline phosphate solution, 5 ml. neutral methylene blue solution and 15 ml. chloroform.
2. Shake gently for 30 seconds. Excessive agitation may cause emulsion troubles. Allow the layers to separate and draw the chloroform layer into another separating funnel containing 110 ml. distilled water and 5 ml. acid methylene blue solution. Rinse the first separating funnel with 2.0 ml. chloroform and add the rinsing to the second separating funnel.
3. Shake the second separating funnel for 30 seconds and allow the layers to separate. Pass the chloroform layer through a small funnel plugged with cotton wool (moistened with chloroform) into a 50 ml. volumetric flask. Rinse the separating funnel with 2.0 ml. chloroform and add the rinsing to the volumetric flask.
4. Repeat the extraction with two further 10 ml. portions of chloroform and collect the extracts in the 50 ml. volumetric flask and make up to the mark with chloroform. Measure the optical density at 650 nm.
5. Prepare standards simultaneously as described above covering the range from 20 to 200μg surfactant. Conduct a blank also. Measure the absorbance at 650 nm (or with suitable orange filter) using a light path of 1 cm. Plot a calibration curve. Read the number of micrograms of surfactant in the sample equivalent to the observed optical density.

Express the result in mg surfactant as Manoxol OT or as the reference material used per litre of the sample.

For samples containing sulfide: Place a suitable volume of sample in a separating funnel and add 10 ml. alkaline phosphate solution followed by 2 ml. hydrogen peroxide (20 volume). Stand for five minutes and dilute to 110 ml. with water. Add 15 ml. chloroform and 5 ml neutral methylene blue solution and proceed as described above from Step No. 2.

CHAPTER - 70

TANNIN AND LIGNIN

The effluent from pulp and paper industries and tanning industries contribute tannin and lignin to water courses. Tannin is also used in the internal treatment of boiler waters to prevent scale formation. Hence the blow down waters contains appreciable quantities of tannin.

REDUCTION METHOD

Principle

Determination of tannin and lignin is based upon the capacity of them to reduce tungstophosphoric and molybdophosphoric acids to produce a blue color.

Reagents

1. Tannin-lignin reagent: Dissolve 25g. sodium tungstate dihydrate. Na_2WO_4. $2H_2O$, 5g. molybdophosphoric acid (also known as phosphomolybdic acid) $20MoO_3.2H_3PO_4.48H_2O$ and 12.5 ml. 85% phosphoric acid, H_3PO_4 in 200 ml. distilled water. Reflux the mixture for 2 hrs. Cool well and make up to 250 ml. with distilled water.

2. Sodium carbonate solution, (saturated): Dissolve 100 g. sodium carbonate, Na_2CO_3 in 250 ml. warm distilled water and dilute to 500 ml. Keep it in a rubber–stoppered bottle.

3. Stock tannin or lignin solution: The sample under investigation may contain either tannin or lignin compounds. According to the substance present in the sample, standard solutions have to be prepared. Dissolve 1.000 g. tannic acid or lignin compound (which is known to be present in the sample) in distilled water and make up to 1000 ml. in a volumetric flask.

4. Standard tannin (or lignin) solution: Pipet 10.0 ml. of the above stock solution and make up to 1 litre in a volumetric flask.

1.00 ml. = 0.01 mg tannin (or lignin)

Procedure

1. Place 50.0 ml. clear sample in a 100 ml. Nessler tube.

2. Add 2 ml. tannin-lignin reducing agent and mix well.
3. After 5 minutes add 10 ml. sodium carbonate solution and stir well.
4. Allow 10 minutes for maximum color development.
5. Simultaneously prepare standards by using 1.0, 2.0, 3.0, 4.0, 6.0, 8.0, 10.0 ml. standard tannin (or lignin) solution and adding the reagents in order as mentioned above.
6. Prepare a blank also.
7. Compare the color of the sample with that of the standards visually. To make photometric comparison, use a light path of 1 cm. or longer at a wavelength of 700nm.

Report the result in general terms as 'Tannin like'or 'Lignin like' or simply as 'hydroxylated aromatic' compounds.

CHAPTER - 71
TEMPERATURE

Temperature measurements are usually made with mercury filled centigrade thermometer. At present digital thermometers are available for temperature measurement. The reading should be made by dipping the thermometer in the sample (water or waste water). Sufficient time should be allowed before constant reading is obtained. Normally the important temperature is that at the time of sampling. The temperature should be expressed to the nearest degree centigrade.

Surface waters vary considerably in temperature between winter and summer. In well supplies, however the temperature is more equable the year around. Regarding springs, the higher the temperature the deeper the source from which it emerges.

Temperature readings are more useful in industrial water supplies for process use, or heat-transmission calculations. Temperature measurements are also useful in detecting unsuspected source of pollution. In limnologic studies temperature at various depths is an important parameter.

CHAPTER - 72

TURBIDITY

Turbidity is an important parameter for characterizing water quality. It is an important control parameter in assessing filter bed efficiencies and to check the effectiveness of backwashing of filter beds. It is an expression of optical property of a sample (water or waste water) containing insoluble substances which cause light to be scattered rather than transmitted in straight lines. The amount and angular distribution of this scattered light is governed not only by the quantity of the insoluble substances but also by their size, shape and refractive index. In most of the waters turbidity is due to colloidal and extremely fine dispersions. Suspended matter such as clay, silt, finely divided organic and inorganic matter, plankton and other microscopic organisms also contribute to turbidity.

Turbidity measurements are useful in various ways. The sudden appearance of turbidity in a water supply is an indication of pollution particularly if the source is near to drains, cesspools, ditches or manured grounds. Water from recently constructed wells usually contains suspended mineral matters such as sand, chalk etc., but they become clear after some time in use. Waters containing iron salts when first drawn will be bright, but they rapidly become turbid on contact with air, due to the formation of more highly oxidized and insoluble salts.

At water works, the turbidity measurements are useful in discovering imperfect filtration. They are also used to determine the effectiveness of treatment produced with different chemicals and the dosages needed. Thus they aid in the selection of the most effective and economical chemical to use.

Turbidity is an important consideration in public water supplies because of the following reasons.

1. Aesthetically turbidity is objectionable.
2. Filtration of water is rendered more difficult when turbidity increases. Since turbidity shortens the filter runs, satisfactory operation of filter beds becomes impossible.
3. There are chances for the pathogenic organisms to be enclosed in the turbidity causing particles and they may not be exposed to the disinfectant.

Turbidity is also undesirable in waters for ice making, beverages and brewing, textiles, pulp and paper, steam boilers and turbine operation and other industrial processes.

Methods of measurement: Instrumental methods are the widely employed methods for the measurement of turbidity. They are preferable to visual methods because of their greater precision, sensitivity and wide range of applicability.

INSTRUMENTAL METHOD

The turbidity can be measured either absorptiometrically or nephelometrically. Nephelometers are more sensitive than absorptiometers, and are widely used for the determination of low turbidities of potable water. It should be noted that instruments of different design may give different result for the same sample, because of the variations in light source, angle of light acceptance of the detector etc. Thus comparisons of results obtained with different instruments should be made with caution.

Preparation of standard suspensions: Prepare turbidity-free water first, by passing distilled water through a membrane filter (with a pore size below 100μm). Discard the first 200 ml and use the further volumes for the preparation of solutions and dilutions.

1. Stock turbidity suspension

(a) Dissolve 1.00g. hydrazine sulfate, $(NH_2)_2.H_2SO_4$ in distilled water (free from turbidity) and make up to 100 ml in a volumetric flask.

(b) Dissolve 10.00g. hexamethylene tetramine, $(CH_2)_6N_4$ in distilled water (free from turbidity) and makeup to 100 ml in a volumetric flask.

(c) Pipet 5.0 ml hydrazine sulfate solution and 5.0 ml hexamethylene tetramine solution in a 100 ml volumetric flask. Allow to stand for 24 hr at 25±3°C. Then make up to the mark with turbidity-free distilled water and shake well. This is the stock turbidity suspension.

1.0 ml = 400 NTU (Nephelometric Turbidity Units)

Prepare the component solutions and suspension monthly.

2. Standard turbidity suspension: Pipet 10.0 ml stock turbidity suspension into a 100 ml volumetric flask and make up to the mark.

1.0 ml = 40 NTU

Prepare weekly.

3. Working turbidity suspension: Dilute standard suspensions as required, with turbidity-free distilled water. Prepare weekly.

Procedure

Follow the manufacturers' operating instructions. Prepare a series of working turbidity suspensions covering the range up to 40 units and measure the reading. Prepare a calibration curve relating instrument reading to turbidity. Find out the turbidity of the sample either directly from the instrument reading or from the calibration graph. The calibration curve should be checked from time to time.

The following points should be borne in mind while determining the turbidity of the sample.

1. Shake the sample well to disperse the solids thoroughly. Pour the sample into the turbidimeter tube after all the bubbles have disappeared.
2. If the sample is very cold, allow it to reach the room temperature before testing.
3. If the turbidity exceeds 40 units, dilute the sample appropriately with turbidity-free distilled water.
4. If it is necessary to correct for color, filter a portion of the well shaken sample through a membrane filter (pore size – 0.45 to 0.5μm). Measure the turbidity of the filtrate and subtract from the turbidity of the unfiltered sample. The resulting value is the true turbidity of the sample.

Express the results in terms of NTU after rounding of them to the nearest value as given in the following table.

Results obtained NTU	Record to the nearest NTU
0 – 1.0	0.05
1 – 10	0.1
11 – 40	1
41 – 100	5
101 – 400	10
401 - 100	50

CHAPTER - 73

VOLATILE ACIDS

This is an important parameter to be tested for effluents subjected to anaerobic digestion, especially for distillery and sugar factory effluents. During anaerobic digestion, the complex materials are converted to low molecular weight compounds. Among them, short chain fatty acids such as acetic, propionic, butyric acids are important components. Valeric, isovaleric and caproic acids are also important though not to a greater extent. All these low molecular weight fatty acids are termed as volatile acids as they are distillable at atmospheric pressure.

Applications of determination of Volatile Acids

1. It provides information concerning the anaerobic degradation of the material in the digester and indicates environmental conditions suited for the optimum activity of methane producing bacteria.
2. Volatile acids determination in conjunction with pH measurements are highly useful in controlling the environmental conditions during the initiation of anaerobic digestion for better production of methane.
3. The onset of unfavorable conditions or the accumulation of acids could be detected immediately by the determination of volatile acids which is not possible with pH measurement alone as it will take longer periods for an indication in the change in pH. Volatile acids if accumulate in the anaerobic digester may react with bicarbonates and destroy the buffering capacity of the system and produce more carbon dioxide resulting in a gradual drop of pH. Under these conditions methane producing bacteria which are sensitive to pH and temperature changes cannot multiply and cannot convert the fatty acids to methane. This condition is detected by the volatile acids determination.
4. Determination of volatile acids can be carried out within a short time and at least expense.

Methods available for the determination of volatile acids

1. Distillation and titration with a standard alkali (Both direct distillation and steam distillation procedures are described below).
2. Gas-chromatography/Column partition chromatography.

3. Colorimetric Method.

Direct distillation and Steam Distillation: Direct distillation procedure is the preferred one as it is rapid and sufficiently accurate for most practical purposes. Recovery of the volatile acids is reasonable (about 70%) and results at an accuracy of ± 50 mg/l could be obtained.

Sulfuric acid is added to the sample for the release of volatile acids. Hence it is very important that distillation has to be stopped as soon as the appropriate amount of distillate is collected. If the distillation is carried out indefinitely, sulfuric acid would be reduced to sulfur dioxide which combines with water to form sulfurous acid and yield abnormally high values. Rate of distillation is also an important factor in direct distillation.

Most of the difficulties encountered in direct distillation are eliminated in the steam distillation procedure. The distillation may be continued until the acids are recovered completely. There are no chances for the concentration of the sample and sulfuric acid in the flask and the results obtained are more accurate. The major disadvantage with this procedure is requirement of more time and complexity of apparatus.

DIRECT DISTILLATION METHOD

Apparatus Required: Distillation Apparatus (See Figure 43-1)

1. Distillation flask: 500 ml capacity.
2. Condenser: About 75 cm.
3. Receiving flask.

Reagents

1. Sulfuric acid 1+1: Add slowly 100 ml sulfuric acid to 100 ml distilled water in a beaker and cool.

2. Sodium hydroxide solution, 0.1N: Prepare by diluting standardized 1N.NaOH solution. (See Chapter – 10 "ACIDITY")

3. Phenolphthalein indicator solution: Dissolve 500mg. phenolphthalein in 50 ml ethyl or isopropyl alcohol and add 50 ml distilled water. Add sodium hydroxide solution (0.02N) dropwise until a faint pink color persists.

Procedure

1. Centrifuge[*] 200 ml of effluent sample for 5 minutes to separate the sludge and liquor.
2. Combine the supernatant liquor in a beaker.
3. Place 100 ml of the supernatant liquor in the distillation flask.
4. Add 100 ml distilled water and a few glass beads or antibumping granules.
5. Add 5 ml 1+1 H_2SO_4 and mix thoroughly.
6. Connect the flask with the condenser and start distillation. The rate of distillation is 5 ml/minute.
7. Collect 150 ml distillate and immediately titrate with 0.1N.NaOH using phenolphthalein indicator. The end point is the appearance of pink color that persists for a short time.

Calculation

$$mg/l\ Volatile\ acids\ (as\ acetic\ acid) = \frac{ml.\ NaOH \times 6000}{ml.\ sample} \times \frac{100}{70}$$

It is assumed that only about 70% acids are recovered (this factor may vary form 68 to 85%). But in order to find out the actual amount of volatile acids present, the factor (100/70) is used in the above calculation.

STEAM DISTILLATION METHOD

Apparatus required: (See Figure 73-1)

1. Steam generator: 1 litre capacity with a steam bypass valve.
2. Distillation flask: 500 ml capacity.
3. Condenser: 75cm long.
4. Bent adaptor tube.
5. Receiving flask with a soda-lime guard tube to prevent the entry of air.

Assemble the apparatus as shown below.

[*] The sample is centrifuged to remove sludge which may decompose and release other organic acids. The sample may also be taken directly without centrifugation, but due care should be taken to prevent excess heating of the walls of the distillation flask above the level of the mixture in the flask. Preferably electric heaters may be employed.

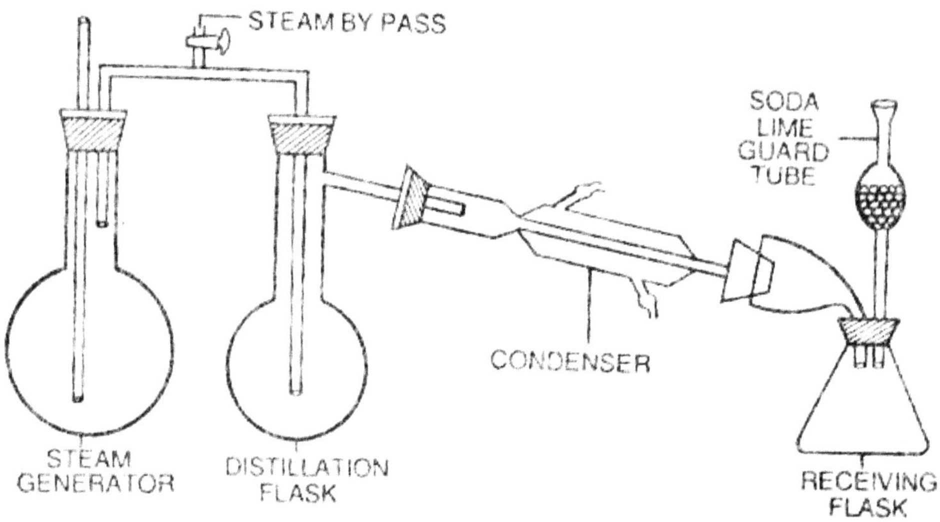

Fig. 73-1. Steam Distillation apparatus for the determination of Votaltile Acids

Reagents

All the reagents prepared for direct distillation procedure. In addition prepare the following reagents.

1. **Ferric chloride solution:** Dissolve 21g. ferric chloride $FeCl_3.6H_2O$ in 250 ml distilled water. Keep in tightly stoppered bottle.
2. **Diatomaceous – Silica filter aid.**
3. **Magnesium sulfate crystals, $MgSO_4.7H_2O$.**

Procedure

1. Place 200 to 500 ml or appropriate amount of the sample in a beaker and adjust the pH to 3.5 with 1 + 1 H_2SO_4.
2. Add 0.6 ml ferric chloride solution per each 100 ml sample.
3. Add 50g. filter aid and mix thoroughly.
4. Place a filter paper on the Buchner funnel and coat it with a thin layer of filter aid.
5. Filter the sample solution prepared above, through the funnel by suction in to an evaporating dish or basin.

6. Wash the residue thoroughly with distilled water.
7. Adjust the pH of the filtrate in the basin to about 11 with sodium hydroxide solution. Concentrate to about 150 ml preferably by keeping over a water bath. Cool in a refrigerator.
8. Transfer the cooled filtrate quantitatively to the distillation flask and adjust the pH to 4 with 1 + 1 H_2SO_4.
9. Add $MgSO_4$ to slight excess of saturation ($MgSO_4$ is added to avoid the effects of varying concentrations of dissolved solids and mineral acid during distillation).
10. Heat the distillation flask with a small flame until rapid evolution of acid commences.
11. Start steam distillation at a slow rate (i.e.) 200 ml distillate has to be collected in about 20 to 25 minutes. Continue distillation at a faster rate and collect a total volume of 600 ml distillate.
    (Foaming during distillation can be avoided by the control of steam rate).
12. Titrate the distillate with 0.1N.NaOH using phenolphthalein indicator.

Calculation

$$mg/l \text{ Volatile acids (as acetic acid)} = \frac{ml.\ NaOH \times 6000}{ml.\ sample}$$

CHAPTER - 74

ZINC

Zinc is commonly found in small quantities in domestic water supplies and industrial waters due to the corrosion of galvanized iron and brass in condensing, cooling and distributing systems. Zinc may also enter into a water supply through the discharge of industrial effluents such as galvanizing wastes, zinc plating wastes, viscose rayon wastes, and rubber factory wastes.

Zinc has no known adverse physiological effects upon man. In fact, it is an essential and beneficial element in human nutrition. From aesthetic considerations, however, high concentrations (above 5 mg/l) of zinc in domestic water are undesirable. A bitter astringent taste is produced by zinc. Concentrations of zinc above 30 mg/l give water a milky appearance and causes a greasy film on boiling. It is now reported that even 5 mg/l can cause a greasy film on boiling.

Two methods for the determination of zinc using dithizone as the complexing agent are given. The first method is applicable to treated and unpolluted waters. The second method is applicable to waste waters and industrial effluents where the interferences are eliminated by the use of bis-2-hydroxy -ethyl dithio carbamate.

DITHIZONE METHOD (MIXED COLOR METHOD)

Principle

Zinc reacts with dithizone (diphenyl thiocarbazone) to form a coordinate compound which when extracted into carbon-tetrachloride is red in color and is used to measure the zinc content.

Interferences

About twenty metals react with dithizone to form colored complexes. Among them, copper, cobalt, cadmium, lead, nickel and tin are important. However they are present only in small quantities in waters and these interferences can be overcame by adjusting the pH of the solution to 4.0 to 5.5 and by the addition of sufficient amount of sodium thiosulfate. If the interfering ions are present in large quantities dithiozone method using bis-2hydroxy - ethyl dithio-carbamate may be employed. Another important factor in the determination of zinc is the purity of reagents and glassware. The reagents should be free from zinc. Since glassware contribute zinc they must be thoroughly cleansed with sodium - citrate solution before use. Dithizone and zinc dithizonate are readily decomposable in strong light and it is advisable to conduct the experiment in subdued light.

Reagents

Prepare all the reagents using zinc-free distilled water which can be prepared by redistilling the distilled water in all Pyrex distillation apparatus.

1. **Hydrochloric acid, approximately 0.02 N:** Dilute 1.0 ml conc.HCl to 600 ml with zinc- free distilled water.
2. **Zinc stock solution:** Dissolve 100mg. 30-mesh zinc metal in about 1 ml of 1 + 1 HCl. Make up to 1000 ml in a volumetric flask with zinc-free distilled water.

$$1.0 \text{ ml} = 100\mu g \text{ Zn}$$

3. **Zinc working solution:** Pipet 10.0 ml of zinc stock solution in to a 1000 ml volumetric flask and make up to mark with zinc-free distilled water.

$$1.0 \text{ ml} = 1\mu g \text{ Zn } (0.001 \text{mg. Zn})$$

Prepare freshly just before use.

4. **Dithizone stock solution:** Dissolve 25mg. dithizone (diphenyl thiocarbazone) in 250 ml carbon tetrachloride. Place it in an amber bottle and keep in a refrigerator.
5. **Dithizone working solution:** Dilute 1 volume of dithizone stock solution with 9 volumes of carbon tetra chloride. Prepare just before use.
6. **Acetate buffer solution**
   (a) **Sodium acetate solution, approximately 2N:** Dissolve 68g. sodium acetate,. $NaC_2H_3O_2.3H_2O$ in little zinc-free distilled water and dilute to 250 ml.
   (b) **Acetic acid, 1+ 7:** Add 25 ml glacial acetic acid to 175 ml zinc-free distilled water.

   Mix 100 ml of sodium acetate solution and 100 ml of acetic acid solution. Extract with 10 ml portions of dithizone stock solution until the last extract remains green. Remove the traces of dithizone by extracting with carbon tetra chloride.
7. **Sodium thiosulfate solution:** Dissolve 25g. sodium thiosulfate. $Na_2S_2O_3.5H_2O$ in 100 ml zinc-free distilled water. Make it free from zinc by extracting with dithizone stock solution and remove the traces of dithizone by extraction with carbon-tetrachloride.
8. **Sodium citrate solution:** Dissolve 20g. sodium citrate, $Na_3C_6H_5O_7.2H_2O$ in 180 ml zinc-free distilled water. Purify by dithizone extraction. Use this solution for final cleaning of glassware.

Procedure

1. Place an aliquot of the sample with zinc content below 0.008 mg in a 50 ml beaker. Adjust the volume of the sample to 10 ml either by evaporation or dilution. Adjust the pH* to 2-3 with HCl. Transfer to a 100 ml separating funnel.
2. In to a series of 100 ml separating funnels, place 1.0, 2.0, ...8.0 ml of zinc working solution and dilute to 10 ml with zinc-free distilled water. Include a separating funnel containing 10 ml zinc-free distilled water as the blank.
3. To the blank, standards and sample and the following reagents in order with mixing after each addition;
   (i) 5.0 ml acetate buffer
   (ii) 1.0 ml sodium thiosulfate solution, and
   (iii) 10.0ml dithizone working solution.
4. Shake the funnel vigorously for about four minutes. Allow the layers to separate and draw the CCl_4 layer in to the absorption cell after drying the stem of the funnel with filter paper strips or cotton swab.
5. Measure the red color of the zinc dithizonate at 535 nm (use green filter if filter photometer is used) after setting the photometer at 100% transmittance with the blank. [If the green color of the unreacted dithizone is measured, set the photometer with the blank at 10% transmittance at a wavelength of 620 nm (red filter)]. Prepare a calibration curve. Find out the mg equivalent of zinc to the observed optical density. Visual comparison can also be made with CCl_4 extracts using Nessler tubes.

The various colors obtained with zinc standards are as follows:

Zinc in mg	Color
0.000	Green
0.001	Blue
0.002	Blue violet
0.003	Blue violet
0.004	Red violet
0.005	Red violet
0.006	Red violet
0.007	Violet
0.008	Violet

* If the sample is preserved by the addition of acid , remove the acid by evaporating suitable volume of the sample to dryness in a silica dish and bring the pH to 2-3. Do not neutralize the excess acid with sodium hydroxide or ammonia solution.

DITHIZONE METHOD USING BIS-2-HYDROXY - ETHYL DITHIOCARBAMATE (MONO COLOR METHOD)

This method is applicable to polluted waters, and waste waters.

Reagents

1. **Methyl red indicator solution:** Dissolve 0.1g. methyl red sodium salt and dilute to 1000 ml with distilled water.
2. **Sodium citrate solution:** Dissolve 10g. sodium citrate $Na_3C_6H_5O_7.2H_2O$ in 90 ml zinc-free distilled water. Purify by dithizone extraction.
3. **Conc. ammonia solution**
4. **Potassium cyanide solution:** Dissolve 5g. potassium cyanide in 100 ml zinc-free distilled water.
5. **Acetic acid, conc.**
6. **Bis (2-hydroxy ethyl dithio) carbamate solution:** Dissolve 4.0g. diethanol amine and 1 ml carbon disulfide, CS_2 in 40 ml methyl alcohol. Prepare freshly.
7. **Dithizone stock solution:** Dissolve 50mg. dithizone in a little carbon tetra chloride and make up to 200 ml with CCl_4.
8. **Dithizone working solution:** Dilute 50 ml stock dithizone solution to 250 ml with CCl_4.
9. **Sodium sulfide solution I:** Dissolve 3.0g. $Na_2S.9H_2O$ or 1.65g. $Na_2S.3H_2O$ in 100 ml zinc-free distilled water.
10. **Sodium sulfide solution II:** Dilute 4.0 ml of the above sodium sulfide solution – I to 100 ml with zinc-free distilled water. Prepare just before use.
11. **Zinc stock solution:** See 'Mixed Color Method'.
12. **Zinc working solution:** See 'Mixed Color Method'.

Procedure

1. Pipet in to a series of 125 ml separating funnels, appropriate volumes of zinc working solution to cover the range from 0.0 upto 40μg zinc. Adjust the volume to 20 ml. Include a separating funnel containing 20 ml zinc- free water as the blank.
2. Place in a separating funnel an aliquot of the acid digested sample containing 10-40μg Zn and adjust the volume to 20 ml.
3. To the blank, standards and sample add 2 drops of methyl red indicator and 2.0 ml sodium citrate solution. If the indicator is not yellow at this point, add conc.ammonia

solution dropwise until the indicator just turns yellow. Add 1 ml potassium cyanide solution. Add conc.acetic acid drop by drop until the solution becomes neutral peach in color.

4. Add 5 ml CCl_4 to the separating funnel and extract the methyl red. Discard the yellow $CC1_4$ layer.
5. Add 1 ml bis (2-hydroxyethyl)dithiocarbamate solution. Shake well and add 10 ml working dithizone solution. Draw off the CCl_4 layer in to another separating funnel. Extract the aqueous solution in the first separating funnel with 5 ml portions of working dithizone solution until the last extract remains green. Combine all the CCl_4 extracts.
6. To the combined CCl_4 extracts in the separating funnel add 10 ml sodium sulfide solution II. Shake well and discard the aqueous layer. Wash the CCl_4 layer with further 10 ml portions of sodium sulfide solution II until the unreacted dithizone has completely been removed as shown by the colour of the aqueous layer either colorless or very pale yellow.
7. Draw off the CCl_4 layer into a 50 mol volumetric flask after removing the water adhering to the stem of the funnel by means of cotton swab. Make upto the mark with CCl_4.
8. Determine the absorbance of the blank, standards and sample at 535 nm. Plot a calibration curve. Find out the mg equivalent of zinc to the optical density.

Samples containing excessive amounts of Copper, Lead and Cadmium: If the aliquot taken for the estimation contains more than 2 mg of copper, lead and cadmium either separately or jointly, then they should be precipitated as sulfides. Place the aliquot in a 100 ml beaker and adjust the volume to 20 ml. Bring the acidity to 0.5N by adding 6N.HNO_3 (or ammonia solution as necessary). Pass H_2S to the cold solution for five minutes and filter through a sintered crucible. Wash the precipitate with hot water. Remove H_2S from the filtrate by boiling.

Cool and transfer to a 125 ml separating funnel. Proceed from Step No. 3 of the procedure described above.

CHAPTER - 75

PROBABLE COMBINATIONS

(CHECKING THE CORRECTNESS OF ANALYSIS)

After all the saline constituents have been determined, the probable combination of the ions may be worked out. The 'probable combination' almost represent the form in which the saline constituent are most probably taken up and dissolved in water. Calculation of probable combinations is also a means of checking the correctness of analysis. The sum total of the probable combinations should almost be equal to that of the solids dried at 180°C. If there is any marked difference, the analytical results have to be checked up.

The following hints would be useful in computing the probable combination:

(1) Combine the ions in the order of insolubility of salts.

Order of insolubility: Cations

$$Ca^{2+} > Mg^{2+} > Na^{+} > K^{+}$$

Order of insolubility: Anions

$$CO_3^{2-} > SO_4^{2-} > Cl^{-} > NO_3^{-}$$

(2) Some of the forms of the saline constituents present in water on evaporation and drying may be changed to other form in the residue. (e.g) $Ca(HCO_3)_2$ will be changed to $CaCO_3$. $NaHCO_3$ changes to Na_2CO_3.

(3) Most of the water of crystallization of compounds will be lost. The remaining water of crystallization should be included in the calculation. Calcium sulfate loses all of its water of crystallization at 110°C, while magnesium sulfate retains 2 molecules of water. Calcium chloride also loses part of its water of crystallization and becomes $CaCl_2.2H_2O$. Magnesium chloride loses $2H_2O$ and becomes $MgCl_2.4H_2O$ at 100°C.

(4) It is assumed that silica exists in free state and is expressed as SiO_2.

(5) Round off final summed up figure to one decimal.

Following are the conversion factors used in the calculation of saline constituents.

TABLE – 75.1

CONVERSION FACTORS FOR SOME SALINE CONSTITUENTS IN WATER

From	To	Multiply by
$CaCO_3$	$MgCO_3$	0.84
$CaCO_3$	Ca	0.4
Ca	$CaCO_3$	2.5
$CaCO_3$	Mg	0.24
Mg	$CaCO_3$	4.16
$CaCO_3$	$NaHCO_3$	1.68
$CaCO_3$	CO_3	0.60
$CaCO_3$	HCO_3	1.22
CO_3	$CaCO_3$	1.66
CO_3	$MgCO_3$	1.4
CO_3	$NaHCO_3$	1.4
Ca	$CaSO_4$	3.4
SO_4	$CaSO_4$	1.42
Mg	$MgSO_4$	5.0
SO_4	$MgSO_4$	1.25
Na	Na_2SO_4	3.09
SO_4	Na_2SO_4	1.48
K	K_2SO_4	2.23
SO_4	K_2SO_4	1.81
Ca	$CaCl_2$	2.77
Cl	$CaCl_2$	1.56
Mg	$MgCl_2$	3.95
Cl	$MgCl_2$	1.34
Na	NaCl	2.54
Cl	NaCl	1.65
K	KCl	1.91
Cl	KCl	2.1
Ca	$Ca(NO_3)_2$	4.1
NO_3	$Ca(NO_3)_2$	1.32
Mg	$Mg(NO_3)_2$	6.17
NO_3	$Mg(NO_3)_2$	1.19
Na	$NaNO_3$	3.69
NO_3	$NaNO_3$	1.37
K	KNO_3	2.59
NO_3	KNO_3	1.63
N	NO_3	4.43
NO_3	N	0.23

Problem 75 – 1

The results of examination of a well water is given below. Work out the probable combination of the constituents.

Calcium hardness (as $CaCO_3$), mg/l = 308
Magnesium hardness (as $CaCO_3$), mg/l = 272
Total hardness (as $CaCO_3$), mg/l = 580
Iron (as Fe), mg/l = 0.1

Total alkalinity (as $CaCO_3$), mg/l = 316
Sulfate (as SO_4^{2-}), mg/l = 119
Chloride (as Cl^-), mg/l = 204
Nitrate (as NO_3^-), mg/l = 35.5
Dissolved silica (as SiO_2), mg/l = 60

Total dissolved solids, dried at 180°C, mg/l = 1080

STEP – 1

Calcium hardness is lower than alkalinity.

∴ All calcium is present as $CaCO_3$ (i.e) **308 mg/l**

STEP – 2

$$\begin{aligned} &\textit{Total alkalinity} &= 316 \\ &\textit{Calcium hardness} &= 308 \\ &\therefore \textit{Remaining alkalinity} &= 316 - 308 = 8 \end{aligned}$$

This remaining alkalinity is present as $MgCO_3$.

100 pats of $CaCO_3$ is equivalent to 84 parts of $MgCO_3$.

∴ 8 parts of $CaCO_3$ is equivalent to $= \frac{84}{100} \times 8$ = **6.72 $MgCO_3$**.

STEP – 3

Magnesium hardness (as $CaCO_3$) = 272

In this, 8 parts is utilized for the formation of $MgCO_3$. The remaining is present as non-carbonate hardness.

$$272 - 8 = 264$$

This magnesium would present as $MgSO_4$ alone or $MgSO_4$ and $MgCl_2$.

96 parts of SO_4 would combine with magnesium to form 120 parts of $MgSO_4$.

∴ 119 parts of SO_4 would form $= \frac{120}{96} \times 119$ = **148.75 $MgSO_4$.**

$MgSO_4 = 148.75$

$SO_4 = 119.00$

$\therefore Used\ Mg = 148.75 - 119.00 = 29.75$

Calculate water of crystallization of $MgSO_4$ as shown below.

$MgSO_4$ - $2H_2O$

120 36

148.75 $\frac{36}{120} \times 148.75$ = **44.6**

STEP – 4

Magnesium hardness (as $CaCO_3$) remaining after the formation of $MgCO_3$= 264. Convert this magnesium hardness to Mg.

$CaCO_3$ - Mg

100 24

264 $\frac{24}{100} \times 264$ = 63.4 Mg

Mg used for the formation of $MgSO_4$ = 29.75.

∴ Remaining Mg = 63.4 – 29.75 = 33.65 mg.

The remaining Mg would combine with chloride to form $MgCl_2$.

Mg - $MgCl_2$

24 95

33.65 $\frac{95}{24} \times 33.65$ = **133.2 $MgCl_2$.**

Calculate the water of crystallization of $MgCl_2$ as shown below.

$MgCl_2$ - $4H_2O$

95 72

133.2 $\frac{72}{95} \times 133.2$ = **100.95.**

STEP – 5

$MgCl_2 = 133.20$

$Mg = 33.65$

$\therefore Used\ Cl = 133.20 - 33.65 = 99.55$

Total Cl^- = 204.00

Used Cl^- = 99.55

$\therefore$ *Remaining* Cl^- = 204.00 - 99.55 = 104.45

This remaining chloride would combine with sodium to form sodium chloride.

Cl - NaCl

35.5 58.5

104.45 $\frac{58.5}{35.5} \times 104.45$ = **172.20 NaCl.**

STEP – 6

Now the remaining anion is only nitrate. It would be present as sodium nitrate.

62 parts of NO_3 would combine with sodium to form 85 parts of $NaNO_3$.

$\therefore$ 35.5 parts of NO_3 would combine with sodium as follows:

NO_3 - $NaNO_3$

62 85

35.5 $\frac{85}{62} \times 35.5$ = **48.67 $NaNO_3$.**

Now all these constituents have to be summed up. It is assumed that silica exists in free state and it has to be summed up as such with the worked out probable combination.

Constituents		Water of Crystallization
$CaCO_3$	= 308.00	
$MgCO_3$	= 6.72	
$MgSO_4$	= 148.75	+ 44.6
$MgCl_2$	= 133.20	+ 100.95
NaCl	= 172.20	
$NaNO_3$	= 48.67	
Silica	= 60.00	
	877.54	+ 144.55 = **1023.09**

= 1023

The theoretically obtained value 1023 is in well agreement with the value obtained practically (1080).

Since the quantity of iron is insignificant, it is not brought into calculation. However if present in significant concentrations, it has to be calculated as Fe_2O_3.

Problem 75 – 2

Work out the probable combination of a sample of bore well water, the results of which are given below.

Calcium hardness (as $CaCO_3$), mg/l = 68
Magnesium hardness (as $CaCO_3$), mg/l = 88
Total hardness (as $CaCO_3$), mg/l = 156

Total alkalinity (as $CaCO_3$), mg/l = 364
Sulfate (as SO_4^{2-}), mg/l = 122
Chloride (as Cl^-), mg/l = 52
Nitrate (as NO_3^-), mg/l = 26.5
Dissolved silica (as SiO_2), mg/l = 70

Total dissolved solids, dried at 180°C, mg/l = 730

STEP – 1

In this water, total alkalinity is higher than total hardness. Therefore all the calcium and magnesium are present as $CaCO_3$ and $MgCO_3$ respectively.

STEP – 2

All calcium hardness is present in as $CaCO_3$. (i.e) **68**.

STEP – 3

All magnesium hardness is present as $MgCO_3$ and calculated as shown below.
100 parts of $CaCO_3$ is equivalent to 84 parts of $MgCO_3$.
∴ 88 parts of $CaCO_3$ (Mg hardness) is equivalent to

$CaCO_3$ - $MgCO_3$

100 84

88 $\frac{84}{100} \times 88$ = **73.92 $MgCO_3$.**

STEP – 4

The excess alkalinity over total hardness would present as sodium carbonate.

Total Alkalinity–Total Hardness

364–156 = 208

50 parts of $CaCO_3$ is equivalent to 53 parts of Na_2CO_3.

$\therefore$ 208 parts of $CaCO_3$ is equivalent to

$CaCO_3$ - Na_2CO_3

50 53

208 $\frac{53}{50} \times 208 =$ **220.48 Na_2CO_3.**

Since calcium and magnesium have been exhausted, all the anions now would combine with sodium to form sodium salts.

STEP – 5

96 parts of SO_4 would combine with Na to form 142 parts of Na_2SO_4.

$\therefore$ 122 parts of SO_4 would combine with Na as follows:

SO_4 - Na_2SO_4

96 142

122 $\frac{142}{96} \times 122 =$ **180.45 Na_2SO_4.**

STEP – 6

35.45 parts of Cl would combine with Na to form 58.5 parts of NaCl.

$\therefore$ 52 parts of Cl would combine with Na as follows:

Cl - NaCl

35.45 58.5

52 $\frac{58.5}{35.45} \times 52 =$ **85.81 NaCl.**

STEP – 7

62 parts of NO_3 would combine with Na to form 85 parts of $NaNO_3$.

$\therefore$ 26.5 parts of NO_3 would combine with Na as follows:

NO_3 - $NaNO_3$

62 85

26.5 $\frac{85}{62} \times 26.5$ = **36.33 $NaNO_3$.**

All these constituents have to be summed up now.

Constituents		Water of Crystallization
$CaCO_3$	= 68.00	
$MgCO_3$	= 73.92	
Na_2CO_3	= 220.48	
Na_2SO_4	= 180.45	
NaCl	= 85.81	
$NaNO_3$	= 36.33	
Silica	= 70.00	

734.99 + No water of crstallization = **734.99**

= 735

This value obtained theoretically (735) is in well agreement with the practically obtained value (730mg/l).

Problem 75 - 3

Work out the probable combination of a sample of bore well water, the results of which are given below.

Calcium hardness (as $CaCO_3$), mg/l = 416
Magnesium hardness (as $CaCO_3$), mg/l = 276
Total hardness (as $CaCO_3$), mg/l = 692
Iron (as Fe), mg/l = 0.1

Total alkalinity (as $CaCO_3$), mg/l = 350
Sulfate (as SO_4^{2-}), mg/l = 47.2
Chloride (as Cl^-), mg/l = 304
Nitrate (as NO_3^-), mg/l = 77
Dissolved silica (as SiO_2), mg/l = 100

Total dissolved solids, dried at 180°C, mg/l = 1290

STEP – 1

Calcium hardness value is higher than alkalinity. Hence calcium hardness equal to the alkalinity is present as $CaCO_3$ and the excess over the alkalinity would be $CaSO_4$.

$CaCO_3$ = **350**

STEP – 2

Calcium hardness–Total alkalnity

414–350 = 66 $CaCO_3$.

Convert this 66 mg/l $CaCO_3$ to Ca by multiplying with 0.4.

$66 \times 0.4 = 26.4$ mg Ca

96 parts of SO_4 would combine with Ca to form 136 parts of $CaSO_4$.

∴ 47.2 parts of SO_4 would combine with Ca to form

SO_4	**- $CaSO_4$**
96	136
47.2	$\frac{136}{96} \times 47.2 =$ **66.96 $CaSO_4$.**

STEP – 3

$$CaSO_4 = 66.96$$
$$SO_4 = 47.20$$
$$\therefore \textit{Used Ca} = 66.96 - 47.20 = 19.76$$

Calcium remaining after the formation of $CaCO_3$ = 26.4 mg
Calcium used for the formation of $CaSO_4$ = 19.76 mg
∴ Calcium remaining after the formation of $CaSO_4$ = 26.4 – 19.76 = 6.64 mg Ca
This remaining calcium will be present as calcium chloride.

STEP – 4

40 parts of Ca would combine with 70.9 parts of Cl^- to form 110.9 parts of $CaCl_2$.

∴ 6.64 parts of Ca would combine with Cl^- to form $CaCl_2$ as follows:

Ca	**- $CaCl_2$**
40	110.9
6.64	$\frac{110.9}{40} \times 6.64 =$ **18.41 $CaCl_2$.**

Water of crystallization

$CaCl_2$ - $2H_2O$

110.9 36

18.41 $\frac{36}{110.9} \times 18.41 =$ **5.97.**

STEP - 5

$CaCl_2 = 18.41$

$Ca = 6.64$

$\therefore Cl = 18.41 - 6.64 = 11.77$

Total chloride present = 304.00
Chloride used for the formation of $CaCl_2$ = 11.77
∴ Remaining chloride = 304.00 – 11.77 = 292.23

Magnesium hardness (276 mg/l) is expressed in terms of $CaCO_3$.
Convert it into Mg by multiplying with 0.24.

$276 \times 0.24 = 66.24$ mg Mg

24 parts of Mg would combine with 70.9 Cl^- to form 94.9 parts of $MgCl_2$.
∴ 66.24 parts of Mg would combine with Cl^- to form

Mg - $MgCl_2$

24 94.9

66.24 $\frac{94.9}{24} \times 66.24 =$ **261.92 $MgCl_2$.**

Water of crystallization

$MgCl_2$ - $4H_2O$

94.9 72

261.92 $\frac{72}{94.9} \times 261.92 =$ **198.5.**

STEP - 6

$MgCl_2 = 261.92$

$Mg = 66.24$

$\therefore Cl = 261.92 - 66.24 = 195.68$

Chloride remaining after the formation of $CaCl_2$ = 292.23 mg

Chloride used for the formation of $MgCl_2$ = 195.68 mg
∴ Remaining chloride = 292.23 – 195.68 = 96.55 mg

All these chloride would present as NaCl.
35.45 parts of Cl^- would combine with 23 parts of Na to form 58.45 parts of NaCl.
∴ 96.55 parts of Cl^- would combine with Na to form NaCl as shown below.

Cl - NaCl
35.45 58.45
96.55 $\frac{58.45}{35.45} \times 96.55 =$ **159.19 NaCl.**

STEP - 7
62 parts of NO_3 would combine with 23 parts of Na to form 85 parts of $NaNO_3$.
∴ 77 parts of NO_3 would combine with Na to form $NaNO_3$ as shown below.

NO_3 - $NaNO_3$
62 85
77 $\frac{85}{62} \times 77 =$ **105.56 $NaNO_3$.**

STEP – 8
Since iron is present in negligible concentration, it may be omitted from the calculation.
It is assumed that silica exists in free state. It has to be summed up as such with the found out probable combinations.
Now the probable combinations may be represented as follows:

Constituents		Water of Crystallization
$CaCO_3$	= 350.00	
$CaSO_4$	= 66.96	
$CaCl_2$	= 18.41	+ 5.97
$MgCl_2$	= 261.92	+ 198.5
NaCl	= 159.19	
$NaNO_3$	= 105.56	
Silica	= 100.00	
	1062.04	+ 204.47 = **1266.51**

= 1267

This value (1267) is very near to the practically obtained value (1290).

Problem 75 – 4

The results of examination of an open well water is as follows:

Calcium hardness (as $CaCO_3$), mg/l	*= 68*	*Total alkalinity (as $CaCO_3$), mg/l*	*= 408*
Magnesium hardness (as $CaCO_3$), mg/l	*= 308*	*Sulfate (as SO_4^{2-}), mg/l*	*= 33.1*
Total hardness (as $CaCO_3$), mg/l	*= 376*	*Chloride (as Cl^-), mg/l*	*= 40*
Iron (as Fe), mg/l	*= - 0 -*	*Nitrate (as NO_3^-), mg/l*	*= 3.9*
		Dissolved silica (as SiO_2), mg/l	*= 70*

Total dissolved solids, dried at 180°C, mg/l = 535

Work out the probable combinations of the constituents.

STEP – 1

The value of calcium hardness is lower than alkalinity value and hence all calcium would present as $CaCO_3$.

∴ Calcium carbonate (as $CaCO_3$) = **68 mg/l**

STEP – 2

Since total hardness is lower than total alkalinity, all magnesium would present as $MgCO_3$. Magnesium is expressed in terms of $CaCO_3$. Hence convert it into $MgCO_3$ by multiplying with 0.84.

308×0.84 = **258.72 $MgCO_3$**.

STEP - 3

The alkalinity value is higher than total hardness, and the value in excess over the hardness indicates the presence of bicarbonates of sodium and potassium. Usually potassium occurs in trace quantities and hence all the bicarbonate remaining is calculated as $NaHCO_3$; but at the drying temperature (more than 100°C) the sodium bicarbonate ($NaHCO_3$) would be converted to sodium carbonate (Na_2CO_3).

$$Alkalinity - Total\ hardness = 408 - 376 = 32\ mg\ CaCO_3.$$

Convert this excess alkalinity (32 mg/l) into sodium carbonate by multiplying with 1.06.

$$32 \times 1.06 = 33.92\ Na_2CO_3.$$

STEP – 4

Now the anions SO_4^{2-}, Cl^- and NO_3^- have to be accounted for. Since all calcium and magnesium combined with CO_3, the cations available are sodium and potassium. As already indicated, potassium occurs only in trace quantities and hence the remaining anions are calculated in combination with sodium only.

i) 96 parts of SO_4^{2-} would combine with 46 parts of Na^+ to form 142 parts of Na_2SO_4.

$\therefore$ 33.1 parts of SO_4^{2-} would combine with Na to form

$= \frac{142}{96} \times 33.1$ = **48.96 Na_2SO_4**.

ii) 35.45 parts of Cl^- would combine with 23 parts of Na^+ to form 58.45 parts of NaCl.

$\therefore$ 40 parts of Cl^- would combine with Na to form

$= \frac{58.45}{35.45} \times 40$ = **65.95 NaCl**.

iii) 62 parts of NO_3^- would combine with 23 parts of Na^+ to form 85 parts of $NaNO_3$.

$\therefore$ 3.5 parts of NO_3^- would combine with Na^+ to form

$= \frac{85}{62} \times 3.9$ = **5.35 $NaNO_3$**.

Now all the cations and anions are balanced.

The water sample contains no iron. So it does not come to the picture. It is assumed that silica exists in free state and it has to be summed up as such with the worked out probable combination.

Constituents		Water of Crystallization	
$CaCO_3$	= 68.00		
$MgCO_3$	= 258.72		
Na_2CO_3	= 33.92		
Na_2SO_4	= 48.96		
NaCl	= 65.95		
$NaNO_3$	= 5.35		
Silica	= 70.00		
	550.90 +	No water of crstallization	= 550.90
			= 551

The theoretically calculated value (551) is in well agreement with the value obtained practically (535).

CHAPTER - 76

DETERMINATION OF STEAM PURITY

Steam purity refers to the amount of solid, liquid or vaporous contamination in the steam. Steam purity is normally expressed in terms of solids content. Precise system control is required for the operation of modern superheated steam turbines. Solids in the steam leaving a boiler can deposit in the superheater and turbines resulting in serious damage. For this reason, close control of steam purity is essential.

METHODS OF STEAM PURITY MEASUREMENT

A number of methods are available for measuring steam purity. Among them, the most important are:

(i) Conductivity method
(ii) Sodium ion Analysis method
(iii) Anion Analysis method.

CONDUCTIVITY METHOD

Measurement of specific conductance is the most commonly used method. It is a rapid and reasonably accurate method. When boiler water is carried over in steam, the dissolved solids content of the boiler water contaminates the steam with the resultant increase in the conductivity of the steam sample. This increase in conductivity is measured and steam purity is thus assessed. The major disadvantage with this method is the contribution to the conductivity by the dissolved gases such as carbondioxide and ammonia. 1 ppm ammonia is reported to increase the conductivity by 8.0 to 9.0 μs (micro siemens) and 1 ppm carbondioxide by 5.0 μs. Thus a sample containing less than 1 ppm dissolved solids with a conductivity in the range of 1.0 – 2.0 μs will show abnormally high conductivity in the presence of ammonia and carbondioxide. In order to obtain a proper measure of dissolved solids, the concentration of these gases in steam have to be measured first and the conductivity equivalent to such concentration have to be found out and deducted from the measured conductivity value. The obtained value is the conductivity of solids in steam from which steam purity can be assessed.

Alternatively, the steam may be degassed using suitable equipments, prior to measurement of conductivity. Such degassing equipments are available readily at recent times.

SODIUM ION ANALYSIS METHOD

This method is based on the measurement of sodium in steam sample. Because water treatment practices are such that most soluble chemical constituents of boiler feed water are converted to sodium salts, most solids present in steam are sodium salts (with minor amounts of calcium, magnesium, iron and copper). Sodium constitutes approximately one third of the total solids in most boiler water (i.e.) if 1 ppm sodium is found, then the solids content of the steam can be regarded as 3 ppm (approximately).

Sodium is measured using sodium ion analyzers. These instruments are reported to operate in the concentration range of 0.1 ppb to 100000 ppm of sodium ion with a sensitivity of 0.1 ppb. Sodium ion analysis method is extremely useful as a continuous monitoring method as well as for field testing.

The concentration of sodium in steam can also be determined using flame photometers. These instruments are much more sensitive to quick changes in operating conditions and detects peaks in solids concentration. They are capable of accurate measurement in low ppb range despite quick variations. When these technics are followed, samples should be collected only in special laboratory – prepared polyethylene bottles and due care should be exercised to protect against contamination.

ANION ANALYSIS METHOD

Anion Analysis is also gaining prominence at recent times as a method to assess steam purity. These instruments measure the concentration of anions such as chlorides, sulfates and acetates etc., in the steam and provides a means of steam purity assessment.

It must be remembered that whatever may be the method of analysis, samples must be truly representative of the steam being generated in order to ensure accurate analysis. When the sampling procedures are not followed properly, the steam purity evaluation is of little or no value.

Dissolved solids are carried into the steam through the entrained boiler water and their concentration vary with the pressure of the boiler. The recommended upper limits are indicated below: for boilers operating below 400 psi, the steam may contain upto 1 ppm of solids, at 600 psi, 0.3 – 0.4 ppm and in superheated steam for high-pressure turbines, the concentration of solids should not exceed 30 ppb.

PART III

MICROBIOLOGICAL ANALYSIS

CHAPTER - 77

MICROBIOLOGICAL ANALYSIS OF INDUSTRIAL WATERS - AN INTRODUCTION

Microbiological examination is regarded as the main component of analysis of water which provides unique information on water quality. The chief function of this examination is to detect the potentially dangerous fecal pollution and the presence of organisms that cause diseases. While this is the case with drinking and domestic waters, industrial waters are subjected to microbiological examination to detect organisms that may contaminate or interfere in process or spoil the food products or develop unwanted growth in the water-carrying structures such as pipes and conduits. There are a wide variety of such organisms present in industrial waters. The most prominent organisms are listed below.

1. Coliform bacteria
2. Proteolytic bacteria
3. Lipolytic bacteria
4. Thermophilic bacteria
5. Gelatin liquefying bacteria
6. Iron bacteria
7. Sulfate-reducing bacteria
8. Sulfur-oxidizing bacteria
9. Yeast and mould
10. Slime-forming organisms

The analytical procedure for the detection and enumeration of these organisms are described in separate chapters. This apart, test for the total number of microbiological organisms is also given as a separate chapter.

TOTAL BACTERIAL COUNT

The number of bacteria present in water is taken into account while assessing the quality of water for industrial use as it gives an idea on the quality of water which is determined by a test termed 'Standard Plate Count'. This test measures the number of microbial organisms present in a water irrespective of the type and hence gives a first hand information on the hygienic status of a water. The test is also useful in assessing the efficacy of certain water treatment processes, namely, coagulation, filtration and disinfection. It also gives an early indication of pollution. For example, if any sudden increase in the microbial count is observed with unpolluted waters, it indicates the access of soil, dust or other pollutants into water sources.

Most industries require water that contain only a few number of microorganisms. Food processing industries need water showing a standard plate count below 50; fermentation industries require water with a total bacterial count below 100. The number of organisms shall not exceed 100 per ml for a water used in the manufacture of rayon. For tannery also, the limit is 100 per ml. For test procedure see Chapter - 82 "STANDARD PLATE COUNT"

COLIFORM BACTERIA

This test is carried out on industrial waters intended mainly for food processing and fermentation. The presence of coliform organisms indicate fecal contamination. Hence absence of coliform organisms in water is an essential requisite for food processing industries including dairies, bakeries, canneries, confectionaries, meat packing and also for fermentation industries. Besides indicating fecal contamination, coliform organisms are also reported to retard fermentation. Coliform organisms are the best lactose fermenters and the lactic acid produced by lactose fermentation would diminish the yeast activity.

The coliform or total coliform group include all of the gram-negative, non-sporing, rod shaped bacteria that ferment lactose in 24 to 48 hrs at 37°C with the production of acid and gas, under aerobic conditions.

There are two types of coliform bacteria – fecal and non fecal califorms. The species Escherichia Coli is exclusively of fecal origin. Other coliforms, even though occur in feces, they are normal inhabitants of soil and water. They are assumed of fecal origin unless shown to be otherwise. Hence in the absence of E.coli, any positive result for the coliform test is regarded as an indication of fecal contamination of water. Separate test procedures are given for total coliforms (fecal + non-fecal) and for fecal coliforms. See Chapter – 83 COLIFORM BACTERIA (TOTAL COLIFORMS)" and Chapter – 84 "FECAL COLIFORMS (ESCHERICHHCOLI)".

PROTEOLYTIC AND LIPOLYTIC BACTERIA

Proteolytic bacteria are a group of bacteria that hydrolyze proteins. The prorteolytic bacteria are primarily the aerobic spore-formers (bacillus). Another group of bacteria namely, Lipolytic bacteria hydrolyzes fats and lipids. Hence waters containing these bacteria are undesirable for food processing industries. For Test Procedure, see Chapters – 85 and 86 respectively.

THERMOPHILIC BACTERIA

Thermophilic bacteria grow well at the temperatures used in pasteurization, especially when the low-temperature method is followed. These organisms are usually found in milk and are able to survive to pasteurization and can grow at such temperatures. They have an optimum range of 50 to 52°C. The thermophilic bacteria are found in soil and improperly cleaned milking machines and on utensils and pasteurizing equipments. If the utensils and equipments are not thoroughly washed, these organisms collect in the milk films and act as foci for seeding new lots and deteriorate milk. For Test Procedure, see Chapter – 87.

GELATIN LIQUEFYING BACTERIA

There is a group of bacteria capable of transforming a gel in to liquid and they are collectively termed as "Gelatin Liquefying Bacteria". Their presence is detrimental in waters intended for gelatin and glue manufacture, photographic film manufacture and food processing industries. Enterobacter liquefaciens and Enterobacter cloacae are the important organisms among this group. For Test Procedure, see Chapter – 88.

IRON BACTERIA

The iron bacteria are of the most important types of nuisance organisms in water. They are generally filamentous and typically found in fresh water. Unlike algae, they are capable of growing in dark and hence they are also seen in pipe lines, reservoirs, stagnant surface waters, wells and springs. They transform soluble compounds of iron to insoluble compounds (ferric hydroxide) which may be deposited in a sheath around the organism (Sphaerotilus) or secreted so as to form stalks or ribbons attached to the cell (Gallionella) or deposit on pipes. The large amount of brown slime so produced will impart a reddish tinge (and an unpleasant odor) to water rendering it unsuitable for drinking and other industrial uses. It is especially troublesome in water intended for boiler feed and cooling. Heavy growths of these organism at times block the passage of water in pipes, valves nozzles and other parts of water distribution systems, and reduce the flow rate considerably. Their presence may cause pitting and tuberculations in pipes.

In many cooling water systems, iron and manganese bacteria growth causes serious problems necessitating treatment. Some of these growth are matted and fibrous in appearance, some other are dense and resemble inorganic scale. They frequently break loose in large masses and clog nozzles, pumps and recirculating systems etc., as indicated already. This apart, the decaying masses often cause bad tastes and odors in water. Further, iron bacteria are reported to coat the filter media and ion exchange resins impeding the operation and resulting in a poorly filtered / softened water. For Test Procedure, see Chapter – 89.

SULFATE-REDUCING BACTERIA

Sulfate-reducing bacteria are widely distributed in nature and are common in soils, lake waters, cooling waters, sea water, sewage and marine mud. Sulfate reducing bacteria are micro organisms capable of reducing sulfate (or other forms of oxidized sulfate) to hydrogen sulfide. The hydrogen sulfide is a corrosive agent and attack severely the metallic structures and concrete with the formation of 'black water'. These are normally seen at dead ends of mains and the black color is due to ferrous sulfide. Sulfate reducing bacteria are reported to be the most common organism found in the slimes that plug up tube of cooling systems. They form a coating over the surface the tubes as slime, acts as an insulator, resulting in reduced heat transfer.

The most common species are *Desulfovibrio desulfuricans, Desulfovibrio vulgaris and Desulfovibrio salexigens*. Another sulfate-reducing bacteria, namely *Dt. nigrificans* (which is thermophilic) is often found in industrial cooling waters. For Test Procedure, see Chapter – 90.

SULFUR-OXIDIZING BACTERIA

Sulfur-oxidizing bacteria are a group of organisms which oxidize elemental sulfur to sulfate or hydrogen sulfide to sulfur. Waters containing sulfur oxidizing bacteria are frequently seen with thread - like growths. These waters often contain elemental sulfur in finely divided colloidal form. This sulfur may be due to the oxidation of hydrogen sulfide by bacteria in the presence of dissolved oxygen. Their presence is highly undesirable in industrial water. The sulfuric acid and hydrogen sulfide produced during their metabolism is destructive to metallic and concrete structures. *Thiobacillus thioparus* and *Thiobaciullus - thiooxidants* are the more common forms belonging to this group. For Test Procedure, see Chapter – 91.

YEAST AND MOULD

Yeasts are unicellular fungi which occurs as egg shaped or spherical cells by simple budding. They form characteristic creamy colonies on culture. They are easily differentiated from bacterial colonies by their relatively larger size and morphological features. The size of yeasts vary considerably ranging from 1 to 5 micro meter (or more) in length. Moulds are filamentous fungi and they form true mycelia and reproduce by the formation of different types of spores.

Yeast and mould in water are highly undesirable as they tend to deteriorate the quality of the final product or interfere in the process. Pulp and paper industry, textile industry, rayon manufacturing industry are some of the industries which virtually need water absolutely free from yeast and mould.

For Test Procedure, see Chapter - 92

SLIME-FORMING ORGANISMS

Many micro-organisms are capable of producing gummy or mucilaginous materials, either as capsular structures or as an extra cellular excretion products. This group includes a wide variety of organisms. Among them the most important ones are iron bacteria, sulfate-reducing bacteria, certain other higher bacteria (myxobacters) and some fungi. The presence of these slime-forming organisms is especially undesirable in water intended for paper mills and food processing industries. Besides, the slimes formed inside the pipes and tubes impede heat transfer, particularly in cooling towers. Hence cooling waters should be free from slime forming organisms. The slimes also act as a base for the profused growth of other bacteria and certain algae. To eliminate these organisms, normally waters are treated with chlorine or other suitable disinfectants.

These slimes also coat resin particles of ion exchangers and block the surface of the resin resulting in a severe loss of exchange capacity. The bacterial slimes may be destroyed by chlorination of the feed water or by the application of bleach liquors along with regenerant solution. Slimes formed over the resins may also be removed by applying warm caustic soda solution to the resin. After applying caustic soda, the softener should be allowed to stand for several hours to ensure complete removal of slimes and then backwashed to remove the dead slime from the bed.

For Test Procedure, see Chapter - 93

CHAPTER - 78

EQUIPMENTS, GLASSWARE AND OTHER ITEMS REQUIRED FOR MICROBIOLOGICAL ANALYSIS

In this chapter the essential equipments, glassware, other items and media required for bacteriological examination are listed out.

EQUIPMENTS

1. **Analytical Balance.**
   Accuracy 1 mg.
2. **Autoclave with pressure gauge.**

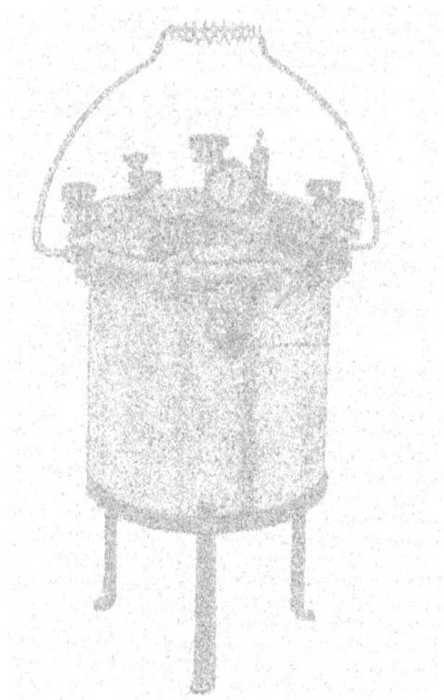

Fig 78-1. Autoclave with pressure gauge

3. **Hot air Sterilizer** of adequate size to prevent crowding of the interior.
4. **Steam –sterilizer:** (Koch (or) Arnold type)
5. **Circular water** bath made up of stainless steel with concentric rings.

6. Membrane Filtration Apparatus – Complete Set with oil – free Pump

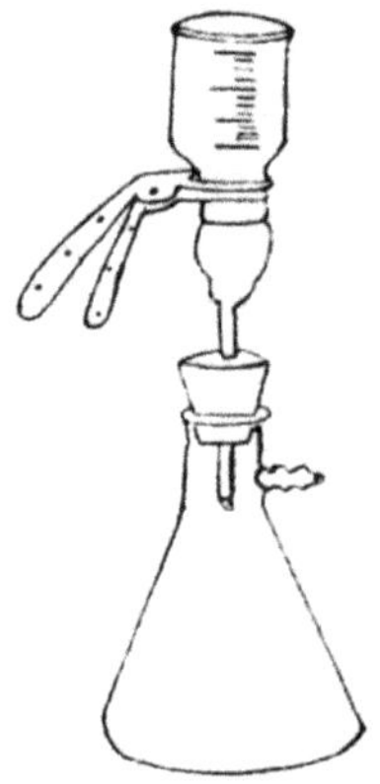

Fig 78–2. Membrane Filtration Apparatus

7. Bacteriological Incubator with thermostat control and glass doors inside. The incubator must control temperature within specified tolerance limits of the tests.

Fig 78–3. Bacteriological Incubator

8. Serological Water bath, made up of stainless steel to incubate cultures under examination at 44°C.

9. Colony Counter - Transistorised digital colony counter that provideds good visibility and magnification, with guide plate grid marked.

Fig 78-4. Colony Counter

10. Microscopes: Stereo microscope 1 No for identifiying and counting the colonies when membrane filter technique is used and microscope (ordinary) 1 No with an oil immersion lens of magnification (x 100).

11. Refrigerator: To store media and reagents.

12. LPG gas cylinders, with regulator, burner and LPG stove.

GLASS WARE

1. Sampling bottles

These are used for sampling raw waters and samples after each stage of treatment. These bottles should be made up of good quality glass, free from alkali. The bottle should have a ground glass stopper with an overlapping rim to protect the mouth from the dust and contamination by figures during handling.

Capacity: 125 ml , 250 ml , 500 and 1000 ml.

Screw capped bottles are also satisfactory provided bacteriostatic or nutritive compounds are not produced from the liners or caps. Plastic bottles can also be used , but they have to withstand the sterilizing conditions.

2. Dilution bottles or tubes

Borosilicate or other good quality glass bottles with screw caps and inert liners. Bottles / tubes with graduation on the side is preferable.

Capacity: Bottles - app 150 ml. Tubes - app 50 ml.

3. Reagent bottles

Made up of borosilicate glass, plain, narrow-mouth with interchangeable flat head stopper.

Capacity: 125 ml, 250ml and 500 ml.

4. Pipets

Mohr type pipets are found to be useful. Generally 10 ml and 1 ml pipets find more use in bacteriological examination. Pipets with straight sided unbroken tips, and clearly marked graduations should be used. Pipets with damaged tips should be discarded. Before sterilization each pipet should be plugged with non-absorbent cotton at the mouth end to prevent contamination by organisms from the mouth. (No 10 ml pipet should have an error of ±1% and no 1 ml pipet should have an error of more than ±2%)

5. Petridishes

App OD × height = 100 mm × 17 mm

Petridishes with beaded edges have to be used. The bottom of the dishes shall be free from bubbles and scratches, and shall be flat, so that the thickness of the medium would be uniform throughout the plate.

6. Test tubes

Test tubes with the following dimensions are used for dilution and culturing.

App. length x OD

- 150 mm x 25 mm
- 150 mm x 18 mm
- 150 mm x 16 mm
- 100 mm x 12 mm

For the preparation of dilutions, the tubes of size 150 mm × 25 mm are to be used. For the test for coliforms and other organisms for which 10ml double-strength medium is required, the most satisfactory size of the tube is 150 mm × 18 mm. Where 5m single strength medium is required , the tubes of size 150 mm × 16 mm are suitable. For carbohydrate fermentation tests (sugar reactions) the tubes of 100 mm x 12 mm are satisfactory.

7. Durham Tubes

These are small glass tubes closed at one end. They are placed in inverted position in tubes containing liquid medium to find out the production of gas during reaction. Size of Durham tubes to be used with 5 and 10 ml volumes of medium is 35 mm×8 mm (length × OD), and for the 50 ml volumes 75 mm × 10 mm. For tubes containing sugar medium (for the test for fermentation of carbohydrates), Durham tubes of size 25 mm × 3 mm is preferred.

8. Flasks

Made up of glass , boiling , florence ,flat bottomed with the following capacities are required for the preparation of media and for inoculating high volumes of water into the medium.

- 2000ml
- 1000ml
- 500ml
- 250ml
- 100ml

9. Beakers

Glass with spout, Griffin, low form of various capacities are needed for dissolution and general use.

- 2000ml
- 1000ml
- 500ml
- 400ml
- 250ml
- 100ml

10. Measuring cylinders

Glass / polypropylene (translucent) / TPX (glass like transparency)

Measuring cylinders with broad base are preferable. The following capacities are required.

- 1000ml
- 500ml
- 250ml
- 100ml
- 50ml
- 10ml

11. Funnels

Glass, Plain, 60° angle, dia 100 mm and 75 mm. Buchner funnel is also required for filtration of certain media.

12. Glass Slides

For microscopical examination (Gram – staining)

13. Glass Rods

App. length × dia.

200 mm × 9 mm with a small button at one end

14. Glass tubes

App. length × dia.

200 mm × 4 mm

MISCELLANEOUS ITEMS

1. Petridish case

With lid and retriever. Made up of stainless steel.

App length × dia.

25cm × 12 cm.

2. Pipet case

With lid, made up of stainless steel.

App length × dia.

40 cm × 5 cm.

3. Inoculating Loops and Needles

Length of the wire 6.5 cm, dia of the loop 3 mm.

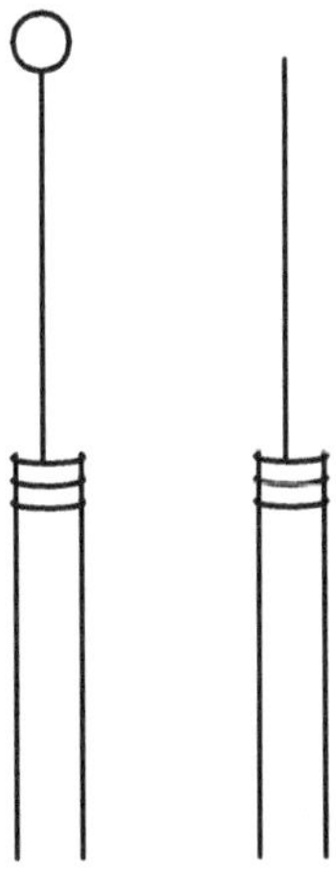

Fig 78-5. Inoculating Loop and Needle

4. Nichrome Wire: (24 SWG - 0.56 mm)

5. Test tube Basket

Stainless steel /Aluminium / PVC/ Polythene.

Different sizes are required .

15 cm × 15 cm × 15 cm

15 cm × 10 cm × 10 cm

6. Test tube stand (Rack),Poly carbonate.

7. Kahn racks: To hold tubes to keep them at 44°C. Made up of stainless steel.

8. Thermometers: Temperature ranges

0 – 55° C

0 – 110°C

0 – 300°C

9. Cotton – Absorbent

10. Cotton – Non absorbent in different colors

11. Slip on caps (Aluminium or plastic that can withstand autoclaving)

MEDIA AND REAGENTS

Dehydrated Media

1. Formate Ricinoleate Broth
2. Mac Conkey's Broth (Double Strength)
3. Mac Conkey's Broth (Single Strength)
4. Medium for Thiobacillus thiooxidans (Sulfur Medium)
5. Medium for Thiobacillus thioparus (Thiosulfate – oxidizing Medium)
6. Membrane Endo Agar (M – Endo Agar)
7. Membrane Fecal Coliform Medium (M – FC Medium)
8. Modified Formate Lactose Glutamate Medium (Double Strength)
9. Modified Formate Lactose Glutamate Medium (Single Strength)
10. Potato Dextrose Agar (for Yeast and Mould)
11. Sulfate – reducing Medium (Starkey's Medium)
12. Tributyrin Agar
13. Yeast Extract Agar

Bacteriological Ingredients and Chemicals

If dehydrated media are not available or if it is intended prepare the media in the laboratory itself the following bacteriological ingredients and chemicals are required.

1. Agar
2. Ammonium chloride
3. Ammonium dihydrogen phosphate
4. Ammonium sulfate [$(NH_4)_2 SO_4$]
5. Amyl alcohol (or) Iso amyl alcohol
6. Aniline blue
7. Basic fuchsin
8. Beef extract
9. Bile salt
10. Bile salt No.3
11. Bromocresol purple
12. Bromothymol blue

13. Calcium chloride ($CaCl_2.2H_2O$)
14. Casitone
15. Dipotassium hydrogen phosphate (K_2HPO_4)
16. Ethyl alcohol
17. Ferric chloride ($FeCl_3.6H_2O$)
18. Ferric Citrate (scales)
19. Gelatin
20. Glucose
21. Hydrochloric acid, con (Sp.gr 1.18)
22. Isopropyl alcohol
23. L – Arginine monohydro chloride
24. L - Aspartic acid
25. L - Cystine
26. Lactose
27. Magnesium sulfate ($MgSO_4.7H_2O$)
28. Meat extract
29. Methyl red
30. Mohr's salt (Ferrous ammonium sulfate)
31. Neutral red
32. Nicotinic acid
33. Pantothenic acid
34. Paradimethyl amino benzaldehyde
35. Peptone
36. Phenol red
37. Potassium chloride
38. Potassium dihydrogen phosphate (KH_2PO_4)
39. Proteose peptone No.3
40. Ringer Tablet or Powder
41. Rosolic acid
42. Skimmed milk powder
43. Sodium bicarbonate
44. Sodium chloride
45. Sodium citrate

46. Sodium desoxycholate
47. Sodium formate
48. Sodium glutamate
49. Sodium hydroxide pellets
50. Sodium lactate
51. Sodium lauryl sulfate
52. Sodium ricinoleate
53. Sodium sulfate
54. Sodium sulfite
55. Sodium thiosulfate ($Na_2S_2O_3.5H_2O$)
56. Thiamin (Aneurin hydrochloride)
57. Thiopeptone
58. Tributyrin
59. Tryptone
60. Tryptose
61. Yeast extract
62. α-naphthol

CHAPTER - 79

LABORATORY TECHNICS

Before proceeding into the bacteriological examination of water, the analyst must be conversant with the general laboratory practices and technics involved in bacteriological examination. They are dealt under the following heads.

1. Cleaning of Glassware
2. Preparation of Media
3. Sterilization
4. Preparation of Dilutions
5. Inoculation
6. Incubation
7. Plating on Solid medium
8. Colony Picking
9. Good House Keeping and General Instructions

CLEANING OF GLASSWARE

In microbiological examination cleanliness of glassware is vital in securing valid results. All the glassware are cleaned by soaking overnight in 1 + 9 hydrochloric acid, and rinsing up in tap water followed by a final rinse with distilled water. Alternatively, a good laboratory grade detergent may be used.

PREPARATION OF MEDIA

Dealt elaborately in Chapter – 80 "PREPARATION OF MEDIA AND REAGENTS".

STERILIZATION

'Sterilization of Media' and 'Sterilization of Glassware' are important steps involved in bacteriological examination. Sterilization of media is discussed in detail in Chapter - 77 "PREPARATION OF MEDIA AND REAGENTS". Sterilization of glassware is given below.

The hot air oven is used to sterilize glassware such as petridishes, pipets, sample bottles and flasks.

Petri dishes

Place the clean petridishes in the petridish case (made up of stainless stell) and sterilize in the hot air-oven at 160°C for 1 hour.

Pipets

Place a pad of cotton* at the bottom of a pipet case (stainless steel). Then place 10ml/1ml pipets inside. Also place a pad of cotton in the lid, and close the case, and sterilize in the hot air-oven at 160°C for 1 hour. It is to be mentioned that a small cotton plug should be thrust into the mouth of pipet before sterilization.

Sample Bottles

Before sterilization of sample bottles, a strip of paper of about 75mm by 10mm should be inserted between the stopper and the neck of the bottle to prevent jamming of the stopper and cracking of the glass on cooling. Atleast half of the length of the paper should project on the outside of the bottle so that when the stopper is taken out of the bottle, the paper can be removed without contaminating the rim by fingers. It is most important to wrap the bottles completely with kraft paper (or aluminium foil). The bottles are sterilized at 160°C for 1 hour by placing them in hot air oven.

The following precautions have to be observed before sterilization in hot air oven.

(i) The glassware must be absolutely dry.

(ii) The oven is not hot while articles are placed inside. The temperature should be steadily increased to the specified temperature and maintained for the stipulated period for effective sterilization.

(iii) The articles have to be taken away from the oven only after they have cooled down.

(iv) Temperature above 160°C is highly undesirable, as higher temperature results in the charring of the kraft paper wrapped over the bottle and cotton plug inserted in the mouth of the pipets.

---

* Cotton is placed to protect the pipet from damage.

Alternatively sample bottles and dilution bottles may be sterilized in an autoclave at 121°C (15 lb/in^2) for 20 minutes.

Inoculating Needles and Loops

These are sterilized by keeping them in the flame of the burner to redness. After sterilization, the needles and loops should be allowed to air cool otherwise it may kill the bacteria under examination.

PREPARATION OF DILUTIONS

After the sample has been shaken thoroughly (as described above), one or more tenfold dilutions have to be made as shown in Figure 79-1 depending upon the nature of the sample. Dilution is made in the tubes containing 9 ml Ringer solution.

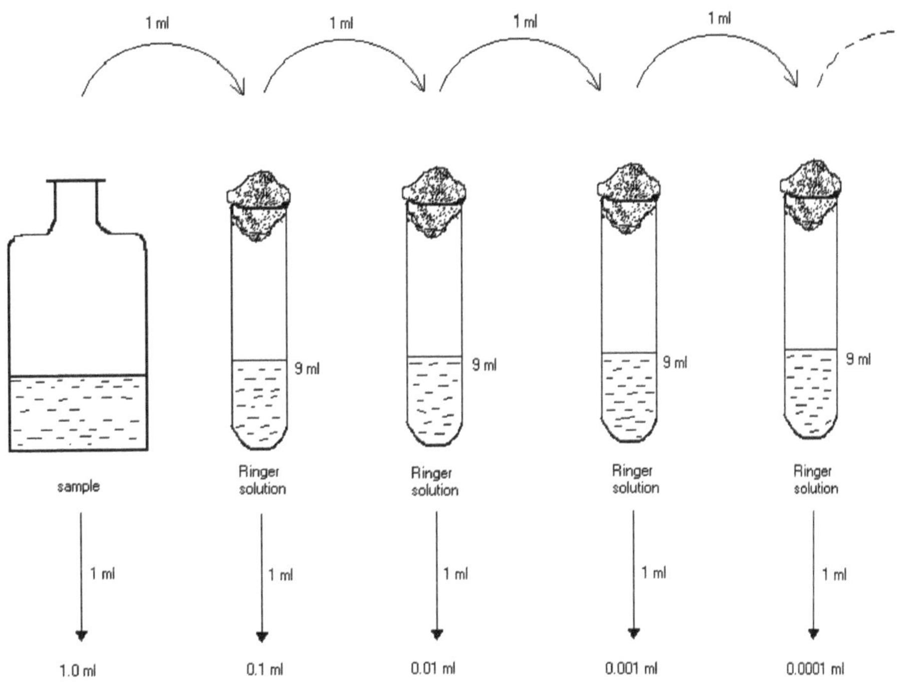

Fig 79-1. Diagrammatic representation of the Dilution technic

Ringer solution (Quarter strength) may be prepared using Ringer tablet or powder available commercially. The specified quantities of the tablet or powder has to be dissolved completely in the specified volume of distilled water. Then the solution is placed in 9 ml quantities in test tubes of size 150 mm x 25 mm and auto claved at 121°C for 20 minutes.

INOCULATION

The term 'inoculation' means the introduction of water samples or bacterial cultures into sterilized media under aseptic conditions.

The following guidelines would be useful in achieving aseptic conditions.

(i) All the glassware should be sterilized.

(ii) Do not touch any part of the container, pipet etc., which will come in contact with the sample or culture.

(iii) Flame the top of the culture tube immediately after removal of the cap (or cotton plug).

(iv) Do not raise the lid of the petridish longer than the height required.

Volume of water samples are transferred using pipets to culture medium. The pipet should be lightly flamed along the entire length by passing through the flame of the burner, prior to drawal of water from the sample bottle. The transfer of culture from one tube to a fresh tube (termed as 'subculturing') is achieved with the aid of a bacteriological loop. The internal diameter of the loop is normally 3mm. A loopful of the culture holds approximately 5μl and contains sufficient bacteria for subculturing. The loop is sterilized before and after use by heating to redness the entire length of the wire, in the flame of the burner. Similarly, a straight wire (without loop) is used to inoculate a semi solid medium for the motility test. The straight wire is dipped in the grown up culture and stabbed deep into the middle of the semi solid medium in the tube.

INCUBATION

After inoculation of the sample, the tubes / petridishes have to be placed in the incubator to provide the desired growth temperature for the organism for the specified period. The petridishes should be placed in the inverted position in the incubator. Most of the organisms grow well at 37°C. Thermostatically controlled air incubators are used for keeping at 37°C.

PLATING ON SOLID MEDIUM

To obtain pure culture of the organism, a loopful of culture is transferred from the tube containing broth and placed over the solid medium specified for the organism) in a petridish. Then streak the inoculum over the plate by touching the surface of the agar lightly with the loop. Either simple-streak is made as shown in Fig. 79–2 or cross-streak as shown in Fig. 79–3.

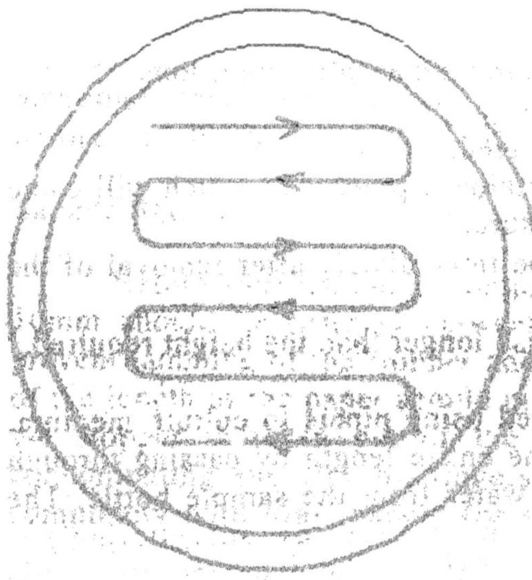

Fig 79-2. Simple streak

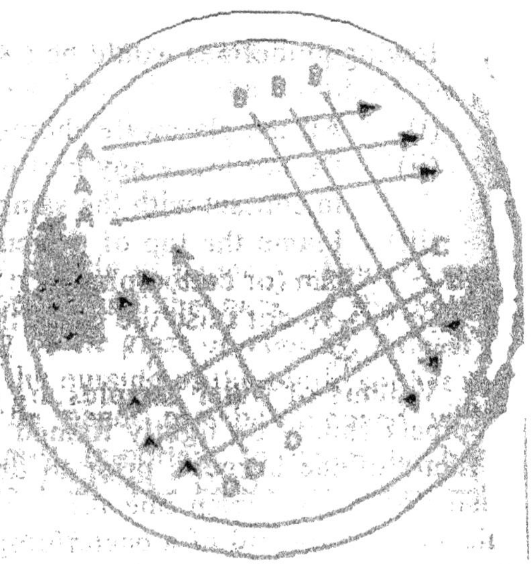

Fig 79-3. Cross streak

Place a loopful of culture at one end of the plate and spread out as well. Make more distinct streaks (AAA) with sterilized loop. Make further cross-streaks (BBB), (CCC) and (DDD) as shown in the figure using separate loops for each streak. Whatever may be the method of streaking, the resulting culture should show a number of discrete colonies.

COLONY PICKING

On incubation for a specified period colonies grown on the solid media. A few of the colonies seem to be characteristics of the organism should be picked up using a bacteriological loop and transferred to the broth in the tube aseptically.

In case, where colonies are small that could not be picked up using bacteriological loop and distance between adjacent colonies is very small, a straight wire (similar to bacteriological loop but without the loop end) may be used for picking colonies.

GOOD HOUSE KEEPING AND GENERAL INSTRUCTIONS

The laboratory intended for microbiological examination should be free from dust. High standards of cleanliness should be maintained in work areas. Benches should be cleaned everyday and work areas swabbed with methylated spirit before starting the work. Floors should be wet-mopped with water containing a suitable disinfectant.

All used glassware items such as petridishes, test tubes, slides etc., should be placed in a bin and disinfected at 121°C (15lb/in^2) for 20 minutes.

Used pipets need to be disinfected by placing them in a plastic tube containing suitable disinfectant solution and keeping them overnight in the solution. Then they should be thoroughly washed with tap water and finally rinsed with distilled water.

CHAPTER - 80

PREPARATION OF MEDIA AND REAGENTS

The preparation of culture media and reagents plays a vital role in the bacteriological examination of water. Now a days, most of the media are available readily in dehydrated form. Use of these commercial dehydrated media involves only weighing and rehydration, thus totally eliminates the laborious procedures to be carried out in the preparation of media. Hence purchase and use of commercially available good quality dehydrated culture media are strongly recommended.

Follow the manufacturers' instruction for rehydration and sterilization. However in case of non-availability of dehydrated media, they can be prepared in the laboratory itself. Before proceeding into the preparation of media, the basic components used and principles of media preparation and sterilization are discussed briefly.

The basic ingredients of bacteriological media are peptone, agar and bile salts. Needless to say water is the most important constituent.

Wherever the word water is used in the preparation of media, it denotes distilled water or deionised water, which is free from traces of dissolved metals and bactericidal or inhibitory compounds. Where peptone is mentioned, any standard bacteriological peptone will be satisfactory unless a particular grade is specified. Most peptones form clear solutions. However, some peptones may form a precipitate when alkali is added. Media containing these peptones need to be filtered.

Agar is not a nutrient. It is added in order to solidify the liquid medium. Superior quality agars are available at recent times. A satisfactory gel strength can be obtained with 0.8% to 2% of agar in the medium depending on the quality of the agar. Hence the quantity of agar mentioned in the media may be slightly modified to obtain a satisfactory gel strength. Most of the commercially available agars are excellent in character in that they form only clear solutions and do not form any precipitate when mixed with other ingredients or on sterilization. If such agars are not available and agars of poor quality need to be used, the medium should be filtered.

GENERAL INSTRUCTIONS FOR THE PREPARATION OF BACTERIOLOGICAL MEDIA

In the preparation of liquid media, ingredients are dissolved, pH adjusted, tubed and sterilized. Then stored in refrigerator or kept as stock solutions. Separate precautions are not necessary.

While preparing solid media, it is generally advisable to dissolve first the soluble ingredients such as the salts, peptone, beef extract etc., The agar is added in the last. For dissolving the agar, it is preferable to use water bath than direct heating to avoid the possibility of concentration and sticking to the bottom of the flask.

pH adjustment

All micro-organisms have a particular 'reaction pH' at which growth is optimal. Hence, in order to secure the best growth, adjustment of the pH (which controls the reaction) should be made as accurately as possible. There are two methods of pH measurement. One is the method of using electrically operated pH meters. The other method involves the use of indicator dyes that change the color at different pH values. Given below in the Table 80-1 are the common indicators that are generally employed and their color changes.

TABLE 80-1
INDICATORS AND THEIR pH RANGES

S. No.	Indicators	pH Range	Color change
1	Thymol Blue (Acid range)	1.2 - 2.8	Red to Yellow
2	Bromophenol Blue	2.8 - 4.6	Yellow to Violet
3	Bromocresol Green	3.6 - 5.2	Yellow to Blue
4	Methyl Red	4.4 - 6.2	Red to Yellow
5	Bromocresol Purple	5.2 - 6.8	Yellow to Violet
6	Bromothymol Blue	6.0 - 7.6	Yellow to Blue
7	Neutral Red	6.8 - 8.0	Red to Yellow
8	Phenol Red	6.8 - 8.4	Yellow to purple pink
9	Cresol Red	7.2 - 8.8	Yellow to Violet red
10	Thymol Blue (Alkaline range)	8.0 - 9.6	Yellow to Blue
11	Phenolphthalein	8.3 - 10.0	Colorless t Violet red
12	Thymolphthalein	9.3 - 10.5	Colorless to Blue

The solution of these indicators and discs indicating their color changes are readily available. To compare the colors with the use of these discs, a comparator. has to be purchased.

To adjust the pH, take 10ml of medium (of which the pH has to be adjusted) in both the comparator tubes. Place the tubes in the two slots provided in the comparator. Place the color disc of the corresponding indicator in the hinged door of the comparator and set the disc to the required pH by rotating it. Now one of the tube appears colored and the other will not exhibit any color. Add sodium hydroxide (0.1N or 0.05N) [or hydrochloric acid (0.1N or 0.05N)] dropwise until medium in the tube acquires the color. Find out the volume of sodium hydroxide (or hydrochloric acid) required for adjusting the pH. Now the total volume of the alkali (or acid) required for pH adjustment is calculated from the following formula and added to the total volume of the medium.

$$\text{Amount of alkali or acid required to adjust the whole of the medium} = \frac{\text{ml (NaOH or HCl)}}{10} \times \text{Total volume of medium}$$

Instead of dye indicators, pH meter can be used to measure the pH values and to determine the quantity of alkali or acid required.

Filtration of Media and Agars

Liquid media can be filtered through filter paper or by using Buchner funnel. Melted agar is filtered through cotton while it is still hot (80° to $90^{\circ}C$).

Tubing of Media

Media is filled into test tubes by using glass funnels into which liquid medium is filled up and poured into tubes through a rubber tubing fitted with a tapered delivery tube and pinch cock to control the flow of medium. While filling the medium, close the mouth of the funnel using a large dia petridish.

If Durham fermentation tubes are to be placed inside the tubes, they should be placed before filling the tubes with the medium. Otherwise air will be entrapped and mimic like a positive.

After filling with media the tubes have to be closed by plugging the mouth with good grade non-absorbent cotton. The plug should be inserted to a depth of 3 to 4 cm in the tube and about 2cm should project outside. It should fit firmly but not so tightly otherwise removal of cotton plug from the tube will become difficult. At recent times, aluminium caps and caps madeup of polymers are available. They can be used instead of cotton-wool plugs. Solid media may be distributed in test tubes and closed with cotton wool plugs or slip-on metal caps.

STERILIZATION

Sterility is achieved by employing dry heat or moist heat or by filtration. Glassware and other articles are sterilized by dry heat. See Chapter – 76 "LABORATORY TECHNICS".

Media is sterilized preferably by moist heat. However certain liquids cannot be sterilized by moist heat as they are heat sensitive. Such components are sterilized by filtration.

Sterilization by moist heat

Most heat is preferred over dry heat for sterilization of liquids. Auto- clave is used in the laboratory for moist-heat sterilization. Steam and pressure leads to effective sterilization as it has good penetrating power and coagulates microbial protoplasm. The relationship between the pressure of steam and temperature is shown in Table 80-2.

TABLE 80-2
RELATIONSHIP BETWEEN STEAM PRESSURE AND TEMPERATURE

Pressure		* Temperature °C
Lb/in^2	Kg/Cm^2	
0	0	100
5	0.35	108.3
10	0.70	115.5
15	1.05	121.6
20	1.41	126.6
25	1.76	130.5
30	2.11	134.4

* Values for pure steam. It is correct only if all air is evacuated from the sterilizing chamber. Since a mixture of air and steam at a given pressure gives a lower temperature than given by pure steam only.

The sterilized media after cooling should be stored in refrigerator. Refrigerated media should be allowed to reach room temperature before use.

CHAPTER - 81

SAMPLING OF WATER FOR MICROBIOLOGICAL ANALYSIS

The success of bacteriological examination mainly depends on sampling. It is important that a sample should be representative of the water to be examined. Utmost care is essential to avoid extraneous contamination during collection. Experienced personnel may be employed for collection of samples to avoid any error.

The methods of sampling described below are applicable to industrial water supplies from sources such as well, rivers, streams, lakes, reservoirs and pipelines and conduits, processing tanks and vats, spray ponds, cooling towers and from each stage of water treatment.

Sampling Bottles

Sterilized glass bottles (made of good quality glass - free from excessive alkali) should only be used. The bottle should have ground-glass stopper with an overlapping rim to protect the lip of the bottle from contamination of particulate matter in the air and from contamination caused by fingers during removal of the stopper. Alternatively screw capped bottles may be used. Normally aluminium screw caps are used. However the inner side of it should be lined with silicone rubber which will withstand repeated sterilization at 160°C. The liners/cap should not release any nutritive/bacteriostatic compound. Plastic bottles withstanding sterilization at 160°C can also be used.

The size of the bottle depends upon the organisms for which tests are carried out and testing procedures. The minimum capacity of the bottle is 300ml. However higher capacity (1000ml) may be required for membrane filtration technic.

Preparation of bottles

Sample bottles have to be cleaned thoroughly by soaking overnight in 1+9 hydrochloric acid and rinsing in tap water followed by a final rinse with distilled water. Then it has to be sterilized. Before sterilization, a strip of paper of about 75mm by 10mm should be inserted between the stopper and neck of the bottle to prevent jamming of the stopper and cracking of the glass on cooling. Atleast half of the length of the paper should project on the outside of the bottle so that when the stopper is taken out of the bottle, the paper can be removed without contaminating the rim by fingers. It is most important to wrap the bottles completely with kraft paper (or aluminium foil). The bottles are sterilized at 160°C for one hour by placing them in hot air oven. If the water to be sampled contains or is likely to contain chlorine/chloramine, place 0.1ml of 3% solution of sodium thiosulfate in the bottle before sterilization. Alternatively, sample bottles may be sterilized in an autoclave at 121°C for 20 minutes.

Method of collection

The bottle should be kept stoppered until the moment it is required for filling. Do not rinse the bottle with sample. Remove the string which fixes the cover with bottle. Remove the stopper and keep it in one hand and collect sample and stopper immediately. Until the completion of collection of water, the stopper and neck of the bottle should not be allowed to touch any thing. If the sample is to be collected from a spring or a surface water course, remove the cover also. After sampling replace the same in its original position on the bottle.

Order of taking sample

When a sample is being collected for various examination (namely chemical, bacteriological and microscopical examination) from the same sampling point, the sample for bacteriological examination should be collected first.

Sampling from open wells

Collect the sample as described above using a weighted sampling bottle with string and rope. If the well is provided with an electrical pump, the delivery point of water should be sterilized by flaming it using a pad of cotton wool or asbestos rope soaked in alcohol* for about 10minutes. Then allow water to waste for about 10minutes. The water should then be allowed to flow directly in to the bottle from delivery point.

Sampling from a drilled source (bore well)

Collect the sample as described above by flaming the mouth of the delivery tube allowing the water to waste for sometime and then collecting in the bottle.

* cotton wool or asbestos soaked in alcohol when burnt will not deposit soot.

Sampling from springs and surface water course

In order to obtain a representative sample, the sampling point must be located in such a way that it is not too near the bank, or too far.

While taking the sample, the cover and stopper of the bottle should be removed and held in one hand. Keeping the base with the other hand, the bottle should be plunged neck downwards below the surface for a distance of about one foot. The bottle should be then inverted and filled so that the neck points towards water current. When completely full, the bottle should be brought rapidly above the surface and immediately stoppered. Where it is not possible to collect the sample directly, the sample may be obtained by lowering into water a specially weighted sampling bottle*. To lower the bottle into water, use a rope the one end of which is tied to the string attached to the neck of the bottle firmly (the string is tied to the bottle even before sterilization)

Sampling from open vats or tanks

As described above

Sampling from a sample line or tap

Allow the water to run from the sample line or tap for at least 5 minutes or long enough to flush the entire part of the system. Turn off the sample outlet. Flame the outside of the outlet using a pad of cotton wool or asbestos rope soaked in alcohol, until the whole tap is unbearably hot to the touch. The tap should be cooled by allowing water to run to waste for a few seconds.

Use appropriate sample bottle for collection (i.e) if the water to be sampled contains chlorine, the bottle containing sodium thiosulfate has to be used. If does not contain any chlorine, use bottle that does not contain sodium thiosulfate.

Remove the stopper from the sample bottle. Hold the bottle by the bottom to avoid touching the neck. Quickly hold the bottle under the flowing water (gentle stream) and fill it to three fourths of its volume. Replace the stopper immediately.

---

* Instead, stainless steel jug (round-bottomed) the inside of which is previously sterilized by igniting methylated spirit can be used.

Sampling at various stages of treatment

The ducts/canals carrying water after each treatment have to be provided with taps to facilitate drawal of water for analysis. These taps should be sterilized by flaming as described above and after allowing the water to waste for about 10 minutes, the sample has to be collected in the bottle.

Preservation of sample

Do not add any preservative to the sample. The sample should be examined as soon as possible after collection. If the sample is held more than 2 hrs before examination, it has to be stored in a refrigerator at a temperature not more than 10°C. (If the sample is to be transported, it has to be kept in an iced container). In no case, the interval between collection and analysis of the sample must be more than 6 hrs.

Labelling

The sample has to be labelled with the following information.

(i) Sample Number
(ii) Date and Time of collection
(iii) Source of sample
(iv) Exact point of sampling
(v) Temperature of the sample during collection
(vi) Signature of sample collector.

Information to be submitted along with the sample

In addition to labelling, other information relevant to sample (such as treatment it has undergone - filtration, chlorination or other treatment) have also to be furnished along with the sample. For other details see Chapter – 5, "SAMPLING OF WATER FOR CHEMICAL ANALYSIS" in Part – II.

CHAPTER - 82

STANDARD PLATE COUNT (TOTAL BACTERIAL COUNT)

POUR PLATE METHOD

Among the methods available for enumeration of viable bacteria the pour plate method is widely employed and more useful, as it reveals to certain extent the nature of bacteria also.

The pour plate method is a direct method of determination.

Summary of Test

1.0ml or dilutions of 1.0ml of the sample (if it is known to contain more bacterial organisms) are grown on specified agar media and the colonies are counted. The preferred medium is yeast extract agar.

Glassware

(i) Sterilized petridishes – bottom plate 100mm dia, side wall height minimum 15mm. Should be free from scratches and bubbles and be flat.

(ii) Sterilzed pipet 1 ml capacity.

Media and Reagents

1. Yeast extract agar

Dehydrated form of this agar is readily available in the market. If not available, prepare the agar as per the procedure given below.

Yeast extract	....	3 g
Peptone	....	5 g
Agar	....	15 g
Distilled water	....	1000 ml

Dissolve yeast extract and peptone in water. Adjust the pH to 7.3. Add the agar and steam to dissolve. Distribute in 15 ml quantities in tubes of size 150mm × 18 mm and autoclave at 115°C (10 lb/in^2) for 20 minutes. The final pH should be 7.2. Store it in a refrigerator.

2. Ringer Solution (Quarter Strength): Ringer tablets are readily available commercially. If not available, it can be prepared as follows:

Sodium chloride	....	2.25 g
Potassium chloride	....	105 mg
Calcium chloride	....	120 mg
Sodium bicarbonate	....	50 mg
Distilled water	....	1000 ml

Dissolve all the salts in distilled water. Place 9 ml quantities in tubes of size 150mm × 25mm and autoclave at 121°C (15 lb/in^2) for 20 minutes.

Procedure

Melting of Agar

Place the required number of agar tubes in a steel jug containing water. The level of water in the jug should be higher enough to atleast 3cm above the level of agar in the tube. Heat the jug until the agar melts completely. Allow to cool it in the atmosphere until a temperature of 45° to 50°C is reached.

During cooling , dilutions of the sample have to be prepared and volumes of sample are to be placed in the petridishes.

Preparation of dilutions

Normally 1.0ml sample is taken for standard plate count test. In case if dilutions are to be made, follow the procedure given in Chapter - 79, "LABORATORY TECHNICS".

Raw waters and samples collected from stages of treatment may need dilution before proceeding with the test.

Placing sample in the dish

Shake the bottle vigorously 25 times in order to distribute the bacteria evenly throughout the water. Remove the stopper and flame the mouth. Withdraw 1.0 ml of the sample from the bottle using a 1ml sterilized pipet and place it in the sterile petridish, lifting the lid just enough to permit the operation as shown in figure 82-1. Close the lid immediately as soon as the sample is placed.

(As indicated above, raw waters and samples collected from stages of treatment may need dilution before proceeding with the test. In these cases withdraw 1.0ml of 1 in 10 dilution, 1 in 100 dilution using a 1.0ml sterilized pipet and place it in the petridish as described above and proceed. Finally multiply the count by the dilution factor).

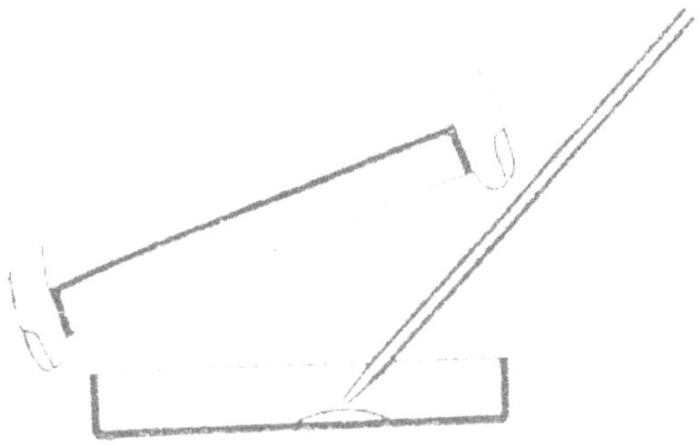

Fig 82-1. Placing the sample in petridish

Mixing Agar with the sample

When the molten agar reaches 45°C, raise the lid of the dish as little as possible and pour the agar into it quickly (See figure 82-2). See that the agar is poured not over the sample but at a little distance. Pouring over the sample may kill some bacterial cells. Also see the lid is held just above the base to minimize air-borne contamination.

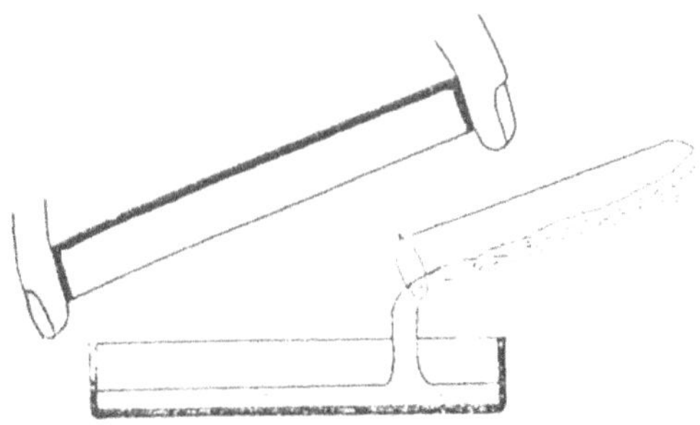

Fig 82-2. Pouring the molten agar

Keeping the plate flat on the table, mix the agar and the sample by moving the plate 5 times in each of the following directions (Fig 82-3).

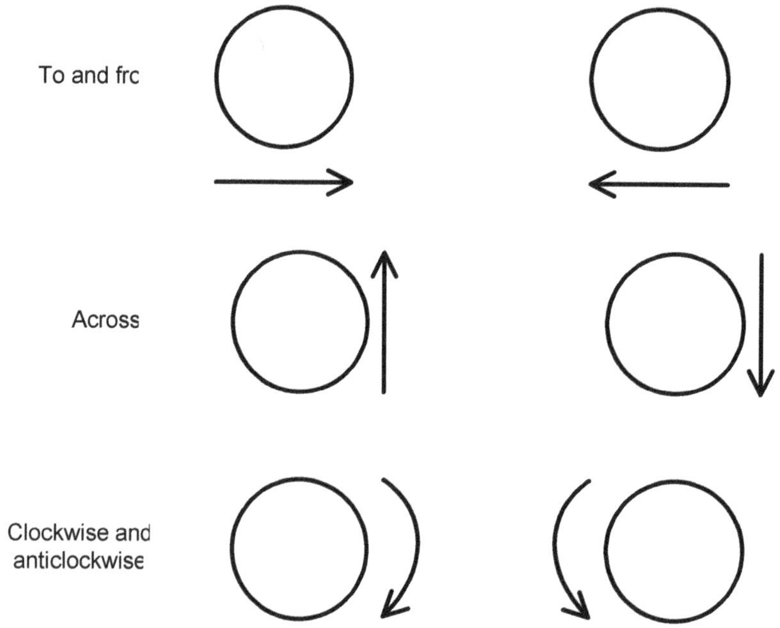

Fig 82-3. Mixing of Agar with the sample

While mixing the agar with the sample, care should be taken not to splash the agar on the lid or out of the dish. Complete the mixing within 15 seconds. Keep the plates undisturbed on the table until the agar is completely solidified.

Incubation

Keep the plates in the inverted position in the incubator. Incubate at 37°C for 24 hrs.

Counting the colonies

The plates kept for incubation are removed from the incubator after 24 hrs and the colonies are counted. Colonies may be counted manually (if necessary with the aid of a magnifying glass), or using colony counter.

The best plates to count are those with 30 – 300 colonies. Count all the colonies carefully. A plate with less than 30 colonies may be counted if inoculated with undiluted water. Each spreader should be counted as 1 colony. If half of the area of the plate is covered with spreader, discard it. In certain cases, even, dilutions show colonies more than 300. In these cases, the plate is divided into sectors and colonies in the sector are counted. If dilution is made multiply by the dilution factor. Record the results only to two significant figures.

Reporting results

The colony should be reported not in terms of the number of organisms per ml but as the number of colonies developing on the medium. The report should invariably include the agar used, incubation temperature and period.

Number of colonies per ml on yeast extract agar (at 37° after 24 hrs).

The present day trend is to express the total bacterial count in terms of 'Colony Forming Units' (CFU) per ml.

CHAPTER - 83

COLIFORM BACTERIA

(TOTAL COLIFORMS)

Coliform bacteria in water can exist in different densities and are determined by growing the cells in liquid or solid medium. The widely employed methods are, 1) MPN Method and 2) Membrane Filtration Method. MPN method for assessing the density of coliforms is described below taking into account its advantages over other methods.

MPN METHOD FOR TOTAL COLIFORMS

MPN method (Most Probable Number) is popularly termed as Multiple Tube method. Sometimes it is also referred as Dilution method. This method employed for the enumeration of coliforms, yields good results.

Summary of Test

In the MPN method, measured volumes of sample (50 ml, 10 ml, 1 ml or one or more dilutions of it) are transferred to bottles and / or tubes containing a liquid differential medium. On incubation at specified temperature for specified period, the organisms will either produce acid or gas or both, which is considered as positive. The positive tubes are counted and the probable number of coliform organisms present in the sample is derived from the MPN Table.

Media and Reagents

1. Modified Formate Lactose Glutamate medium (Double Strength)
2. Modified Formate Lactose Glutamate medium (Single Strength)
3. Formate Ricinoleate broth
4. Ringer solution (Quarter strength). See Chapter – 82, "STANDARD PLATE COUNT".

These media in dehydrated form are readily available commercially. Follow the manufacturers' instructions for rehydration and sterilization. If dehydrated media are not available, they can be prepared in the laboratory as detailed below.

1. Modified Formate Glutamate Lactose Medium (Double Strength)

Lactose	....	20 mg
Sodium glutamate	....	12.7 mg
L - Arginine monohydrochloride	....	40 mg
L - Aspartic acid	....	48 mg
L - Cystine	....	40 mg
Sodium formate	....	0.5 mg
Dipotassium hydrogen phosphate (K_2HPO_4)	....	1.8 mg
Ammonium chloride	....	5 g
Magnesium sulfate ($MgSO_4.7H_2O$)	....	0.2 mg
Calcium chloride ($CaCl_2.2H_2O$)	....	20 mg
Ferric citrate scale	....	20 mg
Thiamin (Aneurin hydrochloride)	....	2 mg
Nicotinic acid	....	2 mg
Pantothenic acid	....	2 mg
Bromocresol purple (1% ethanolic solution)	....	2 ml
Distilled water	....	1000 ml

Dissolve all the ingredients in water (excepting bromocresol purple) and adjust the pH to 6.8 or slightly above, so that the final pH after sterilization is 6.7. Now add 2 ml of bromocresol purple. Mix well and distribute in 50 ml quantities in 150 ml capacity bottles, 10 ml volumes in tubes of size 150 mm ×18mm each containing Durham fermentation tubes. Autoclave at 115°C (10 lb/in^2) for 10 minutes.

2. Modified Formate Lactose Glutamate medium (Single Strength)

The single strength medium is conveniently prepared by 1:1 dilution of double strength medium with sterile distilled water. Place 5 ml quantities in tubes of size 150 mm ×16 mm provided with Durham fermentation tubes and autoclave at 115°C (10 lb/in^2) for 10 minutes.

3. Formate Ricinoleate Broth

Peptone	....	5 g
Lactose	....	5 g
Sodium formate	....	5 g
Sodium Ricinoleate	....	1 g
Distilled water	....	1000 ml

Mix the ingredients, steam to dissolve, adjust pH to 7.6. Distribute in 5 ml quantitites in tubes of size 150 mm × 16 mm provided with Durham fermentation tubes. Sterilize at 115°C (10 lb/in^2) for 10 minutes.

Procedure

Inoculation and Incubation

Invert the sample bottle several times and shake for thorough mixing of the sample. Flame the mouth of the bottle and pour out a little of the water. Again shake the bottle well. Now transfer required volumes of the sample to bottles/tubes. If the sample is unpolluted water or treated water, transfer 50 ml of the sample into the bottle containing modified formate glutamate lactose medium (double strength). Add 10 ml volumes of the sample into five tubes containing same medium double strength. Transfer 1 ml volumes to five tubes containing same medium of single strength. For raw water and polluted waters, the following volumes may be taken, each in five tubes.

10ml, 1ml, 0.1ml (omit 50ml volume in this case)

If the water is highly polluted, dilution of the sample is necessary. Dilution of the sample has to be made as per the procedure given in Chapter – 79 "LABORATORY TECHNICS".

The neck of the bottles and media tubes should be flamed prior to the transfer of the sample to avoid any contamination. During the preparation of dilution, use separate pipets for each dilution. While pipeting, it should be ensured that the whole of the inner surface of the pipet should be thoroughly wet which is achieved by sucking up water several times. Similarly excess water adhering to the outer surface of the pipet should be removed by touching inside the bottle / tube.

The inoculated bottles and tubes are then incubated at 37°C for 48 hrs. The tube in which gas is entrapped in the inner Durham tube is considered as positive. The positive tubes are then subjected to confirmed test as described below.

Using a 3 mm loop, transfer the culture from the positive tubes to formate ricinoleate broth in tubes. Incubate at 37°C for 48 hrs. Any amount of gas produced and entrapped in the Durham tube constitutes a positive confirmed test. Count the number of positive tubes.

Deriving the number of coliforms

When the scheme-one 50 ml, five 10 ml and five 1.0 ml is employed, the MPN is derived from Table – 83.1. For example, the positives are seen in one 50 ml, in three 10 ml and in two 1 ml tubes (i.e) 1-3-2, the MPN of coliform is derived from the Table as 14.

If samples are inoculated in volumes of five 10 ml, five 1 ml and five 0.1 ml, MPN of coliforms is found out from Table 83–2. For example if 4 tubes in 10 ml, 3 tubes in 1 ml and 2 tubes in 0.1 ml are found as positive then look for the corresponding value to 4-3-2 in the Table 83–2. The MPN is 40. Incase dilutions of the samples are used, multiply by the dilution factor. For example, if 0.01ml volume is used for inoculation, multiply the MPN value by a factor of 10. If 0.001 ml volume is used, multiply by 100 and so on.

Reporting the results

Report the result as MPN of Coliform per 100 ml sample.

TABLE 83–1

MPN TABLE – I (FOR VOLUMES OF 50ml, 10ml AND 1ml)

Quantity of water put up in each tube →	50ml	10ml	1ml	Most probable number of organisms in 100 ml water
Number of tubes used →	1	5	5	
Number of tubes giving positive reaction →	0	0	1	1
	0	0	2	2
	0	1	0	1
	0	1	1	2
	0	1	2	3
	0	2	0	2
	0	2	1	3
	0	2	2	4
	0	3	0	3
	0	3	1	5
	0	4	0	5
	1	0	0	1
	1	0	1	3
	1	0	2	4
	1	0	3	6
	1	1	0	3
	1	1	1	5
	1	1	2	7
	1	1	3	9

Quantity of water put up in each tube →	50ml	10ml	1ml	Most probable number of organisms in 100 ml water
Number of tubes used →	**1**	**5**	**5**	
Number of tubes giving positive reaction →	1	2	0	5
	1	2	1	7
	1	2	2	10
	1	2	3	12
	1	3	0	8
	1	3	1	11
	1	3	2	14
	1	3	3	18
	1	3	4	20
	1	4	0	13
	1	4	1	17
	1	4	2	20
	1	4	3	30
	1	4	4	35
	1	4	5	40
	1	5	0	25
	1	5	1	35
	1	5	2	50
	1	5	3	90
	1	5	4	160
	1	5	5	180+

Note: The most probable numbers from 0 to 20 are correct to the nearest unit; above 20 are correct to the nearest 5.

TABLE 83–2

MPN TABLE – II (FOR VOLUMES OF 10ml, 1ml and 0.1ml)

Quantity of water put up in each tube →	10ml	1ml	0.1ml	Most probable number of organisms in 100ml water
Number of tubes used →	5	5	5	
Number of tubes giving positive reaction →	0	0	1	2
	0	0	2	4
	0	1	0	2
	0	1	1	4
	0	1	2	6
	0	2	0	4
	0	2	1	6
	0	3	0	6
	1	0	0	2
	1	0	1	4
	1	0	2	6
	1	0	3	8
	1	1	0	4
	1	1	1	6
	1	1	2	8
	1	2	0	6
	1	2	1	8
	1	2	2	10
	1	3	0	8
	1	3	1	10
	1	4	0	11
	2	0	0	5
	2	0	1	7
	2	0	2	9
	2	0	3	12
	2	1	0	7
	2	1	1	9
	2	1	2	12
	2	2	0	9
	2	2	1	12
	2	2	2	14
	2	3	0	12
	2	3	1	14
	2	4	0	15
	3	0	0	8
	3	0	1	11

Quantity of water put up in each tube →	**10ml**	**1ml**	**0.1ml**	**Most probable number of organisms in 100ml water**
Number of tubes used →	**5**	**5**	**5**	
Number of tubes giving positive reaction →	3	0	2	13
	3	1	0	11
	3	1	1	14
	3	1	2	17
	3	1	3	20
	3	2	0	14
	3	2	1	17
	3	2	2	20
	3	3	0	17
	3	3	1	20
	3	4	0	20
	3	4	1	25
	3	5	0	25
	4	0	0	13
	4	0	1	17
	4	0	2	20
	4	0	3	25
	4	1	0	17
	4	1	1	20
	4	1	2	25
	4	2	0	20
	4	2	1	25
	4	2	2	30
	4	3	0	25
	4	3	1	35
	4	3	2	40
	4	4	0	35
	4	4	1	40
	4	4	2	45
	4	5	0	40
	4	5	1	50
	4	5	2	55
	5	0	0	25
	5	0	1	30
	5	0	2	45
	5	0	3	60
	5	0	4	75
	5	1	0	35
	5	1	1	45

Quantity of water put up in each tube →	10ml	1ml	0.1ml	Most probable number of organisms in 100ml water
Number of tubes used →	5	5	5	
Number of tubes giving positive reaction →	5	1	2	65
	5	1	3	85
	5	1	4	115
	5	2	0	50
	5	2	1	70
	5	2	2	95
	5	2	3	120
	5	2	4	150
	5	2	5	175
	5	3	0	80
	5	3	1	110
	5	3	2	140
	5	3	3	175
	5	3	4	200
	5	3	5	250
	5	4	0	130
	5	4	1	170
	5	4	2	225
	5	4	3	275
	5	4	4	350
	5	4	5	425
	5	5	0	250
	5	5	1	350
	5	5	2	550
	5	5	3	900
	5	5	4	1600
	5	5	5	1800 +

Note: The most probable number from 0 to 20 are correct to the nearest unit; from 20 to 200 are correct to the nearest 5; from 200 to 500 are correct to the nearest 25; above 500 are correct to the nearest 50.

In addition to the count of coliform organisms, if identification of coliform bacteria is required, the following procedure of differentiation of coliform organisms may be carried out. This differentiation tests not only identify the organisms but also constitutes a test for fecal coliforms too.

DIFFERENTIATION OF COLIFORM ORGANISMS (IMViC Tests)

Differentiation tests popularly known as IMViC tests alongwith fermentation at 44°C are carried out to identify the organism present among the coliforms. The abbreviation is expanded below.

I – Indole production Test
M – Methyl red Test
V – Voges–Proskauer Test
C – Citrate utilization Test

To carry out these differentiation tests, typical colonies are picked up, inoculated into tubes containing MacConkey's broth single strength, and incubated at 37°C for 24 hrs. Then the above biochemical tests are carried out. Before proceeding with the test procedures, preparation of test solutions and indicators are described.

Media and Reagents

1) Peptone Water
2) Kovac's reagent
3) Glucose – phosphate broth
4) Methyl red indicator
5) α – naphthol (alcoholic) solution, 5%
6) Koser's citrate medium
7) MacConkey's broth (Single strength)

The above media and reagents are readily available commercially. They can also be prepared in the laboratory as detailed below.

1) Peptone Water

See Chapter – 84, "FECAL COLIFORMS (ESCHERICHIA COLI)"

2) Kovac's Reagent

See Chapter – 84, "FECAL COLIFORMS (ESCHERICHIA COLI)"

3) Glucose – phosphate broth

Peptone	....	5g
Dipotassium hydrogen phosphate	....	5 g
Glucose	....	5 g
Distilled water	....	1000 ml

Add peptone and dipotassium hydrogen phosphate in distilled water and steam until they dissolve. Cool and adjust the pH to 7.5. Add glucose and mix thoroughly. Distribute in 4 ml volumes in tubes of size 100 mm x 12 mm and auto clave at 115°C for 10 minutes.

4) Methyl red indicator

Methyl red	....	0.1 g
Ethyl alcohol	....	300 ml
Distilled water	....	200 ml

Dissolve methyl red in ethyl alcohol and add distilled water to the mixture.

5) α – naphthol (alcoholic) solution, 5%

Dissolve 5 g, α – napththol in 100 ml absolute ethyl alcohol. Store in a refrigerator (5 to 10°C).

6) Koser's Citrate Medium

Sodium chloride	....	5.0 g
Magnesium sulfate ($MgSO_4.7H_2O$)	....	0.2 g
Ammonium dihydrogen phosphate	....	1.0 g
Dipotassium hydrogen phosphate	....	1.0 g
Sodium citrate	....	2.0 g
Bromothymol blue indicator solution (0.2% aqueous solution)	....	40 ml

Dissolve all the ingredients in 1000 ml water. Adjust the pH to 6.8 $\pm$ 0.1 with 1N. sodium hydroxide solution. Distribute in 5 ml quantities in tubes of size 100 mm x 12 mm and sterlize at 121°C for 15 minutes.

7) Macconkey's broth (Single strength)

See Chapter – 84, "FECAL COLIFORMS (ESCHERICHIA COLI)"

Procedure

Place 2 or 3 loopfuls of inoculum from the positive tubes (modified formate glutamate - lactose medium) on a MacConkey's agar plate and streak using an inoculating loop as per the method described under section "Plating on Solid Medium" in Chapter – 79, "Laboratory Technics". Incubate the plate at 37° C for 24 hrs. After incubation, pickup atleast four colonies and subculture them in separate tubes containing MacConkey's broth (single strength) and incubate at 37° C for 24 hrs. After incubation, conduct the following biochemical tests.

Indole Test

Transfer the culture from MacConkey's broth to a tube containing peptone water. Incubate at 37° C for 24 hrs. After incubation, add 0.2 to 0.3 ml Kovac's reagent and shake the tube gently. Formation of red ring in the upper amyl-alcohol layer constitutes a positive indole test. Record the result.

Methyl Red Test

Transfer the culture from MacConkey's broth to a tueb containing glucose phosphate broth. Incubate at 37° C for 3 days. Add five drops of methyl red indicator solution. A distinct red color indicates a positive result, while yellow color indicates a negative result; pink or pale red colors are considered as doubtful result. Record the results.

Voges – Proskauer Test

Transfer the culture from MacConkey's broth to a tube containing glucose – phosphate broth. Incubate at 37°C for 3 days. Add to 1.0 ml of the culture 0.6 ml of 5% α - naphthol (alcoholic solution) and observe for a color change. Development of a pink to crimson color within 5 minutes constitutes a positive test. Development of copper color indicates a negative test. Record the result of the test.

Citrate Utilisation Test

Transfer a loopful of culture to tubes containing Koser's citrate medium, and incubate at 37° C for 3 days. Color change from pale green to blue or bluish green constitutes a positive test. The original pale green color of the medium indicates a negative test. Record the result.

Lactose Fermentation at 44°C

A loopful of inoculum is transferred to tubes containing MacConkey's broth(single strength) and incubated at 44°C (preferably in a water bath) for 24 hrs. Production and accumulation of gas inside the Durham fermentation tube indicates a positive test. Record the result.

With the results of the above five tests, the particular type of coliform organism present in the sample may be found out from the following table.

TABLE 83–1
DIFFERENTIATION OF COLIFORM GROUP

Indole Test	MR Test	VP Test	Test for Citrate utilization	Lactose fermentation at 44° C	Organism
+	+	–	–	+	E.Coli I
–	+	–	–	–	E.Coli II
+	+	–	–	–	E.Coli III
–	+	–	+	–	Citrobacter ferundii I
+	+	–	+	–	Citrobacter ferundii II
–	–	+	+	–	Klebisella aerogenes I
+	–	+	+	–	Klebsiella aerogenes II
–	+	–	–	+	Irregular II
–	–	+	+	+	Irregular VI

MEMBRANE FILTRATION METHOD FOR TOTAL COLIFORMS

Most of the literature indicate that recovery of organisms with disinfected water is less when membrane filtration technic is employed. However, this method enables direct emuneration of coliform organisms. Results can be obtained in 24 hrs and higher volumes of samples can be taken for examination.

Media and Reagents

Membrane Endo Agar (M – Endo Agar)

Dehydrated form of this medium is available readily. It can also be prepared in the laboratory as described below.

Yeast extract	....	1.5 g
Casitone or trypticase	....	5 g
Thiopeptone or thiotone	....	5 g
Tryptose or polypeptone	....	10 g
Lactose	....	12.5 g
Sodium desoxy cholate	....	100 mg
Dipotassium hydrogen phosphate	....	4.375 g
Potassium dihydrogen phosphate	....	1.375 g
Sodium chloride	....	5 g
Sodium lawryl sulfate	....	50 mg
Sodium sulfite	....	2.1 g
Basic fuchsin	....	1.05 g
Agar	....	15 g
Ethanol	....	20 ml
Distilled water	....	980 ml

Mix ethanol with distilled water. Dissolve all the ingredients in this alcoholic solution. Heat to near boiling to dissolve agar. Remove from heat and cool to a temperature below 45°C. **Do not sterlize by autoclaving**. Final pH of the medium should be between 7.1 and 7.3. Store the finished medium in a refrigerator. Discard any unused medium after 4 days.

Procedure

Melt and pour 5 – 6 ml of M – Endo Agar in the bottom of a petri dish (50 x 12 mm) to a depth of 2 to 3 mm and to allow to harden. Shake the sample vigorously about 25 times and filter 250 ml or suitable lower aliquot of the sample through the membrane filter. Rinse the sides of the funnel walls with about 20 – 30 ml of sterile distilled water.

After filtration, using a sterile forceps remove the filter aseptically and place grid side up on the surface of M – Endo agar. Incubate at 37° C for 24 hrs. [Place the dish in the inverted position in the incubator]. Coliforms grow as pink to dark red colonies with a surface metallic sheen. The sheen may vary in size from a small pin-head to the full-size of the colony (covering the entire colony). Count the sheen colonies. For counting, use of a binocular wide-field dissecting microscope with a magnification of 10 – 15 x is recommended. Colonies that lack sheen may be pink, red, white, or colorless and they have to be considered as non-coliforms.

Express the result as number of coliforms per 250 ml or per 100 ml or per unit volume of sample taken for determination.

CHAPTER - 84

FECAL COLIFORMS

(ESCHERICHIA COLI)

The fecal coliforms are part of the total coliform group. Coliform organisms derived exclusively from fecal matter are termed, "fecal coliforms". The other coliform organisms – the normal inhabitants of soil and water are referred as non-fecal coliforms. The major species in the fecal coliform group is 'Escherichia coli'. The presence of it indicates fecal pollution and the possible presence of enteric pathogens.

Fecal coliforms are Gram–negative rods, non-spore forming, and ferment lactose at 44° C within 24 hrs, with the production of acid and gas.

Fecal coliforms can be enumerated by MPN method or membrane filtration method. MPN method is pereferable for enumeration of fecal coliforms in treated water. MPN method is preferable because recovery of these organisms in membrane filter method is reported to be less with treated waters. Subsequently, lower results may be obtained with membrane filtration method. MPN method for the enumeration of fecal coliforms is described in detail. A brief account of membrane filtration is also given, taking into account certain advantages of the method.

MPN METHOD FOR FECAL COLIFORMS

Summary of Test

The test for coliforms is carred out using MacConkey broth. Culture from the positive tubes are inoculated into peptone water and incubated at 44° C for 24 hrs. Production of indole after incubation indicates a positive reaction. The number of positive tubes are counted and with this number, fecal coliform density is obtained from the MPN table.

Media and Reagents

1. Mac Conkey's broth (Double Strength)
2. Mac Conkey's broth (Single Strength)

3. Peptone Water
4. Kovac's reagent

These media in dehydrated form are readily available commercially. Follow the manufacturers' instructions for rehydration and sterilization. If dehydrated media are not available, they can be prepared in the laboratory as detailed below.

1. Mac Conkey's broth – Double Strength

Bile salt	....	10 g
Peptone	....	40 g
Sodium chloride	....	10 g
Lactose	....	20 g
Bromocresol purple (1% alcoholic solution)	....	2 ml
Distilled water	....	1000 ml

Mix all the ingredients in water except lactose and bromocersol purple, steam to dissolve and keep in a refrigerator (4° C) overnight. Filter while still cold, add the lactose to the filtrate and allow to dissolve. Adjust the pH to 7.4 with 1N. sodium hydroxide solution. Add the bromocresol purple indicator solution and mix well. Distribute in 10 ml quantities in test tubes of size 150 mm x 18 mm containing Durham tubes placed in inverted position. Also distribute in 50 ml volumes in bottles of 150 ml capacity. This bottle should also contain a Durham tube of suitable size. Sterlize at 115° C for 15 minutes (10 lb / in^2)

2. MacConkey's Broth – Single Strength

This is prepared in the same way as the double strength MacConkey's broth, using half the quantities of the ingredients for the same quantity of water (1000 ml). Distribute in 5 ml quantites in tubes of size 150 mm x 15 mm containing Durham tubes of suitable size. Sterlize at 115° C for 15 minutes (10 lb / in^2).

3. Peptone Water (For Indole Test)

Tryptone	....	20 g
Sodium chloride	....	5 g

(Tryptone is the preferred substance. However if tryptone is not available, it may be replaced with peptone. Use the same quantity as tryptone).

Dissolve the above ingredients in 1000 ml water and steam it. Cool it and adjust the pH of the solution to 7.5. Tube in 5 ml quantites in 100 mm x 12 mm tubes and autoclave at 121° C for 20 minutes. The pH of the finished metdium should be 7.4 to 7.6.

3. Kovac's Reagent

Paradimethyl amino benzaldehyde	....	10g
Amyl alcohol (or) Iso – amyl alcohol	....	150 ml
Conc. Hydrochloric acid (Sp. gr 1.18)	....	50 ml

Dissolve para–dimethyl amino benzaldehyde in amyl or iso-amyl alcohol (AR Grade – free from organic bases). Add conc. Hydrochloric acid to this mixture. Mix well. Place it in an amber bottle and keep it in a refrigerator.

Procedure

Inoculate one 50 ml volume, five 10 ml volumes of sample into MacConkey's broth double strength and five 1.0 ml into single strength medium.

Incubate at 37° C for 48 hrs. Transfer inocula from the positive tubes (indicated by the production of gas) into tubes containing peptone water. Incubate at 44° C for 24 hrs in a serological water bath. Observe the following precautions.

1) At the time of inoculation, the temperature of the peptone water should be 44° C.
2) Tubes must be placed in the incubator within 30 minutes after inoculation.
3) The depth of water in the water bath should be above the top level of the culture medium in the tubes.

After incubation, to each tube add 0.25 ml Kovac's reagent. Shake the tube gently. Formation of red ring is the positive reaction. Formation of yellow ring indicates negative reaction.

Count the number of positive tubes. Find out the fecal coliform (Escherichia coli) density from the MPN Table – II (Table 83-2, Chapter 83) with the number of positive tuebs counted.

Express the result as MPN of fecal coliforms (Escherichia coli) per 100 ml sample.

MEMBRANE FILTRATION METHOD FOR FECAL COLIFORMS

Though the accuracy of this method is somewhat dubious, it has certain definite advantages which make it more practical than MPN method. The results of membrane filtration tests could be obtained in 24 hrs unlike MPN method for which it takes 72 hrs. Further, the membrane filtration method provides direct enumeration of fecal coliforms without enrichment and subsequent testing. Volumes fo the sample taken for examination may be incrased to desired levels.

Media and Reagents

Membrane Fecal Coliform Medium (M – FC medium)

This medium is available readily in dehydrated form. It can also be prepared in the laboratory as per the procedure given below.

Tryptose	....	10 g
Proteose Peptone No.3 (or Polypeptone)	....	5 g
Yeast extract	....	3 g
Sodium chloride	....	5 g
Lactose	....	12.5 g
Bile Salt No.3 (or bile salt mixture)	....	1.5 g
Aniline blue	....	0.1 g
Rosolic acid solution (for preparation, see below)	....	10 ml
Distilled water	....	1000 ml

Rosolic acid is prepared by dissolving 1 g. rosolic acid in 100 ml 0.2 N. sodium hydroxide solution (0.8 g NaoH / 100 ml).

Add 10 ml rosolic acid solution to 1000 ml distilled waer. Then dissolve the ingredients in it and heat to near boiling. Remove from the heat and cool to below 50° C. Final pH should be 7.4

Do not sterlize by autoclaving.

Procedure

Saturate the sterile absorbent pad with 2 – 3 ml of M – FC medium and place in the bottom of the petri plate.

Shake the sample vigorously about 25 times and filter 250 ml or suitable lower aliquots through the membrane filter. Rinse the sides of the funnel walls with about 20 – 30 ml of sterile distilled water. Using a sterile forceps, remove the filter from the filter base aseptically and place grid side up on the pad saturated with M-FC medium. Incubate at 44° C for 24 hrs. if it is placed in a water bath for incubation, it has to be kept (inverted) in sealed water proof plastic bags submerged in the water bath.

After 24 hrs, remove the dishes from the incubator and count the blue colonies typical of fecal coliforms. Pinpoint blue colonies should also be counted. Non-fecal coliforms grow as grey or buff or colorless colonies and should not be counted. The colonies may be counted using a microscope of 10 – 15x and a fluorescent lamp.

Express the result as number of fecal coliforms per 250 ml or per 100 ml or per unit volume of sample taken for determination.

CHAPTER - 85

PROTEOLYTIC BACTERIAL COUNT

The method for the detection and enumeration of proteolytic bacteria is similar to that of standard plate count, but the growth medium is yeast extract to which skimmed milk is added. The incubation temperature remains the same (37°C) while the incubation period is 48 hrs. The proteolytic bacteria hydrolyse proteins and will show clear zones of proteolysis around the colony which are counted.

Apparatus Required

As in Chapter – 82, "STANDARD PLATE COUNT".

Media and Reagents

1. **Yeast extract agar**: Dehydrated form of this agar is readily available. If not available, prepare as per the procedure given in Chapter - 82 "STANDARD PLATE COUNT"

2. **Ringer solution, Quarter strength**: Ringer tablets are readily available commercially. If not available, prepare as per procedure given in Chapter - 82 "STANDARD PLATE COUNT"

3. **Skimmed milk:** Place 2 ml of skimmed milk in tubes of size 100 × 12 mm and plug with cotton wool. Sterilize it by auto claving at 121°C (15 lb/in^2) for 15 minutes.

4. **Dil. Hydrochloric acid:** Mix 20 ml conc. HCl with 180 ml distilled water.

Procedure

Follow the same procedure as described for "Standard plate count". Place 1.0 ml or decimal dilutions of the sample in the petridish. But, just before pouring the yeast extract agar, add 1 ml of skimmed milk to the agar in the tube and mix thoroughly by churning action. Now pour the skimmed milk-agar in to the plate and mix with the sample in the plate thoroughly.

Incubate at 37°C for 48± 3 hrs. Remove the plate from the incubator and count colonies with zones of proteolysis. (In order to differentiate a weak acid clearing by the acid – producing organisms from proteolysis, the plate may be flooded with dil.hydrochloric acid (1 + 9). If clear zone remains, it confirms proteolysis.

Report the result as Proteolytic Bacterial Count per ml.

CHAPTER - 86
LIPOLYTIC BACTERIAL COUNT

The method for the detection and enumeration of lipolytic bacteria is similar to that of standard plate count, but tributyrin agar is used instead of yeast extract agar. The incubation temperature is 30°C and incubation period is 72 hrs. The lipolytic bacteria hydrolyse lipids and colonies of these organisms show clear zones around themselves.

Apparatus required

As in Chapter – 82, "STANDARD PLATE COUNT".

Media and Reagents

Tributyrin Agar

Yeastrel (or meat extract)	....	3 g
Peptone	....	5 g
Agar	....	15 g
Tributyrin	....	2 g
Distilled Water	....	1000 ml

Dissolve yeastrel (or meat extract) and peptone in 1000ml water by steaming. Add agar to the yeastrel – peptone mixture and mix well. Then add 2g. tributyrin and mix thoroughly. Adjust the pH value to 7.5 by the addition of hydrochloric acid or sodium hydroxide after predetermination in a small portion using phenol red as indicator. Place in tubes in 10ml quantities and sterilize by steaming at 100° C for 30 minutes for three successive days.

Procedure

Follow the same procedure as described for "Standard Plate Count". But instead of yeast extract agar pour 10 ml of tributyrin agar into the plate and mix thoroughly with the sample. Incubate at 30° C for 72 $\pm$ 3 hrs. Remove the plates from the incubator. Count the colonies with well defined clear zones of lypolysis extending atleast 1 mm from the edge of the colony.

Report the result as Lipolytic Bacterial Count per ml.

CHAPTER - 87

THERMOPHILIC BACTERIAL COUNT

Carry out the procedure as described for "Standard Plate Count". But instead of 10 ml nutrient agar, 15 to 18 ml has to be used. The plates have to be incubated at 55°C. The incubation period is 48 ± 3 hrs.

Report the colony count as Thermophilic Bacterial Count per ml.

CHAPTER - 88

GELATIN LIQUEFYING BACTERIA

Summary of Test

The test procedure is similar to that of Standard Plate Count. 1ml of the sample is placed in petridish and mixed with gelatin medium and incubated at 20°C for 21 days. A positive reaction is indicated by the liquefaction of the medium around the colonies.

Apparatus

Incubator: Adjusted to 20.0° ± 0.2°C.

Media and Reagents

Gelatin medium

Beef extract	....	3.0g
Peptone	....	5.0g
Gelatin	....	120g
Distilled water	....	1000ml

Dissolve all the ingredients in water and keep it on a water bath for complete dissolution. Cool to about 50°C and distribute 10–15ml volume in test tubes. Sterilize in the autoclave at 121°C (15lb / in^2) for 20 minutes. The pH of the medium shall be 6.8.

Procedure

Follow the same technique as described in Chapter – 82, "STANDARD PLATE COUNT".

Shake the sample bottle vigorously inorder to distribute the bacteria evenly throughout the water. Place 1.0 ml of the sample in sterilized petridish (100mm dia). Add the molten gelatin medium (at 30°C) observing all precautions and mix thoroughly by rotation and to and fro movement of the dish placed on a flat table. Place the petridish in the incubator at 20 ± 0.2°C. Examine everyday for 7 days for liquefaction of the medium around the colonies. The dish should be examined as soon as it is taken from the incubator (if the atmospheric temperature is above 20°C the whole medium in the plate would thaw and convert to a liquid within a short time). If the plate does not show any signs of liquefaction it has to be kept for another 14 days and examined daily (it is preferable to inoculate 5 dishes for each sample to detect the presence of these organisms) to avoid any ambiguity.

CHAPTER - 89
IRON BACTERIA

The iron bacteria are one of the most important types of nuisance organisms in water. This group comprises a number of organisms capable of withdrawing iron present in water. The presence of iron bacteria in water is highly undesirable as they cause pitting and tuberculations in pipes. Besides, their profused growth will block the passage of water in pipes, conduits etc., Iron bacteria can also discolor water and cause undesirable odors and tastes. The principal iron bacteria are *Leptothrix ohracea, Gallionella, Ferruginea* and *Crenothrix polyspora.* These are also the most common microorganisms causing trouble in water supplies.

Procedure

Centrifuge the sample and place the residue on a glass slide and spread it evenly. Examine the glass slide under a microscope for the presence of iron bacteria. The figures of some of the prominent forms of iron bacteria are given below.

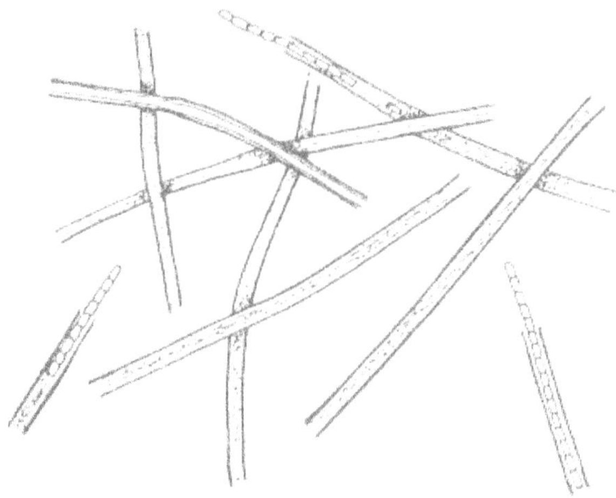

Fig 89-1. Leptothrix Ochracea

Fig 89-2. Gallionella Ferruginea

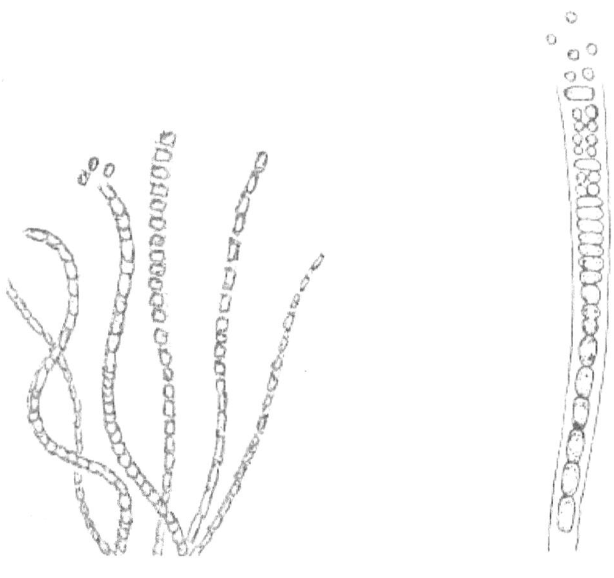

Fig 89-3. (a) Crenothrix Polyspora (b) Filament of Crenothrix Polyspora

CHAPTER - 90

SULFATE-REDUCING BACTERIA

Summary of Test

Detection of sulfate-reducing bacteria is made by mixing suitable volumes of sample of water in the liquid medium (Starkey's medium) followed by incubation at 22 – 30°C under anaerobic conditions for 21 days. A positive reaction will be shown by blackening of the medium. Sulfate reduction produces sulfide which reacts with ferrous iron in the medium to form ferrous sulfide which turns the medium black.

Glassware

Culture bottles: 60ml, glass stoppered bottles, sterilized in hot air sterilizer.

Media and Reagents

Sulfate-reducing medium (Starkey's medium):

Sodium lactate	....	3.5g
Ammonium chloride	....	1.08g
Potassium dihydrogen phosphate	....	0.59g
Magnesium sulfate	....	2.0g
Sodium sulfate	....	0.5g
Sodium chloride	....	0.1g
Mohr's salt (Ferrous Ammonium sulfate)	....	0.4g
Calcium chloride	....	0.1g
Distilled water	....	1000ml

Dissolve all the ingredients in distilled water and sterilize in the autoclave at 121°C ($15 lb/in^2$) for 20 minutes. Keep the medium in bulk in a refrigerator. It can be transferred to bottles whenever necessary.

Procedure

Fill the culture bottle to half the capacity with the recently prepared Starkey's medium. Add the sample to fill the bottle completely and stopper excluding air. Place a small beaker inverted over the bottle to avoid contamination. Incubate at 20° – 30°C.

Examine daily for 3 weeks. If sulfate reducing bacteria is present in the water sample, the medium will become black. Conduct a control test using sterilized water in place of sample for comparision.

CHAPTER - 91
SULFUR-OXIDIZING BACTERIA

Summary of Test

The common organisms belonging to sulfur-oxidizing bacteria are Thiobacillus thioparus and Thiobacillus thio oxidans. Separate test procedures for these organisms are given below. The tests are based on the principle that the former organism oxidizes thiosulfate to sulfur while the later oxidizes sulfur to sulfate.

Media and Reagents

1. Medium for Thiobacillus thioparus (Thiosulfate oxidizing medium)

Sodium thiosulfate ($Na_2S_2O_3.5H_2O$)	....	10.0g
Dipotassium hydrogen phosphate (K_2HPO_4)	....	2.0g
Magnesium sulfate ($MgSO_4.7H_2O$)	....	0.1g
Calcium chloride ($CaCl_2.2H_2O$)	....	0.1g
Ammonium sulfate $(NH_4)_2SO_4$	....	0.1g
Ferric chloride ($FeCl_3.6H_2O$)	....	0.02g
Distilled Water	....	1000 ml

Dissolve all the ingredients in distilled water. Place 50ml quantities in conical flasks of 250ml capacity and autoclave at 121°C (15lb/in^2) for 10 minutes. pH of the medium should be 7.8 after sterilization.

2. Medium for Thiobacillus thio oxidans (Sulfur medium)

Potassium dihydrogen phosphate (KH_2PO_4)	....	3.0g
Magnesium sulfate ($MgSO_4.7H_2O$)	....	0.5g
Ammonium sulfate [$(NH_4)_2SO_4$]	....	0.3g
Calcium chloride ($CaCl_2.2H_2O$)	....	0.25g
Ferric chloride ($FeCl_3.6H_2O$)	....	0.02g
Distilled water	....	1000ml

Dissolve all the ingredients in water. The pH of the solution should be 4.8. Place 1g. elemental sulfur in each of 10 numbers of 250ml conical flasks. To each flask add 100ml of the above solution and mix well. Sterilize for 30 minutes in a steam sterilizer for 3 successive days.

Procedure

Test for Thiobacillus thioparus

Inoculate the medium in the 250ml flask with water. Incubate at 30°C for 3 days. If Thiobacillus thioparus bacteria is present, the surface of the incubated medium would be covered with sulfur by the oxidation of thiosulfate.

Test for Thiobacillus thiooxidans

Inoculate the flasks containing the medium (in 250ml flasks) with 100ml or less volume of the sample and incubate at 25 to 30°C for 4 to 5 days. If the Thiobacillus thioixidans is present, sulfur sinks to the bottom, the pH of the medium would be reduced to about 2.0, and the medium will become turbid.

Quantitative enumeration of these organisms is not required. Hence express the result as positive or negative for the test conducted.

CHAPTER - 92

YEAST AND MOULD

Yeasts are unicellular fungi which occur as egg shaped or spherical cells and reproduce by simple budding. They form characteristic creamy colonies on culture. They are easily differentiated from bacterial colonies by their relatively larger size and morphological features. The size of yeasts vary considerably ranging from 1 to 5μm (or more) in length.

Moulds are filamentous fungi and they form true mycelia and reproduce by the formation of different types of spores.

For the detection and enumeration of yeast and moulds, membrane filtration technic is the preferable method.

Media and Reagents

Potato Dextrose Agar

Potato infusion from 200g. of white potatoes	....	1000ml
Dextrose	....	20g.
Agar	....	15 - 20g.
(Final pH of the medium)	....	3.5±0.1

Boil 200g. peeled and sliced potatoes in about 500ml water for 15 minutes or until it becomes soft. Filter through a cotton wool and makeup to 1000ml with water. Mix the dextrose and agar in the potato infusion and heat the mixture to boiling to dissolve the ingredients. Distribute in flasks and autoclave at 121°C (15lb/in^2) for 15 minutes. The pH of the medium now will be around 5.6.

Before use, melt this potato dextrose agar in steam by keeping over water bath. Cool and acidify to pH 3.5 ± 0.1 with a sterile solution of 10% tartaric acid and mix thoroughly.

Note: The medium should never be heated after the addition of acid, since heating along with acid will result in the hydrolysis of agar destroying its solidifying properties.

Procedure

Follow the membrane filtration technic. Use 250 ml sample for filtration. Soak the filter pad with the potato dextrose agar. After filtration, place the filter over the pad, cover with lid and incubate $30^{\circ} \pm 1^{\circ}C$ for 5 days.

After incubation period, observe the filter thoroughly. Yeast colonies are typical in that they are smooth (or wrinkled) moist, elevated or surface colonies. They are relatively larger than bacterial colonies. If there is any doubt, the colonies under question should be subjected to Gram – staining which clearly indicates Gram – positive yeast cells. The colonies can be examined directly under a microscope also. Mould colonies are easily recognized by their profuse growth of hyphae.

Quantitative enumeration of these organisms is not required. Hence express the result as positive or negative for the test conducted.

CHAPTER – 93

SLIME-FORMING ORGANISMS

A variety of microscopical organisms including bacteria are capable of producing slimes which will adhere to the water carrying structures and pose serious problems. These organisms collectively are termed as 'Slime-Forming Organisms'. Iron bacteria and certain fungi are the most common slime forming organisms reported in the literature. The literature most frequently concerns *Sphaerotilus* and occasionally *Leptothrix*. Among the fungi, *Fusarium* and *Mucor* are noteworthy especially when the pH of water is 5.2 to 5.6. However, the fungi most frequently mentioned among the slimes is *Leptomitus*, more specifically *leptomitus lacteus.* The bacterial groups *Pseudomonas, Flavobacterium, Escherichia, Aerobacter* and *bacillus* are also considered as the primary slime formers. Presence of these organisms is especially undesirable in waters used for cooling, condenser system, food industries, paper mill etc.,

Procedure

Slime forming organisms are normally identified by examination under a microscope. Slime forming bacteria are identified by usual techniques such as enrichment in selective media, plating and subjecting to various biochemical reactions, motility and Gram-staining.

Place on a clear glass slide a drop of the sample and spread it evenly. Cover it with a cover glass and examine under a microscope using low power objective. If no organisms are observed, centrifuge the sample and place the residue on glass slide and examine again under microscope for slime forming organisms. The figures of some of the common slime-forming organisms are given in the next page (Fig 93-1).

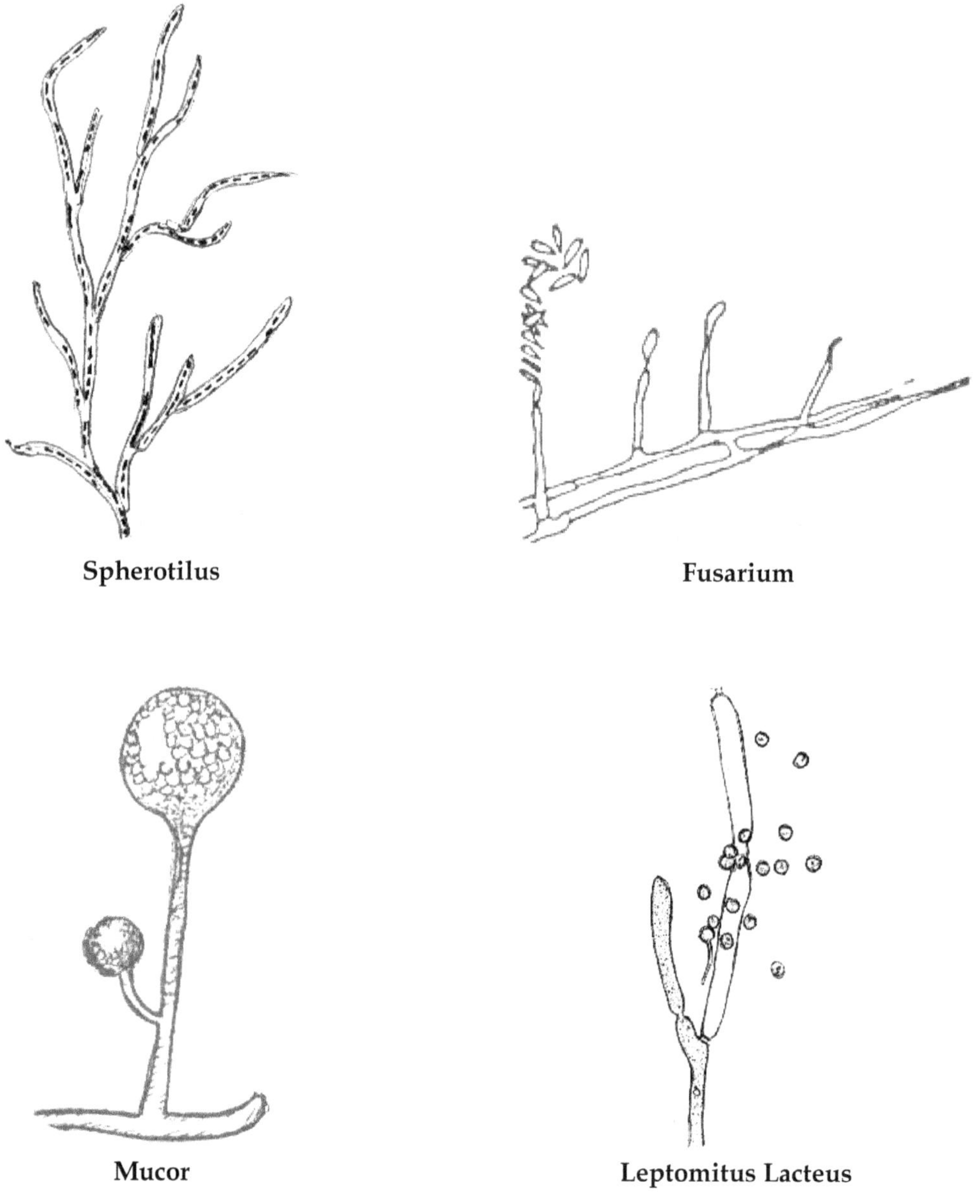

Fig 93-1. Some common slime forming organisms

PART IV

MICROSCOPICAL EXAMINATION

CHAPTER - 94

MICROSCOPICAL EXAMINATION

(BIOLOGICAL EXAMINATION)

Microscopical examination also termed as 'Biological Examination' is mainly intended to assess the raw water quality and the efficacy of various stages of water treatment. It is a tool to detect contamination and to identify and estimate the amount of bacteria (higher forms), algae, mould, protozoa and other planktons. Drinking water is normally subjected to detailed microscopical examination. Since most of the industrial waters are treated to remove all these organisms, normally industrial waters are not examined for biological organisms. However, with respect to cooling waters, (especially once through cooling waters) microscopical examination is very important as some of the algae, and bacteria (such as iron bacteria) are capable of growing in masses and block the passage of water and impede heat transfer. Once through cooling waters are normally not subjected to any treatment (excepting biocidal treatment). In those cases and where the water is not undergone any conventional treatment such as coagulation and flocculation and filtration, biological quality is very important. Many of the organisms (including bacteria) are capable of forming biological slimes and exhibit a profused growth in pipelines and mains and restrict the flow of water. Besides, microscopical examination is useful in various other ways. Since many algae and other organisms impart color and odor to water, microscopical examination is useful in explaining reason for the sudden appearance of color, turbidity and unobjectionable odor, and tastes in water. Microscopical examination is valuable complement to chemical analysis. It is helpful in elucidating the results of chemical analyses, for example, in relating the presence or absence of certain biological forms to oxygen deficiency or super saturation in natural waters. As already stated, it is useful in determining the effectiveness of various treatment stages and to aid in determining the effective chlorine (or suitable biocide) dosage to be applied. The reason for clogging of pipes, screens and filters is indicated by a detailed microscopical examination which helps to design an effective water treatment plant and for trouble-free operation.

Microscopical examination, as the name implies, is observing the water under microscope. Since a drop of water cannot reveal all the organisms, either the bottom sediment in the sample container is taken or a known quantity of water is centrifuged and the residue is subjected to microscopical examination. The flora and fauna of water are numerous and no attempt is made to provide the figures of these organisms. Readers are directed to refer the following publications which serve as source books, in this regard.

1. Edwin Windle Taylor, "The Examination of Waters and Water Supplies", Seventh Edition, 1951, J&A Churchill Ltd, London.

2. H.B. Ward & G.C. Whipple, "Fresh Water Biology" (Ed: W.T. Edmondson), Second Edition, 1966, John Wiley & Sons, New York.

3. American Public Health Association, "Standard Methods for the Examination of Water and Waste Water", 20th Edition.

However, a shortlist of names of the common organisms, their habitat, effect and significance is given below for reference.

TABLE – 94.1

LIST OF SOME COMMON MICROSCOPIC ORGANISMS PRESENT IN WATER

Classification	Name of the organism	Habitat	Effect & Significance
I. PHYTO PLANKTON/ALGAE			
(i) BLUE-GREEN ALGAE (Cyanopyceae)	Agmenellum	Clean waters	Indicator of purification
	Anabaena	Polluted water Impounded sources	Impart color Produce taste and odor Clog filters Toxin producing
	Anacystis	Polluted waters	Causes algal bloom and impart color Produce taste and odor Clog filters Toxin producing
	Gomphosphaeria	Polluted waters	Toxin producing
	Oscillatoria	Polluted waters	Impart color Clog filters

Classification	Name of the organism	Habitat	Effect & Significance
(ii) GREEN ALGAE (Chlorophyceae)			
	Actinastrum	Polluted waters Impounded sources	Impart color
	Ankistrodesmus	Clean water	Indicate clean condition
	Arthrodesmus		
	Closterium	Polluted waters Impounded sources	Clog filters and creates operational difficulties
	Cosmarium	Fresh water sources, sometimes in bogs	
	Coelastrum	Polluted Waters Impounded sources	Impart Color
	Desmidium	Surface waters	
	Euastrum	Polluted waters Impounded sources	Impart color
	Pediastrum	Polluted waters Impounded sources	Impart color
	Scenedesmus	Polluted waters Impounded sources	Impart color
	Selenastrum		
	Staurastrum	Polluted waters	Produce taste and odor
	Tetrastrum		
	Tetraedron	Polluted waters	Impart color
(iii) BACILLARIOPHYCEAE (Diatoms)			
	Anomoneis		
	Asterionella	It is frequent in lakes and fairly large and clear ponds and in hill streams and in reservoirs	Taste and odor producing
		High altitude torrential and temperate waters	Clog filters
	Cocconeis	Clean waters	Indicator of purification
	Cymbella	Rivers and streams Impounded sources	Clog filters and cause operational difficulties
	Coscinodiscus		
	Cyclotella	Clean waters Rivers and streams Impounded sources	Indicator of purity Clog filters and cause operational difficulties
	Diatoma	Rivers and Streams Impounded sources	Clog filters and cause operational difficulties
	Fragilaria	Polluted waters Rivers and Impounded sources	Taste and odor producing Clog filters and cause operational difficulties
	Frustulia	Fresh water sources	

Classification	Name of the organism	Habitat	Effect & Significant
	Gomphonema	Moderately polluted waters	Cause discoloration
	Gyrosigma	Clean water	Indicates clean condition
	Melosira	Rivers and Streams Impounded sources	Indicator of clean water Clog filters and cause operational difficulties
	Meridion	Clean waters	Indicator of purification
	Navicula	Rivers and streams Impounded sources	Clog filters and cause operational difficulties
	Nitzschia	Moderately polluted water	Cause discoloration
	Pinnularia	Clean water	Indicator of purification
	Stauroneis	It is found principally among the ooze at the bottom of lakes and larger – sized ponds.	Cause discoloration
	Stephanodiscus		Cause discoloration
	Synedra	Rivers and Streams Impound sources	Clog filters and cause operational difficulties
	Surirella	Clean waters	Indicator of purification
	Tabellaria	Hill streams High altitude torrential and temperate waters	Taste and odor producing Clog filters
(iv) DINOPHYCEAE			
	Ceratium	Rocky strata, iron bearing and acidic waters	Impart color and fishy taste
	Peridinium	Rocky strata iron bearing and acidic waters	Impart color and fishy taste
(v) CHRYSOPHYCEAE			
	Dinobryon	Clean unpolluted waters Rocky strata, iron bearing and acidic waters	Indicator of clean water. Impart color and fishy taste
	Synura	Clean unpolluted water	Indicator of clean water when present in abundance.
(vi) EUGLENOPHYCEAE			
	Euglena	Polluted waters stagnant pools	Impart color
	Phacus	Polluted waters	Impart color

Classification	Name of the organism	Habitat	Effect & Significance
II. ZOO PLANKTON			
(i) SARCODINA (Amoebas)			
	Actinophrys	Polluted waters	Pollution indicator
	Actinosphaerium		
	Amoeba	Polluted waters	Pollution indicator
	Arcella	Polluted waters	Pollution indicator
	Difflugia	Polluted waters	Pollution indicator
(ii) MASTIGOPHORA (Non-pigmented flagellates)			
	Astasia	Surface of pools and ponds	Impart rich green color to water
	Peranema	Surface of pools and ponds	Impart color
(iii) CILIATA (Ciliated protozoas)			
	Chilodon	Fresh water and salt water	
	Coleps	Highly polluted waters Sewage and activated sludge	Bacteria eaters.
	Colpidium	Highly polluted waters Sewage and Activated sludge	Bacteria eater
	Colpoda	Highly polluted waters Sewage and Activated sludge	Bacteria eater
	Epistylis	Highly polluted waters	
	Euplotes	Highly polluted waters Sewage and Activated sludge	Bacteria eater
	Halteria	Generally found in still ponds in the neighbourhood of organic debris (decaying vegetation)	Indicator of pollution
	Holophrya		
	Lionotus		
	Loxodes	Found generally in the older decaying vegetation.	
	Paramecium	Highly polluted waters Sewage and Activated sludge	Bacteria eater

Clasification	Name of the organism	Habitat	Effect & Significance
	Stentor	Highly polluted waters Sewage and Activated sludge	Bacteria eater
	Strombidium		
	Stylonichia	Found among decaying aquatic plants.	
	Vorticella	Highly polluted waters Sewage and Activated sludge	Bacteria eater
(iv) GASTROTRICHA			
	Chaetonotus	Shallow ponds Beach pools	
(v) ROTIFERA (Rotifers)			
	Anurea	Commonest rotifers found in the surface collections of wide open stretches of water.	
	Brachionus		
	Euchlanis	Prefers to live in clean & clear waters	
	Filinia		
	Keratella		
	Monostyla		
	Philodina	Polluted and algal laden waters	Feed on Algae
	Proales		
	Trichocerca		
(vi) CRUSTACEAE (Crustaceans)			
	Bosmina	Stagnant Polluted waters	Indicator of pollution
	Cyclops	Step wells in tropical countries	Carrier host of guinea worm
	Cypris	In ponds and lakes	
	Daphnia	Stagnant polluted waters	Indicator of pollution
	Diaptomus	Lakes and pools	Feeds on decaying portions of algae and aquatic vegetation which would tend to accumulate and foul the waters. Diaptomus thus usefully aids in sanitation and the cleansing of great lakes and pools.

APPENDIX

TABLE – I
METRIC MULTIPLIERS

There are three basic metric units: meters (which measure length), grams (which measure weight or mass), and litres (which measure volume or capacity). The metric system is based on the number 10 with prefixes to specify the multiple or fraction of the basic unit (of one meter, one gram, one litre). The prefixes commonly used in are:

Prefix	Multiplier
mega (M)	1000,000.0
kilo (K)	1000.0
hecto (h)	100.0
deca (da)	10.0
deci (d)	0.10
centi (c)	0.01
milli (m)	0.001
micro (μ)	0.000001
nano (n)	1×10^{-9}

TABLE – II
PREFIXES USED IN THE METRIC SYSTEM

Prefixes	Meaning	Unit
pico – (one trillionth)	$\frac{1}{1,000,000,000,000} = 10^{-12}$	
nano – (one billionth)	$\frac{1}{1,000,000,000} = 10^{-9}$	
micro – (one millionth)	$\frac{1}{1,000,000} = 10^{-6}$	
milli – (one thousandth)	$\frac{1}{1,000} = 10^{-3}$	
centi – (one hundredth)	$\frac{1}{100} = 10^{-2}$	
deci – (one tenth)	$\frac{1}{10} = 10^{-1}$	meter for length
unit (one)	1	1
deka – (ten)	10	gram for weight or mass
hecto – (one hundred)	$100 = 10^{2}$	
kilo – (one thousand)	$1,000 = 10^{3}$	liter for capacity
mega – (one million)	$1,000,000 = 10^{6}$	
giga – (one billion)	$1,000,000,000 = 10^{9}$	
tera – (one trillion)	$1,000,000,000,000 = 10^{12}$	

TABLE – III
LENGTH

1 Kilometer	= 1000 meters	1 micron	= 0.000001 meter (10^{-6}m)
1 meter	= 100 centimeters	1 millimicron	= 0.000000001 meter (10^{-9}m)
1 meter	= 1000 millimeters	1 Angstrom	= 0.0000000001 meter (10^{-10}m)
1 millimeter	= 0.03937 inch	1 inch	= 2.540 centimeters
1 centimeter	= 0.3937 inch	1 foot	= 30.480 centimeters
1 meter	= 39.37 inches	1 foot	= 0.3048 meter
1 meter	= 3.2808 feet	1 yard	= 91.440 centimeters
1 meter	= 1.09361 yards	1 yard	= 0.9144 meter
1 kilometer	= 0.6214 mile	1 mile	= 1.6093 kilometers.

TABLE – IV
AREA

1 square centimeter	= 0.155 square inch	1 sq. inch	= 6.4516 sq.cms
1 square meter	= 1555.0 sq.inch	1 sq.foot	= 929.0341 sq.cms
1 sq.meter	= 10.764 sq.feet	1 sq.foot	=0.0929 sq.meter
1 sq.meter	= 1.196 sq. yards	1 sq.yard	= 0.8361 sq.meter
1 sq.km	= 0.3861 sq.mile	1 acre	= 43560 sq.ft
1 sq.mile	= 2.590 sq.km	1 acre	= 0.4047 hectare
		1 hectare	= 2.471 acres

TABLE – V
VOLUME

1 kilo litre	= 1000 litres	1 cubic meter	= 1000 litres
1 deci litre	= 0.1 litre	1 cubic meter	= 35.145 cubic feet
1 centi litre	= 0.01 litre	1 cubic feet	= 1.3079 cubic yard
1 milli litre	= 0.001 litre	1 cubic centimeter	= 0.06102 cubic inch
1 micro litre	= 0.000001 litre	1 cubic inch	= 16.3872 cubic centimeter
1 litre	= 1000 milli litres	1 cubic foot	= 0.02832 cubic meter
		1 cubic yard	= 0.7646 cubic meter

TABLE – VI
WEIGHT

1 ton	= 1000 Kilograms	1 microgram	= 0.001 mg (10^{-3} mg or 10^{-6}g)
1 Kilogram	= 1000 grams	1 microgram	= 1000 nanograms (10^{-6} mg or 10^{-9}g)
1 gram	= 1000 milligrams	1 Kilogram	= 2.205 pounds
1 milligram	= 1000 micrograms	1 pound	= 453.6 grams

TABLE – VI
VOLUME CONVERSION FACTORS

From to	cm^3	liters	m^3	in^3	ft^3	yd^3	gal (U.S)	gal (U.K)
cm^3	1	0.001	1×10^{-6}	0.06102	353×10^{-5}	1.3×10^{-6}	2.64×10^{-4}	2.2×10^{-4}
litre	1000	1	0.001	61.02	0.03532	0.00131	0.2642	0.220
m^3	1×10^{-6}	1000	1	6.1×10^{4}	35.31	1.308	264.2	220
in^3	16.39	0.01639	1.64×10^{-5}	1	5.79×10^{-4}	2.14×10^{-5}	0.00433	0.00360
ft^3	283×10^{4}	28.32	0.02832	1728	1	0.03704	7.481	6.229
yd^3	7.65×10^{5}	764.5	0.7646	4.67×10^{4}	27	1	202.0	168.2
gal (U.S)	3785	3.785	0.00379	231.0	0.1337	0.00495	1	0.8327
gas (U.K)	4546	4.546	0.00455	277.4	0.1605	0.00595	1.201	1

TABLE – VIII
PRESSURE UNITS AND CONVERSION FACTORS

From	To	Multiply by
psi (or lb/in^2)	Kg/cm^2	0.07031
psi	atmosphere	0.068046
psi	bar	0.068948
psi	Pa (or) N/m^2	6.8927×10^3
Kg / cm^2	psi	14.223
Kg / cm^2	atmosphere	0.9678
Kg / cm^2	bar	0.9804
Kg / cm^2	Pa (or) N/m^2	9.804×10^4
atmosphere	psi	14.696
atmosphere	Kg/cm^2	1.03325
atmosphere	bar	1.01295
atmosphere	Pa (or) N/m^2	1.01295×10^5
bar	psi	14.5039
bar	Kg/cm^2	1.02
bar	bar	0.9872
bar	Pa (or) N / m^2	1×10^5
Pa (or) N / m^2	psi	1.4508×10^{-4}
Pa (or) N / m^2	Kg/cm^2	1.020×10^{-5}
Pa (or) N / m^2	atmosphere	9.869×10^{-6}
Pa (or) N / m^2	bar	1×10^{-5}

Pressure is also expressed as MN / m^2
1 MN / m^2 = 1 MPa

TABLE - IX
FLOW RATE CONVERSION FACTORS

From to →	*litre/s*	*gal/min*	*ft³/s*	*ft³/min*	*m³/s*	*m³/min*
litre/s	1	15.85	0.03532	2.119	0.001	0.06
gal/min	0.06309	1	0.00223	0.1337	0.000063	0.00379
ft³/s	28.32	448.8	1	60	0.02832	1.699
ft³/min	0.4719	7.481	0.01667	1	0.000472	0.02832
m³/s	1000	15852	35.31	2118.6	1	60
m³/min	16.667	264.2	0.5885	35.31	0.01667	1

TABLE – X
CONVERSION FACTORS FOR TEMPERATURES SCALES

The following formula may be used to convert temperatures from one scale to another

	Temperature wanted in			
Temp. given in	**⁰C**	**⁰K**	**⁰F**	**⁰Rank**
⁰C	C	C + 273.16	1.86 C + 32	1.8C + 491.4
⁰K	K – 273.16	K	1.8K – 459.4	1.8K
⁰F	0.556F – 17.8	0.556F + 255.3	F	F + 459.4
⁰Rank	0.556 Rank – 273.1	0.556 Rank	Rank – 459.4	Rank

TABLE – XI
APPLICATION OF METRIC UNITS

Description	Unit	Symbol
Precipitation run-off	milli metre (1 mm of rain = 1 litre/m²)	mm
River flow	Cubic metre per second	m³/S
Flow in pipes	Cubic metre per second or litre per second	m³/S or l/s
Discharges	Cubic metre per day	m³/d
Usage of water	litre per person per day	l/person/day
Density	Kilogram per cubic metre	Kg/m³
Concentration	milligram per litre or gram per cubic metre	g/m³
B.O.D. loading	Kilogram per cubic metre per day	Kg/m³/d
Hydraulic load per unit area	Cubic metre per square-metre per day	m³/m²/d
Hydraulic load per unit volume	Cubic metre per cubic metre per day	m³/m²/d
Velocity	metre per second	m/s

TABLE – XII
WATER ANALYSIS UNITS
INTER-RELATIONS AND EQUIVALENTS

1 part per million (ppm)	1 milligram per litre
	0.1 part per hundred thousand
	0.07 grain per imperia gallon
	0.0583 grain per U.S. gallon
1 part per hundred thousand (pts/100,000)	10 part per million
	10 milligrams per litre
	0.7 grain per imperial gallon
	0.583 grain per U.S. gallon
1 grain per imperial gallon	14.3 parts per million
	14.3 milligram per litre
	1.43 parts per hundred thousand
	0.833 grain per U.S. gallon
1 grain per U.S. gallon	17.1 parts per million
	17.1 milligrams per litre
	1.71 parts per hundred thousand
	1.2 grains per imperial gallon

TABLE - XIII
HARDNESS CONVERSION FACTORS

Units of Measure	mg/l $CaCO_3$	British gr/gal (Imperial) $CaCO_3$	American gr/gal (US) $CaCO_3$	French Degree Parts/100000 $CaCO_3$	German Degree Parts/100000 $CaCO_3$	meq/l	g/l CaO
mg/l $CaCO_3$	1.0	0.07	0.058	0.1	0.056	0.02	5.6×10^{-4}
British gr/gal (Imperial) $CaCO_3$	14.3	1.0	0.83	1.43	0.83	0.286	8.0×10^{-3}
American gr/gal (US) $CaCO_3$	17.1	1.2	1.0	1.72	0.96	0.343	9.66×10^{-3}
French Degree Parts/100000 $CaCO_3$	10.0	0.7	0.58	1.0	0.56	0.2	5.6×10^{-3}
German Degree Parts/100000 $CaCO_3$	17.9	1.25	1.04	1.79	1.0	0.358	1.0×10^{-2}
meq/l	50.0	3.5	2.9	5.0	2.8	1.0	2.8×10^{-2}
g/l CaO	1790.0	125.0	104.2	179.0	100.0	35.8	1.0

TABLE – XIV
FACTORS FOR INTERCONVERSIONS BETWEEN MILLIGRAM PER LITRE (mg/l) AND MILLIEQUIVALENT PER LITRE (meq/l).

Ion	Multiplying Factors	
	mg/l to meq/l	meq/l to mg/l
Na^{+}	0.04349	22.90
K^{+}	0.02557	39.10
Ca^{2+}	0.04990	20.04
Mg^{+}	0.08226	12.16
Ba^{2+}	0.01456	68.67
Sr^{2+}	0.02283	43.81
Zn^{2+}	0.03060	32.68
Cu^{2+}	0.03148	31.77
Pb^{2+}	0.009653	103.6
Ni^{2+}	0.03047	29.36
Fe^{2+}	0.03581	27.92
Mn^{2+}	0.03640	27.47
Al^{3+}	0.1112	8.994
Fe^{3+}	0.05372	18.62
Cr^{3+}	0.05770	17.33
Mn^{4+}	0.07281	13.73
Cr^{6+}	0.1154	8.666
F^{-}	0.05264	19.00
Cl^{-}	0.02821	35.45
Br^{-}	0.01251	79.91
I^{-}	0.007880	126.9
NO_2^-	0.02174	46.01
NO_3^-	0.01613	62.00
HCO_3^-	0.01639	61.01
OH^{-}	0.05860	17.01
CO_3^{2-}	0.03333	30.00
SO_3^{2-}	0.02498	40.03
SO_4^{2-}	0.02082	48.03
S^{2-}	0.06238	16.03
CrO_4^{2-}	0.01724	58.00
PO_4^{3-}	0.03159	31.66

TABLE - XV
CONVERSION FACTORS FOR GASES

Multiply	By	To obtain
p. p. m. of oxygen	0.698	ml/l
ml/l of oxygen	1.43	p.p.m
p.p.m of carbon dioxide	0.506	ml/l
ml/l of carbon dioxide	1.98	p.p.m

TABLE - XVI
CHEMICAL VOLUMETRIC FACTORS FOR USE IN TITRATIONS

1 ml. 0.01N (N/100) potassium dichromate	= 0.558 mg iron
1 ml. 0.0125N (N/80) potassium permanganate	= 0.25 mg Ca = 0.1 mg oxygen
1 ml. 0.01 N (N/100) potassium permanganate	= 0.2 mg Ca
1 ml. 0.1250N (N/8) sodium thiosulphate	= 1.0 mg oxygen
1 ml. 0.1N (N/10) sodium thiosulphate	= 3.546 mg chlorine = 1.568 mg phenol = 1.734 mg Cr = 3.8697 mg CrO_4
1 ml. 0.0250N (N/40) sodium thiosulphate	= 0.2 mg oxygen = 0.887 mg chlorine = 0.426 mg H_2S = 0.4 mg S = 0.434 mg chromium = 1.589 mg copper = 0.392 mg phenol
1 ml. 0.01250N (N/80) sodium thiosulphate	= 0.1 mg oxygen = 0.443 mg chlorine
1 ml. 0.02N (N/50) H_2SO_4 or HCl	= 1.0 mg $CaCO_3$
1 ml. 0.1N (N/10) NaOH	= 4.8 mg SO_4 = 4.0 mg SO_2
1 ml. 0.02N (N/50) EDTA	= 1.0mg $CaCO_3$
1 ml. 0.01N (N/100) silver nitrate	= 0.54 mg HCN = 0.52 mg CN = 0.3546 mg chloride
1 ml. 0.01428N (N/70) sodium sulphite	= 0.5 mg chlorine
1 ml 1.0N ferrous ammonium sulphate	= 17.425 mg Cr

1 ml conc. H_2SO_4 (Sp. Gr. 1.84/36N) neutralizes 1800 mg alkalinity as $CaCO_3$
1 ml. conc. HCl (Sp. Gr. 1.18/11.6N) neutralizes 580 mg alkalinity as $CaCO_3$
1 g lime (100% CaO) neutralizes 1785 mg acidity as $CaCO_3$
1 g lime (90% CaO) neutralizes 1606 mg acidity as $CaCO_3$

TABLE - XVII
EQUIVALENT WEIGHTS OF COMMONLY USED ACIDS, BASES AND SALTS

Hydrochloric acid	36.46
Sulphuric acid	49
Nitric acid	63
Oxalic acid	63
Acetic acid	60
Sodium hydroxlde	40
Sodium carbonate	53
Sodium bicarbonate	84
Potassium hydroxide	56
Ammonia solution (sp. 0.90)	35
Potassium dichromate	49.03
Potassium permanganate	31.6
Sodium thiosulphate	248.17
Ferrous sulphate	278
Ferrous ammonium sulphate (Mohrs' salt)	392

TABLE - XVIII
SOLUBILITIES OF BICARBONATES, CARBONATES, CHLORIDES, HYDROXIDES AND SULFATES OF CALCIUM MAGNESIUM AND SODIUM

Name	Formula	Solubility, mg/l	
		At 0° C	at 100° C
Calcium bicarbonate	$Ca(HCO_3)_2$	1620	Decompose
Calcium carbonate	$CaCO_3$	15	13
Calcium chloride	$CaCl_2$	336000	554000
Calcium hydroxide	$Ca(OH_2)$	2390	888
Calcium sulfate	$CaSO_4$	1290	1250
Magnesium bicarbonate	$Mg(HCO_3)_2$	37100	Decompose
Magnesium carbonate	$MgCO_3$	101	75
Magnesium chloride	$MgCl_2$	362000	443000
Magnesium hydroxide	$Mg(OH)_2$	17	8
Magnesium sulfate	$MgSO_4$	170000	356000
Sodium bicarbonate	$NaHCO_3$	38700	Decompose
Sodium carbonate	Na_2CO_3	61400	290000
Sodium chloride	$NaCl$	225000	243000
Sodium hydroxide	$NaOH$	370000	970000
Sodium sulfate	Na_2SO_4	33600	210000

TABLE - XIX
CONCENTRATION OF ACIDS AND GASES
Common Commercial Strengths

	Molecular weight	Moles per litre	Grams per litre	Percent by wt.	Sp. Gravity
Acetic acid glacial	60.05	17.4	1045	99.5	1.05
Acetic acid	60.05	6.27	376	36	1.045
		5.75	264	25	1.06
Hydriodic acid	127.9	7.57	969	57	1.70
		5.51	705	47	1.50
		.86	110	10	1.10
Hydrobromic acid	80.92	8.89	720	48	1.50
		6.82	552	40	1.38
Hydrochloric acid	36.5	11.6	424	36	1.18
		2.9	105	10	1.05
Hydrofluoric acid	20.01	32.1	642	55	1.167
		28.8	578	50	1.155
Nitric acid	63.02	15.99	1008	71	1.42
		14.90	938	67	1.40
		13.3	837	61	1.37
Perchloric acid	100.5	11.65	1172	70	1.67
		9.20	923	60	1.54
Phosphoric acid	98.0	14.8	1445	85	1.70
		9.20	923	60	1.54
Sulphuric acid	98.1	18.0	1766	96	1.84
Ammonia solution	17.0	14.8	252	28	0.898
Potassium hydroxide	56.1	13.5	757	50	1.52
		1.94	109	10	1.09
Sodium carbonate	106.00	1.04	110	10	1.10
Sodium hydroxide	40.0	10.1	769	50	1.53

TABLE - XX
AMMONIA SOLUTION
Specific Gravity and Percentage

Sp. Gravity	Percentage Ammonia
0.88	35%
0.885	33%
0.90	28 – 29%
0.91	25%

TABLE - XXI
SOLUBILITY OF OXYGEN IN FRESH WATER

Temperature °C	*°F*	*Dissolved oxygen p.p.m*
0	32.0	14.62
1	33.8	14.23
2	35.6	13.84
3	37.4	13.48
4	39.2	13.13
5	41.0	12.80
6	42.8	12.48
7	44.6	12.17
8	46.4	11.87
9	48.2	11.59
10	50.0	11.33
11	51.8	11.08
12	53.6	10.83
13	55.4	10.60
14	57.2	10.37
15	59.0	10.15
16	60.8	9.95
17	62.6	9.74
18	64.4	9.54
19	66.2	9.35
20	68.0	9.17
21	69.8	8.99
22	71.6	8.83
23	73.4	8.68
24	75.2	8.53
25	77.0	8.38
26	78.8	8.22
27	80.6	8.07
28	82.4	7.92
29	84.2	7.77
30	86.0	7.63

TABLE - XXII
PERIODIC TABLE OF THE ELEMENTS

Key	
1	Atomic Number
H	Symbol
1.008	Atomic Mass

1 IA	2 IIA	3 IIIB	4 IVB	5 VI	6 VIB	7 VIIB	8	9 VIII	10	11 IB	12 IIB	13 IIIA	14 IVA	15 VA	16 VIA	17 VIIA	18 O
1 H 1.008																	2 He 4.003
3 Li 6.939	4 Be 9.012											5 B 10.81	6 C 12.01	7 N 14.01	8 O 16.00	9 F 19.00	10 Ne 20.18
11 Na 22.99	12 Mg 24.31											13 Al 26.98	14 Si 28.09	15 P 30.97	16 S 32.06	17 Cl 35.45	18 Ar 39.95
19 K 39.10	20 Ca 40.08	21 Sc 44.96	22 Ti 47.90	23 V 50.94	24 Cr 52.00	25 Mn 54.94	26 Fe 55.85	27 Co 58.93	28 Ni 58.71	29 Cu 63.54	30 Zn 65.39	31 Ga 69.72	32 Ge 72.59	33 As 74.92	34 Se 78.96	35 Br 79.91	36 Kr 83.80
37 Rb 85.47	38 Sr 87.62	39 Y 88.91	40 Zr 91.22	41 Nb 92.91	42 Mo 95.94	43 Tc (99)	44 Ru 101.1	45 Rh 102.9	46 Pd 106.4	47 Ag 107.9	48 Cd 112.4	49 In 114.8	50 Sn 118.7	51 Sb 121.8	52 Te 127.6	53 I 126.9	54 Xe 131.3
55 Cs 132.9	56 Ba 137.3	57 La 138.9	72 Hf 178.5	73 Ta 181.0	74 W 183.9	75 Re 186.2	76 Os 190.2	77 Ir 192.2	78 Pt 195.1	79 Au 197.0	80 Hg 200.6	81 Ti 204.4	82 Pb 207.2	83 Bi 209.0	84 Po (210)	85 At (210)	86 Rn (222)
87 Fr (223)	88 Ra (226)	89 Ac (237)	104 Unq	105 Unp	106 Unh	107 Uns	108 Uno	109 Une									

58 Ce 140.1	59 Pr 140.9	60 Nd 144.2	61 Pm (147)	62 Sm 150.4	63 Eu 152.0	64 Gd 157.3	65 Tb 158.9	66 Dy 162.5	67 Ho 164.9	68 Er 167.3	69 Tm 168.9	70 Yb 173.0	71 Lu 175.0
90 Th 232.0	91 Pa (231)	92 U 238.0	93 Np (237)	94 Pu (242)	95 Am (243)	96 Cm (247)	97 Bk (247)	98 cF (249)	99 Es (254)	100 Fm (253)	101 Md (256)	102 No (253)	103 Lr (257)

TABLE – XXIII
INTERNATIONAL ATOMIC WEIGHTS 1973

Element	***Symbol***	***Atomic No.***	***Atomic Weight***
Actinium	Ac	89	(227)
Aluminium	Al	13	26.9815
Americium	Am	95	(243)
Antimony	Sb	51	121.75
Argon	Ar	18	39.948
Arsenic	As	33	74.9216
Astaline	At	85	(210)
Barium	Ba	56	137.34
Berkeylium	Bk	97	(247)
Beryllium	Be	4	9.0122
Bismuth	Bi	83	208.9804
Boron	B	5	10.81
Bromine	Br	35	79.904
Cadmium	Cd	48	112.40
Calcium	Ca	20	40.08
Californium	Cf	98	(251)

Element	Symbol	Atomic No.	Atomic Weight
Carbon	C	6	12.011
Cerium	Ce	58	140.12
Cesium	Cs	55	132.9054
Chlorine	Cl	17	35.453
Chromium	Cr	24	51.996
Cobalt	Co	27	58.9332
Copper	Cu	29	63.542
Curium	Cm	96	(247)
Dysprosium	Dy	66	162.50
Einsteinium	Es	99	(254)
Erbium	Er	68	167.26
Europium	Eu	63	151.96
Fermium	Fm	100	(257)
Fluorine	F	9	18.9984
Francium	Fr	87	(223)
Godolinium	Gd	64	157.25
Gallium	Ga	31	69.72
Germanium	Ge	32	72.59
Gold	Au	79	196.9665
Hafnium	Hf	72	178.49
Helium	He	2	4.0026
Holmium	Ho	67	164.9304
Hydrogen	H	1	1.0079
Indium	In	49	114.82
Iodine	I	53	126.9045
Iridium	Ir	77	192.22
Iron	Fe	26	55.847
Krypton	Kr	36	83.80
Lanthanum	La	57	138.9066
Lawrencium	Lr	103	(260)
Lead	Pb	82	207.0
Lithium	Li	3	6.941
Lutetium	Lu	71	174.97
Magnesium	Mg	12	24.305
Manganese	Mn	25	54.9380
Mendelevium	Md	101	(258)
Mercury	Hg	80	200.59
Molybdenum	Mo	42	95.94
Neodymium	Nd	60	144.24
Neon	Ne	10	20.179
Neptunium	Np	93	237.0482
Nickel	Ni	28	58.70
Niobium	Nb	41	92.9064
Nitrogen	N	7	14.0067
Nobelium	No	102	(255)
Osmium	Os	76	190.2
Oxygen	O	8	15.9994
Palladium	Pd	46	106.4
Phosphorus	P	15	30.9738
Platinum	Pt	78	195.09
Plutonium	Pu	94	(244)

Element	Symbol	Atomic No.	Atomic Weight
Polonium	Po	84	(209)
Potassium	K	19	39.098
Praseodymium	Pr	59	140.9077
Promethium	Pm	61	(145)
Protactinium	Pa	91	231.0359
Radium	Ra	88	226.0254
Radon	Rn	86	(222)
Rhenium	Re	75	186.207
Rhodium	Rh	45	102.9055
Rubidium	Rb	37	85.4678
Ruthenium	Ru	44	101.07
Samarium	Sm	62	150.4
Scandium	Sc	21	44.9559
Selenium	Se	34	78.96
Silicon	Si	14	28.086
Silver	Ag	47	107.868
Sodium	Na	11	22.9898
Strontium	Sr	38	87.62
Sulphur	S	16	32.06
Tantalum	Ta	73	180.9479
Technetium	Tc	43	(97)
Tellurium	Te	52	127.60
Teribium	Tb	65	158.9254
Thallium	Tl	81	204.37
Thorium	Th	90	232.0381
Thulium	Tm	69	168.9342
Tin	Sn	50	118.69
titanium	Ti	22	47.90
tungsten	W	74	183.85
Uranium	U	92	238.029
Vanadium	V	23	50.9414
Xenon	Xe	54	131.30
Ytterbium	Yb	70	173.04
Yttrium	Y	39	88.9059
Zinc	Zn	30	65.38
Zirconium	Zr	40	91.22

TABLE - XXIV
CHEMICAL FORMULA, FORMULA WEIGHTS AND EQUIVALENT WEIGHTS

Name	Chemical Formula	Formula Weight	Equivalent Weight
Alum, ammonia	see Ammonium alum		
Alum, filter	see Aluminium sulphate		
Alum, potash	See Potassium oxide		
Aluminium	Al	27.0	9.00
Aluminium chloride	$AlCl_3$	133	44.4
Aluminium chloride	$AlCl_3.6H_2O$	241	80.5
Aluminium hydroxide	$Al(OH)_3$	78.0	26.0
Aluminium oxide	Al_2O_3	102	17.0
Aluminium sulphate	$Al_2(SO_4)_3.I8H_2O$	666	111

Name	Chemical Formula	Formula Weight	Equivalent Weight
Ammonia	NH_3	17.0	17.0
Ammonium ion	NH_4^+	18.0	18.0
Ammonium alum	$Al_2(SO_4)_3.(NH_4)_2SO_4.24H_2O$	906	151
Ammonium chloride	NH_4Cl	53.5	53.5
Ammonium hydroxide	NH_4OH	35.1	35.1
Ammonium sulphate	$(NH_4)_2SO_4$	132	66.1
Barium	Ba	137	68.7
Barium carbonate	$BaCO_3$	197	98.7
Barium chloride	$BaCl_2.2H_2O$	244	122
Barium sulphate	$BaSO_4$	233	117
Bicarbonate ion	HCO3-	61.0	61.0
Bicarbonate of soda	see Sodium bicarbonate		
Blue vitriol	see Copper sulphate		
Calcite	see Calcium carbonate		
Calcium	Ca	40.1	20.05
Calcium bicarbonate	$Ca(HCO_3)_2$	162	81.0
Calcium carbonate	$CaCO_3$	100	50.0
Calcium chloride	$CaCl_2$	111	55.5
Calcium hydroxide	$Ca(OH)_2$	74.1	37.1
Calcium nitrate	$Ca(NO_3)_2$	164	82.1
Calcium phosphate (tri-)	$Ca_3(PO_4)_2$	310	51.7
Calcium sulphate	$CaSO_4$	136	68.1
Calcium sulphate		145	72.6
Calcium sulphate	$CaSO_4.(1/2)H_2O$ (Plaster of Pairs) $CaSO_4.2H_2O$ (gypsum)	172	86.1 3.00
Carbon	C	12.0	
Carbon dioxide (gas)	CO_2	44.0	22.0
Carbonate ion	CO_3^-	60.0	38.0
Carbonic acid	H_2CO_3	62.0	31.0
Caustic potash	see Potassium hydroxide		
Caustic soda	see Sodium hydroxide		
Chalk	see Calcium carbonate		
Chemical lime	see Calcium oxide		
Chile saltpeter	see Sodium nitrate		
Chloride ion	Cl^-	35.5	35.5
Chlorine (gas)	Cl_2	70.9	35.5
Copper	Cu	63.6	31.8
Copper sulphate	$CuSO_4.5H_2O$	250	125
Copperas	See Ferrous sulphate	184	46
Dolomite	$CaCO_3.MgCO_3$		
Dolomitic lime	CaO + various amounts MgO		
Dolomitic lime, hydrated	$Ca(OH_2)$ + various amounts MgO		
Dolomitic limestone	$CaCO_3$ + various amounts $MgCO_3$		
Epsom Salt	See Magnesium sulphate		
Ferric chloride	$FeCl_3$	162	54.1
Ferric chloride	$FeCl_3.6H_2O$	270	90.1
Ferric hydroxide	$Fe(OH)_3$	107	35.6
Ferric oxide	Fe_2O_3	160	26.6
Ferric sulphate	$Fe_2(SO_4)_3$	400	66.6
Ferroso-Ferric hydroxide	$Fe_3(OH)_8$	304	
Ferroso-ferric oxide	Fe_3O_4	232	
Ferrous bicarbonate	$Fe(HCO_3)_2$	178	89.0
Ferrous carbonate	$FeCO_3$	116	57.9

Name	Chemical Formula	Formula Weight	Equivalent Weight
Ferrous hydroxide	$Fe(OH)_2$	89.9	44.9
Ferrous oxide	FeO	71.8	35.9
Ferrous sulphate	$FeSO_4.7H_2O$	278	139
Fluoride ion	F^-	19.0	19.0
Fluorine gas	F_2	38.0	19.0
Fluosilicic acid	H_2SiF_6	144	72.0
Green vitriol	see Ferrous sulphate		
Gypsum	see $CaSO_4.2H_2O$		
Hydrochloric acid	HCl	36.5	36.5
Hydrofluoric acid	HF	20.0	20.0
Hydrofluosilicic acid	See Fluosilicic acid		
Hydrogen ion	H^+	1.01	1.01
Hydrogen gas	H_2	2.02	1.01
Hydrogen sulphide (gas)	H_2S	34.1	17.0
Iodide ion	I^-	127.0	127
Iodide	I_2	254	127
Iron (ferric ion)	Fe^{+++}	55.8	18.6
Iron (ferrous ion)	Fe^{++}	55.8	27.9
Lead	Pb	207	104
Lime	see Calcium oxide		
Lime, chemical	see Calcium oxide		
Lime, dolomitic	see Dolomitic lime		
Lime, hydrated	see Calcium hydroxide		
Lime, slaked	see Calcium hydroxide		
Lime, quick	see Calcium oxide		
Magnesia	see Magnesium oxide		
Magnesite	see Magnesium carbonate		
Magnesium bicarbonate	$Mg(HCO_3)_2$	146	73.2
Magnesium carbonate	$MgCO_3$	84.3	42.2
Magnesium chloride	$MgCl_2$	95.2	47.6
Magnesium hydroxide	$Mg(OH)_2$	58.3	29.2
Magnesium nitrate	$Mg(NO_3)_2$	148	74.2
Magnesium oxide	MgO	40.3	20.2
Magnesium phosphate(tri-)	$Mg_3(PO_4)_2$	263	43.8
Magnesium sulphate	$MgSO_4$	120	60.2
Magnesium sulphate	$MgSO_4.7H_2O$ (Epsom salt)	246	123
Manganese (manganic ion)	Mn^{+++}	54.9	18.3
Manganese (manganous ion)	Mn^{++}	54.9	27.5
Manganese dioxide	MnO_2	86.9	21.7
Manganic hydroxide	$Mn(OH)_3$	106	35.3
Manganic oxide	Mn_2O_3	158	26.3
Manganous bicarbonate	$Mn(HCO_3)_2$	177	88.5
Manganous carbonate	$MnCO_3$	115	57.5
Manganous hydroxide	$Mn(OH)_2$	89.0	44.5
Manganous oxide	MnO	70.9	35.5
Muriatic acid	see Hydrochloric acid		
Nitrate ion	NO_3^-	62.0	62.0
Nitrite cake	see Sodium bisulphate		
Nitric acid	HNO_3	63.0	63.0
Nitrogen	N (valencies 3 and 5)	14.0	
Nitrogen (gas)	N_2	28.0	
Oil of vitriol	see Sulphuric acid		
Oxygen	O	16.0	8.0

Name	Chemical Formula	Formula Weight	Equivalent Weight
Oxygen (gas)	O_2	32.0	8.0
Pearl ash	see Potassium carbonate		
Permanganate of potash	see Potassium permanganate		
Phosphate (ortho ion)	PO_4^{3-}	95.0	31.7
Phosphorus	P (valencies 3 and 5)	31.0	
Phosphorus pentoxide	P_2O_5	142	23.7
Plaster of paris	see $CaSO_4.1/2H_2O$		
Potash	see Potassium carbonate		
Potash alum	see Potassium alum		
Potassium	K	39.1	39.1
Potassium alum	$Al_2(SO_4)_3.K_2SO_4.24H_2O$	949	158
Potassium bicarbonate	$KHCO_3$	100	100
Potassium carbonate	K_2CO_3	138	69.1
Potassium chloride	KCl	74.6	74.6
Potassium hydroxide	KOH	56.1	56.1
Potassium iodide	KI	166	166
Potassium nitrate	KNO_3	101	101
Potassium oxide	K_2O	94.2	47.1
Potassium permanganate	$KMnO_4$	158	31.6
Potassium sulphate	K_2SO_4	174	87.1
Quartz	see Silicon dioxide		
Quicklime	see Calcium oxide		
Sal ammoniac	see Ammonium chloride		
Sal soda	see $Na_2CO_3.10H_2O$		
Salt (common)	see Sodium chloride		
Salt cake	see Sodium sulphate		
Saltpeter	see Potassium nitrate		
Saltpeter, Chile	see Sodium nitrate		
Silica	see Silicon dioxide		
Silicon	Si	28.1	7.03
Silicon dioxide	SiO_2	60.1	
Silver	Ag	108	108
Silver chloride	AgCl	143	143
Silver nitrate	$AgNO_3$	170	170
Soda	see Sodium carbonate		
Soad ash	see Na_2CO_3		
Sodium	Na	23.0	23.0
Sodium bicarbonate	$NaHCO_3$	84.0	84.0
Sodium bisulphate	$NaHSO_4$	120	
Sodium bisulphite	$NaHSO_3$	104	
Sodium carbonate	Na_2CO_3 (Soda ash)	106	53.0
Sodium carbonate	$Na_2CO_3.10H_2O$ (Sal soda)	286	143
Sodium chloride	NaCl	58.5	58.5
Sodium chromate	Na_2CrO_4	162	
Sodium dichromate	$Na_2Cr_2O_7.2H_2O$	298	
Sodium hydroxide	NaOH	40.0	40.0
Sodium nitrate	$NaNO_3$	85.0	85.0
Sodium oxide	Na_2O	62.0	31.0
Sodium phosphates:			
meta (comm. Grade 69.0% P_2O_5)	$NaPO_3$	102	34.0
mono (comm. Grade 58.0% P_2O_5)	NaH_2PO_4	120	40.0
mono (comm. Grade 52.0% P_2O_5)	$NaH_2PO_4.H_2O$	138	46.0
di (comm. Grade 49.0% P_2O_5)	Na_2HPO_4	142	47.3
di (comm. Grade 19.0% P_2O_5)	$Na_2HPO_4.12H_2O$	358	119

Name	Chemical Formula	Formula Weight	Equivalent Weight
tri (comm. Grade 39.5% P_2O_5)	$Na_3PO_4.H_2O$	182	60.7
tri (comm. Grade 18.9% P_2O_5)	$Na_3PO_4.12H_2O$	380	127
Sodium sulphate	Na_2SO_4 (Salt cake)	142	71.0
Sodium sulphate	$Na_2SO_4.10H_2O$ (Glauber's salt)	322	161
Sodium sulphite	Na_2SO_3	126	63.0
Sodium tetrathionate	$Na_2S_4O_6$	270	
Sodium thiosulphate	$Na_2S_2O_3$	158	
Sulphate ion	SO_4^{2-}	96.1	48.0
Sulphide ion	S^{2-}	32.1	16.0
Sulphite ion	SO_3^{2-}	80.1	40.0
Sulphur	S (valencies 2, 4 and 6)	32.1	
Sulphur dioxide	SO_2	64.1	32.0
Sulphur trioxide	SO_3	80.1	40.0
Sulphuric acid	H_2SO_4	98.1	49.0
Sulphurous acid	H_2SO_3	82.1	41.0
Tin	Sn (valencies 2 and 4)	119	
Vitriol	see Sulphuric acid		
Water	H_2O	18	9.00
Zinc	Zn	65.4	32.7

TABLE – XXV
DATA ON CHEMICALS USED IN WATER AND WASTE WATER TREATMENT

Chemical Name, Formula and Formula Weight	Common Name/Trade Name	Available form	Commercial strength	Charact eristic	Solubility gm/100 ml water				Weight/ lb/cu.ft	Storage space tone cu.ft/ton	Shipping containers	Suitable handling materials
					0°C	10°C	20°C	30°C				
Activated Carbon – See Carbon activated												
Aluminium sulfate $Al_2(SO_4)_3.14H_2O$ 594	Alum Filter Alum Sulfate of Alumina	Ivory/Brown colored slab, lump, granule, powder, Also available in liquid form	14.5% to 17.5% Al_2O_3	pH of 1% solution 3.4	60.8	65.3	71.0	78.8	Slab lump (or) ground 57-67 powder-ed 38-45	30-35 45-53	Bags, barrels and in bulk; liquid is transported by Tankers	Dry : Iron, steel Liquid: Lead lined rubber, silicon asphalt
Alum, ammonium $Al_2(SO_4)_3.(NH_4)_2.SO_4.2H_2O$ 906	Ammonia Alum (Sometimes called as crystal alum)	Lump, nut, powder	11% Al_2O_3	pH of 1% solution 3.5	3.9	9.5	15.1	20.0	64-68	30-31	Bags, barrels, boxes, Cartons, Kegs and bulk	Lead lined rubber, silicon Iron, stone-ware.
Alum potassium $Al_2(SO_4)_3.K_2SO_4.24H_2O$ 949	potash alum	lump, crystal	10.7% Al_2O_3	pH 1% solution 3.5	5.7	7.6	11.4	16.6	64-68	30-31	Bags, barrels, boxes, cartons, and kegs	Lead lined rubber, stone-ware.
Bentonite	colloidal clay, volclay, wilkinite	Powder, pallet, mixed sizes	-	-	-	-	-	-	60	-	Bags bulk	Iron steel

Chemical Name, Formula and Formula Weight	Common Name/Trade Name	Availabl e form	Commercial strength	Charact-eristic	Solubility gm/100 ml water				Weight/ lb/cu.ft	Storage space tone cu.ft/ton	Shipping containers	Suitable handling materials
					0°C	10°C	20°C	30°C				
Bleaching power See "chlorinated lime"												
Calcium Carbonate $CaCO_3$ 100	Lime stone, Calcite, Chalk, whiting	Crushed and graded granules; also available in powder form	-	-	-	-	13	-	Granule 100-115 powder 48-71	18-20 28-42	Bags, barrels bulk	
Calcium chloride $CaCl_2$ 111	-	Flake: 77-78% $CaCl_2$ Solid: 73-75% $CaCl_2$ Liquid: 40% $CaCl_2$	-	-	59.5	65.0	74.5	102	60-66	29-31	Flakes: moisture proof bags or drums Solids: drums	-
Calcium hydroixde $Ca(OH)_2$ 74.1	Hydrated lime, slaked lime	Powder (light and dense)	85-99% $Ca(OH)_2$	Hopper agitation required for dry feed of light form.	0.18	0.17	0.16	0.15	25-50	40-80	Bags, barrels, bulk	Rubber, Iron, Steel, cement and asphalt
Calcium hypochlorite $CaOCl_2.4H_2O$- 199	"ATH" "Perchloron" "Pittchlor"	Powder	70% available chlorine	- Generally 1-3% available Cl_2 solution used.	Highly soluble				-	-	Cans and drums	Glass, rubber, stoneware and wood
Calcium oxide CaO 56.1	lime, burnt lime, chemical lime, quick lime, unslaked lime	Lump, pebble ground and powder.	75 to 99% CaO	Slakes with water to form hydrated lime $Ca(OH)_2$ For slaking 1 Kg. lime, 10 litres of water is required	-	-	-	-	Lump 50-65 Ground 50-70 Pulv. 39-71	31-40 29-40 28-52	Bags, barrels and in bulk	Rubber, Iron, steel, cement and asphalt
Calcium sulfate $CaSO_4.2H_2O$ 172	Gypsum	Ground	-	-	0.18	0.19	0.20	0.21	50-60	33-40	Bags, barrels and in bulk	-
Carbon-activated C 12	-	Black granules and powder	-	-	Insoluble				Granular 12-30 Powder 15-28	67-167 72-134	Bags, bulk	Dry iron, steel, wet-rubber, stainless steel.
Caustic soda See Sodium- hydroxide												

Chemical Name, Formula and Formula Weight	Common Name/Trade Name	Available form	Commercial strength	Charact-eristic	Solubility gm/100 ml water				Weight/ lb/cu.ft	Storage space tone cu.ft/ton	Shipping containers	Suitable handling materials
					0°C	10°C	20°C	30°C				
Chlorinated lime $CaO.2CaOCl_2.3H_2O$ 364	Bleaching Powder, chloride of lime	White powder	25-37% available chlorine	Deteriorates on exposure to air	Very soluble				48	–	Drums	Glass, rubber, stone ware, wood
Chlorine Cl_2 70.9	–	Chlorine gas Liquid chlorine (liquid field gas under pressure in steel cylinders	99.8% Cl_2	–	1.46	0.980	0.716	0.562	–	–	Cylinders Tankers, Tank trucks	Dry: Iron Copper, steel, wetgas: glass, hardrubber.
Chlorine dioxide ClO_2 102.9	Chlorine dioxide	Yellow-red gas	26.3% available chlorine	–	–	–	–	–	–	–	Generated at the site itself	Plastic, soft rubber (Hard rubber to be avoided)
Copper sulfate $CuSO_4.5H_2O$ 249.6	Blue vitriol Blue stone	Crystal, lump powder	99% $CuSO_4$	–	19.2	–	26.4	31.2	Crystal 75-90 lump 73-80 Powder 60-64	–	Bags, barrels, drums	Iron, Stainless steel, Silicon, asphalt
Disodium phosphate (see Sodium phosphate)												
Dolomitic lime CaO + MgO (Content of MgO varies)	Dolomitic lime, High magnesium lime	Lump, pebble, ground or powder	58% CaO and 40% MgO	Typical for Silica removal Slakes with water forming $Ca(OH)_2$+ MgO the later Slakes very slowly	–	–	–	–	Lump 50-65 Pebble 60-65 Ground 50-75 Pulv 37-63	31-40 31-34 27-40 32-54	Bags, barrels and in bulk	Wood, Iron steel
Dolomitic lime, hydrated $Ca(OH)_2$ + MgO content of MgO varies	Dolomitic lime, hydrated	Powder	62% $Ca(OH)_2$ and 32% MgO	Typical for silica removal	–	–	–	–	28-52	39-72	Bags, barrels and in bulk	Wood, Iron Steel
Ferric chloride (i) $FeCl_3$ (ii) $FeCl_3.6H_2O$ (iii) $FeCl_3$ (35-45%) solution	Anhydrous ferric chloride Crystal ferric Chloride "Ferrichlor" Chloride of Iron	Green black powder Yellow brown lump Dark brown Syrupy liquid	98% $FeCl_3$ 34% Fe 59-61% $FeCl_3$ 20-21% Fe 37-47% $FeCl_3$ 20-21% Fe	Highly hygroscopic; lumps and Powder should be stored in air tight containers	Completely soluble				–	–	Casks, kegs barrels car-boys trucks, tankers.	Rubber, glass, stoneware synthetic-resins

Chemical Name, Formula and Formula Weight	Common Name/Trade Name	Availabl e form	Commercial strength	Charact-eristic	Solubility gm/100 ml water				Weight/ lb/cu.ft	Storage space tone cu.ft/ton	Shipping containers	Suitable handling materials
					0°C	10°C	20°C	30°C				
Ferric Sulfate $Fe_2(SO_4)_3$ 400	Ferrisul Ferrifloc	Granules Reddish brown powder	Composition from different sources varies 70%-90% $Fe_2(SO_4)_3$ some grades contain $Al_2(SO_4)_3$ also	Mildly hygroscopic coagulated at pH 8.5 – 11.0	Highly soluble, to dissolve in cold water, add 2 parts water to 1 part ferric sulfate with stirring.				60-70	29-34	Bags, barrels, kegs, drum and in bulk	Rubber, plastic, lead ceramic, stainless-steel
Ferrous sulfate $FeSO_4.7H_2O$ 278	Copper as green vitriol	Crystals, granules	55% $FeSO_4$ 20% Fe	Hygro-scopic: cakes on storage. Optimum pH 8.5 – 11.0	28.7	37.5	48.5	60.2	63-66	30-32	Bags, barrels kegs and in bulk	Wood, lead asphalt, concrete
Hydrated lime (See Calcium hydroxide)												
Lime (See Calcium oxide)												
Hydrazine N_2H_4 32.05	–	Colorless oily liquid	–	Powerful reducing agent fuming in air penetrating odor resembling that of ammonia	Miscible with water				–	–	Glass	Glass
Hydrochloric acid HCl 36.45	Muriatic acid	sometime-es colored by traces of iron chlorine and organic matter	38% HCl	Fumes in air pH of 1N HCl is 0.1 0.1N HCl = 1.1 0.01N HCl = 2.02	–	–	–	–	–	–	Bulk in Tankers	Plastic rubber
Meta Sodium-phosphate (See sodium phosphate meta)	–	–	–									
Soda ash (See Sodium carbonate)												
Sodium aluminate $Na_2O.Al_2O_3$ 164	Soda alum	Brown powder liquid (27°Be)	70-80% $Na_2Al_2O_4$ Min. 32% $Na_2.Al_2O_4$	For dry feed hopper agitation required	Complete				50-60	–	Bags, drums tankers	Iron, plastic, rubber, steel
Sodium carbonate Na_2CO_3 106	Soda ash	White powder Extra light, dense	Light soda ash is used in water treatment. 99.4% Na_2CO_3 58% Na_2O	pH of 1% solution 11.3 Hopper agitation required for dry feeding	7.0	12.5	4.5	38.8	23 35 65	–	Bags, barrels bulk [Carloads, trucks]	Iron, rubber, steel

Chemical Name, Formula and Formula Weight	Common Name/Trade Name	Availabl e form	Commercial strength	Charact-eristic	Solubility gm/100 ml water				Weight/ lb/cu.ft	Storage space tone cu.ft/ton	Shipping containers	Suitable handling materials
					$0^\circ C$	$10^0 C$	$20^\circ C$	$30^\circ C$				
Sodium chloride NaCl 58.5	Salt, common salt, rock salt evaporated salt	Crystal powder	98% NaCl	–	35.7	35.8	36.0	36.3	50-70	29-40	Bags, barrels, bulk	Rubber, cement, plastic, bronze
Sodium hydroxide NaOH 40	Caustic soda	Flakes, solid ground, liquid	98.06 NaOH (76% Na_2O) 50-70% solution	Solid Highly hygoscop ic pH of 1% solution 12.9	42.0	51.5	109	119	–	–	Flake : drums Solid : drums Solution: in tankers	Cast iron steel, rubber.
Sodium hypochlorite NaOCl 74.5	Sodium hypochlorite	Light yellow liquid	12-15% available chlorine	–	–	–	–	–	–	–	Carboys, Bulk in tankers	Glass plastics, rubber, ceramics
Sodium hexameta phosphate $(NaPO_3)_6$ 750	'Calgon' glassy phosphate vitreous phosphate	crystal, flakes powder	66% P_2O_5 (unadjusted)	pH of 0.25% solution 6.0-8.3	–	–	–	–	47	–	Bags	Plastics, hard rubber, stainless steel
Sodium phosphate, mono NaH_2PO_4 (anhydrous) 120 $NaH_2PO_4.H_2O$ 138	–				Monohydrate 66.6	 80.4	 98.0	 12.3	Anhydrou s 55-70 Mono hydrate 58-79	29-37 26-35	Bags, barrels, boxes, kegs drums	Cast iron, steel
Sodium phosphate, Disodium – phosphate, Disodium hydrogen phosphate Na_2HPO_4 – 142 (anhydrous) $Na_2HPO_4 .12H_2O$ - 358	Basic sodium phosphate DSP secondary sodium phosphate	Crystal	19.5% P_2O_5	Precipitat es calcium and magnesiu m pH of 1% solution 9	Anhydrous 1.7 Dodeca hydrate 4.2	 3.6 8.9	 7.7 19.3	 20.8 52.4	Anhydrou s 53-62 46-53 Dodeca hydrate	32-39 38-44	Bags, barrels, kegs drum	Cast Iron, steel
Sodium phosphate Trisodium – phosphate $Na_3PO_4.H_2O$ 182 $Na_3PO_4.12H_2O$ 380	Normal sodium phosphate Tertiary sodium phosphate	Crystal coarse Medium standard	19% P_2O_5	pH of 1% solution 11.9	Monohydrate 1.7 Dodeca hydrate 3.5	 4.6 9.5	 12.3 25.5	 22.2 46.4	83-90 56-60	22-24 32-36	Bags, barrels boxes kegs, drum	Cast Iron, steel
Sodium tripoly – phosphate (STPP) polypenta sodium triphosphate $Na_5P_3O_{10}$ – 367.9	–	slightly hygrosco pic granules	–	pH of 1% solution 9.7 to 9.8	20 gm at 25°C and 86.5 gm at 100°C				–	–	–	–
Sodium Silicate $Na_2O.SiO_2$ (variable ratio of Na_2O to SiO_2)	Water glass Silicate of soda	opaque viscous liquid	38° – 40° Be solution	pH of 1% solution 12.3	Miscible in water in all proportions				86.1	–	Drums Tankers	Cast iron rubber, steel
Sodium sulfite Na_2SO_3 126	Sulfite	White crystals powder	23% SO_2	Sulfurous taste and odor	14.0	20.0	27.0	36.0	80-91	23-26	Bags, barrels Kegs and drums	Iron, steel wood

Chemical Name, Formula and Formula Weight	Common Name/Trade Name	Available form	Commercial strength	Characteristic	Solubility gm/100 ml water				Weight/ lb/cu.ft	Storage space tone cu.ft/ton	Shipping containers	Suitable handling materials
					0°C	10°C	20°C	30°C				
Sulfur dioxide SO_2 64.1	Sulfurous acid anhydride	Colorless gas (liquefied gas under pressure in steel cylinders)	99% SO_2	Highly corrosive and irritating gas	22.8	16.2	11.3	7.8	-	-	Steel cylinders, Tankers	Aluminium, brass, stainless steel
Sulfuric acid H_2SO_4 98.1	Oil of vitriol	Liquid	98% 93.2% (66° Be) 77.7% (60° Be)	Approximate pH value of 0.5% solution 1.2	Miscible with water in all proportions				-	-	Bottles, carboys, drums, tankers	Concentr-ated: Iron, steel, Dilute: glass, lead, porcelain rubber.
Tetra Sodium phosphate Pyro $Na_4P_2O_7.10H_2O$ 445.8	Alkaline sodium pyrophosphate	White powder	53% P_2O_5	pH of 1% solution 10.5	-	-	-	0.6	68	-	Bags barrels, Kegs	Cast Iron, steel.

TABLE – XXVI
SIEVE MESH CONVERSION TABLE
SCREEN EQUIVALENTS

Sieve No.	U.S. Standard		Meshes per Inch	Tyler Standard		Meshes per Inch	British Standard	
	Opening			Opening			Opening	
	mm	Inches		mm	Inches		mm	Inches
12	1.68	0.0661	10	1.65	0.065	10	1.68	0.0660
14	1.41	0.0555	12	1.40	0.055	12	1.40	0.0553
16	1.19	0.0469	14	1.17	0.046	14	1.20	0.0474
18	1.00	0.0394	16	0.991	0.039	16	1.00	0.0395
20	0.84	0.0331	20	0.833	0.0328	18	0.853	0.0336
25	0.71	0.0280	24	0.701	0.0276	22	0.669	0.0275
30	0.59	0.0232	28	0.589	0.0232	25	0.599	0.0236
35	0.50	0.0197	32	0.495	0.0195	30	0.500	0.0197
40	0.42	0.0165	35	0.417	0.0164	36	0.422	0.0166
45	0.35	0.0138	42	0.351	0.0138	44	0.353	0.0139
50	0.297	0.0117	48	0.295	0.0116	52	0.295	0.0116
60	0.250	0.0098	60	0.246	0.0097	60	0.251	0.0099
70	0.210	0.0083	65	0.208	0.0082	72	0.211	0.0083
80	0.177	0.0070	80	0.175	0.0069	85	0.178	0.007
100	0.149	0.0059	100	0.147	0.0058	100	0.152	0.006
200	0.074	0.0029	200	0.074	0.0029	200	0.076	0.003
325	0.044	0.0017	325	0.043	0.0017	240	0.066	0.0026

INDEX

A

B

C

D

E

F

G

H

I

J

L

M

N

O

P

R

S

T

REFERENCES

1. Edwin Windle Taylor "THE EXAMINATION OF WATERS AND WATER SUPPLIES" - 6th Edition - 1949, J&A. Churchill Ltd, London.

2. E.B. Sandell "COLORIMETRIC DETERMINATION OF TRACES OF METALS" - 3rd edition - 1959. Inter Science Publishers Inc., New York.

3. Leo.I.Pincus "PRACTICAL BOILER WATER TREATEMENT" - 1962, McGraw Hill Book Co., New York.

4. Louis Meites (ed), "HANDBOOK OF ANALYTICAL CHEMISTRY" - 1963, McGraw-Hill Publishing Company, NewYork.

5. A. I. Vogel, "A TEXT BOOK OF QUANTITIVE INORGANIC ANALYSIS INCLUDING ELEMENTARY INSTRUMENTAL ANALYSIS" - 3rd edition - 1964. ELBS & LongMans Green Co., Ltd. London.

6. H. B. Ward & G. C. Whipple, "FRESH WATER BIOLOGY" (Edited by W.T. Edmonson), 2nd edition - 1966, John Wiley & Sons Inc., New York.

7. Her Majesty's Stationery Office, "REPORTS ON PUBLIC HEALTH AND MEDICAL SUBJECTS NO 71 - BACTERIOLOGICAL EXAMINATION OF WATER SUPPLIES" - 1969, HMSO, London.

8. Association of Official Analytical Chemists, "OFFICIAL METHODS OF ANALYSIS" - 11th edition - 1970 A.O.A.C - Washington DC.

9. American Society for Testing and materials, "ANNUAL BOOK OF ASTM STANDARDS" PART - 23, Water - Atmospheric Analysis - 1972, ASTM. Philadelphia.

10. Her Majesty's Stationery Office, "ANALYSIS OF RAW, POTABLE AND WASTE WATER" - 1972, HMSO, London.

11. D.D. Mara, "BACTERIOLOGY FOR SANITARY ENGINEERS" - 1974, Churchill & Livingstone, Edinburgh & London.

12. Robert Cruickshanks, "MEDICAL MICROBIOLOGY" VOL - I, 12th edition - 1975. Churchill & Livingstone, Edinburgh & London.

13. Robert Cruickshanks, "MEDICAL MICROBIOLOGY" VOL – II, 12th edition – 1975. Churchill & Livingstone, Edinburgh & London.

14. W. Leithe, "THE ANALYSIS OF ORGANIC POLLUTANTS IN WATER AND WASTE WATER" – 1975, Ann Arbor Science Publishers Inc., Michigan.

15. Pelczar, Reid & Chan, "MICROBIOLOGY", 4th edition – 1978. TATA Mc Graw Hill Publishing Co., Ltd., New Delhi.

16. U.S.Environmental Protection Agency, "METHODS FOR CHEMICAL ANALYSIS OF WATER AND WASTES" – EPA-600/4-79-020, U.S. E.P.A., Office of Research and Development, U.S.Government Printing Office, Washington, D.C.

17. Jame W. McCoy, "THE CHEMICAL TREATMENT OF BOILER WATER" – 1981, Chemical Publishing Co., New York.

18. World Health Organisation, "GUIDELINES FOR DRINKING WATER QUALITY", Vol – II, Health Criteria and Other Supporting Information – 1985, WHO, Geneva.

19. R. Anantha Narayanan A.N & C. K. Jayaram Paniker, "TEXT BOOK OF MICROBIOLOGY", 4th edition – 1992, Orient Longman, Madras, (India).

20. World Health Organization, "GUIDELINES FOR DRINKING WATER QUALITY", Vol – I, RECOMMENDATIONS, 2nd edition – 1993, WHO, Geneva.

21. Ganies Bradford Jackson, "APPLIED WATER AND SPENTWATER CHEMISTRY – A LABORATORY MANUAL" – 1993, VanNostrand Reinhold, NewYork.

22. N. Manivasakam, "PHYSICO – CHEMICAL EXAMINATION OF WATER, SEWAGE AND INDUSTRIAL EFFLUENTS" 3rd edition – 1996, Pragati Prakashan, Meerut, (India).

23. American Public Health Association, American Public Health Association, American Water Works Association, Water Environment Federation, "STANDARD METHODS FOR THE EXAMINATION OF WATER AND WASTE WATER" 20th edition – 1998, American Public Health Association, Washington DC.

24. Hach, "WATER ANALYSIS HAND BOOK", 4th edition, Revision 2 – 2003, Hach Company, Colorado, U.S.A.

www.ingramcontent.com/pod-product-compliance
Lightning Source LLC
LaVergne TN
LVHW062111300926
849161LV00010B/2425